M
MARATHON

SEISMIC STUDIES IN PHYSICAL MODELING

International Human Resources Development Corporation   Boston

SEISMIC STUDIES IN PHYSICAL MODELING

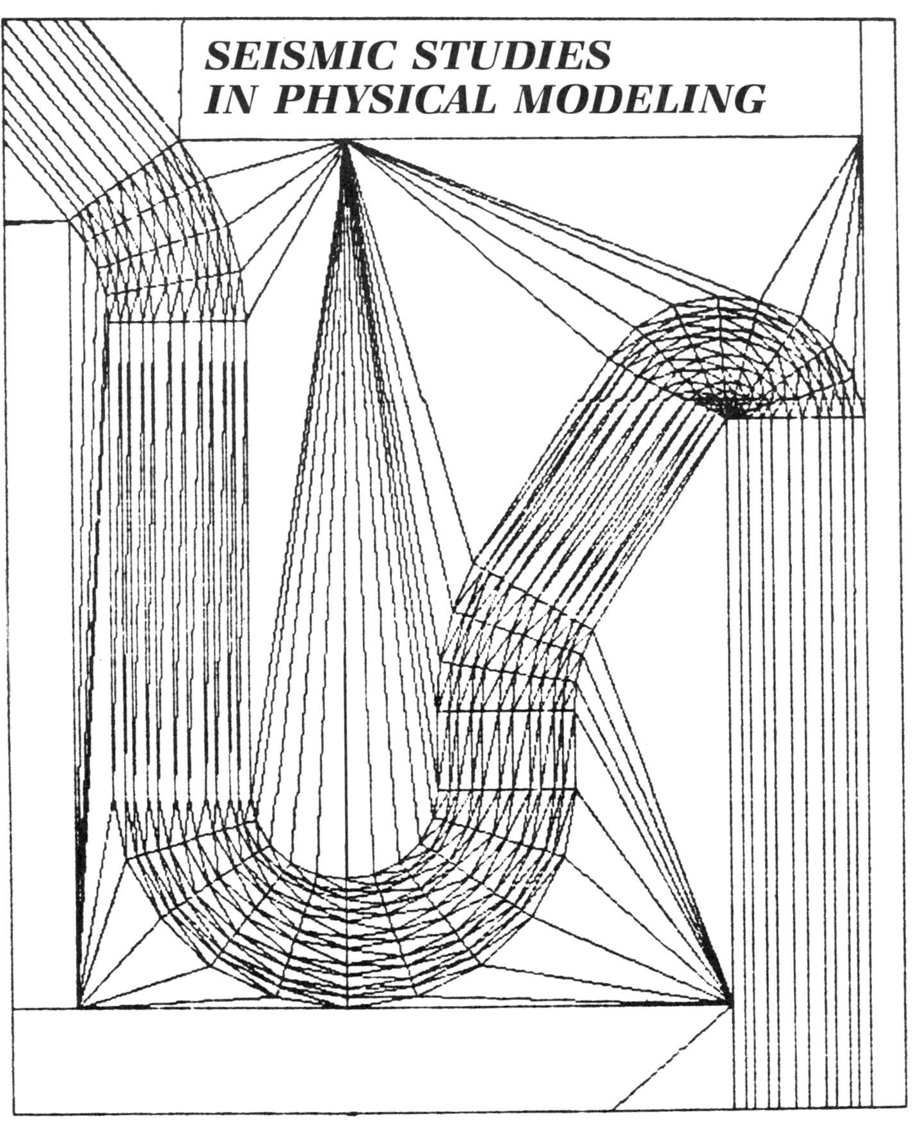

EDITED BY

JOHN A. McDONALD

G. H. F. GARDNER

Seismic Acoustics Laboratory
University of Houston, Texas

FRED J. HILTERMAN

Geophysical Development Corporation
Houston, Texas

Library of Congress Cataloging in Publication Data
Main entry under title:

Seismic studies in physical modeling.

Includes index.
1. Seismic waves—Mathematical models. I. McDonald, John A. (John Andrew), 1931– II. Gardner, G. H. F. III. Hilterman, Fred J.
QE538.5.S428 1983 551.2'2 82-81374
ISBN 0-934634-39-4

Design by Outside Designs

For information address: IHRDC, Publishers, 137 Newbury Street, Boston, MA 02116.

Printed in the United States of America

CONTENTS

LIST OF CONTRIBUTORS

Edip Baysal
Geophysical Development Corporation
Houston, Texas

Lynn L. Chou
Texaco Inc.
Bellaire, Texas

Robert E. Duffy, Jr.
Union Oil of California
Bray, California

G. H. F. Gardner
Seismic Acoustics Laboratory
University of Houston
Houston, Texas

Fred J. Hilterman
Geophysical Development Corporation
Houston, Texas

Dan D. Kosloff
Tel Aviv University
Tel Aviv, Israel

John A. McDonald
Seismic Acoustics Laboratory
University of Houston
Houston, Texas

Thomas R. Morgan
O'Connor Research, Inc.
Denver, Colorado

J. K. Owusu
Petty-Ray Geophysical
GEOSOURCE
Houston, Texas

Thomas A. Smith
N. S. Neidell and Associates
Houston, Texas

SEISMIC STUDIES IN PHYSICAL MODELING

1. PHYSICAL MODELING AT THE SEISMIC ACOUSTICS LABORATORY

John A. McDonald and G. H. F. Gardner

The Seismic Acoustics Laboratory (SAL), a research laboratory at the University of Houston, has been in existence since 1977. The research at this laboratory is aimed at improving the technique of reflection seismology. To this end, it is supported by a consortium of companies, and the results of research are published in half-yearly progress reviews (Table 1.1). By agreement with the sponsors, results are restricted to use only by members of the consortium for six months after first disclosure. After six months, the results are available for publication in the open literature.

It was believed that some of the papers in these semiannual reviews deserved wider dissemination. In all cases, the authors of these papers are now busy professional geophysicists who do not have the time to prepare their papers for the geophysical literature. Therefore, we undertook to select some papers that appeared in the first four years of the SAL (1977–1981), illustrating some of the diverse research benefits that can result from using physical models as an aid to reflection seismology.

The objective of research at the SAL is to investigate seismic data acquisition, processing, and interpretation in order to improve the three-dimensional (3-D) acoustic image of hydrocarbon-bearing rocks. To facilitate this research, a physical seismic-modeling tank was built and theoretical inverse modeling and interpretational techniques have been developed. These papers provide a glimpse of what can be achieved with physical and numerical modeling. The seven papers included in this volume are varied, and yet they give only a suggestion of the possibilities.

Owusu and Gardner demonstrate, for example, that by using a two-step 3-D migration technique, velocities may be determined that are not so dependent on the geometry of reflectors. In a second paper, Owusu and Gardner extend their analysis to one that eliminates both cross-line and in-line dip and curvature effects. Smith and Hilterman discuss diffraction effects that cause waveform distortion within a common-depth-point (CDP) gather. Elsewhere, Smith investigates the use of vertical seismic profiling (VSP) in a modeled oil field and compares it with surface seismic data. Morgan and Hilterman demonstrate the extraction of three-dimensional model parameters by frequency domain imaging.

Table 1.1. *Progress review publication dates*

Volume number	Publication	Month	Year
1	First Year Semi-Annual	March	1978
2	First Year Annual	November	1978
3	Second Year Semi-Annual	May	1979
4	Second Year Annual	November	1979
5	Third Year Semi-Annual	May	1980
6	Third Year Annual	November	1980
7	Fourth Year Semi-Annual	May	1981
8	Fourth Year Annual	November	1981
9	Fifth Year Semi-Annual	May	1982
10	Fifth Year Annual	October	1982

Note: The papers in this volume were published in May 1981 or before.

Chou and associates introduce the concept of "snapshots" for examining wave fields, which have become a continuing research program at SAL under the guidance of Dan Kosloff. And, in an early paper, Duffy makes extensive use of physical modeling in developing exploration methods for coal seams.

These papers present some of the applications of physical and numerical modeling of seismic reflection data; many more are possible, and the results of other research will be published in subsequent volumes.

HISTORICAL VIEW. The use of physical models to study seismic waves dates back almost 60 years. Terada and Tsuboi (1927) and Tsuboi (1927) made a model out of agar-agar, a colloidal gel. They used an electromagnetic chirp as a source and examined such matters as the variation in seismic wave amplitude with depth and the effects of faults and channels on the propagation paths. The second paper (Tsuboi, 1927) studied the dispersion of Rayleigh waves in a two-layer model and the further effects of channels (canals) on the propagation paths.

In the 1950s, several physical modeling systems and experiments were reported. Kaufman and Roever (1951) were probably the first, and they were followed by Northwood and Anderson (1953), Howes, Tejada-Flores, and Randolph (1953), Oliver, Press, and Ewing (1954), Evans et al. (1954), Levin and Hibbard (1955), Clay and McNeil (1955), Hall (1956), Carabelli and Folicaldi (1957), Evans (1959) and Angona (1960), among others.

Kaufman and Roever (1951) set up an apparatus to study transient waves under laboratory conditions. Their results refuted some of the observations by workers studying data from earthquakes and atomic explosions; and Kaufman and Roever identified reflected, refracted, and surface waves over a simple wax model.

Northwood and Anderson (1953) and Howes, Tejada-Flores, and Randolph (1953) went to the extent of using rocklike materials—in one case, concrete, in the other, limestone. Although they used a spark as a source, like Kaufman and Roever, Howes, Tejada-Flores, and Randolph immersed their model in a tank and used water as the propagating medium. Oliver, Press, and Ewing (1954) favored two-dimensional (2-D) models for various reasons, including cost. Their experimental needs, however, were mainly met by discs, as they were studying surface waves. Healy and Press (1960) extended the equipment of Oliver, Press, and Ewing to models in which the velocity and density varied with depth and Rayleigh wave dispersion could be measured.

In the papers of Evans et al. (1954) and Levin and Hibbard (1955) came a realization that by properly scaling a model experiment, wave processes in the real world could effectively be studied. Evans et al. quickly saw the problems that the physical sizes of the source and receiver caused to this kind of scaling. Despite using only a simple two-layer model, Levin and Hibbard (1955) produced surprisingly complex seismograms. Clay and McNeil (1955) compared their results to theoretical solutions with satisfactory results.

Although all the papers mentioned so far were involved with compressional or surface waves, Evans (1959) demonstrated that SH-waves can be generated in a model and that they will produce simple and clear reflections.

Perhaps the first paper describing a model laboratory designed specifically to solve exploration problems is by Angona (1960). In this paper, two-dimensional models were used that had one or more different elastic materials in any desired geometry. The materials employed were plexiglas, copper, aluminum, and steel and they were bonded with epoxy resin. Variations in velocity were achieved by bonding thin sheets of the different materials together; their respective thicknesses allowed models to be made with composite velocities ranging over all the velocities of the materials. This model laboratory is now under the direction of the Dallas Geophysical Laboratory at Southern Methodist University.

Angona's experiments were expanded by Harper (1965), who used models with recesses and extrusions with vertical dimensions less than the wavelength of the pulses. These additions represented simple faults, and some effort was made to explain the resultant seismograms.

In the 1960s, some geophysicists (for example, Silverman, 1969) began to express an interest in holography as a substitute for or addendum to the seismic reflection method. For clarity, the similarities and differences between the methods should be explained. The seismic reflection method depends on the reflection of waves at interfaces between rocks; holography relies also on the scattering of waves at interfaces. The reflection source is impulsive and the holographic source continuous. In theory, holography should be able to produce a three-dimensional image of the subsurface, but in the late 1960s the reflection image was merely a profile. Silverman (1969) proposed the development of "earth holography" and, for demonstration purposes, used a model tank very similar to the one at the SAL. This paper documented an earlier presentation by Farr (1968). Both papers showed that an image can be created of an object immersed in a fluid by the use of coherent sound waves.

The holographic technique did not gain acceptance, and some of the reasons were explained by French, Marcoux, and Matzuk (1973). As they explained it, the holography technique for seismic exploration must be modified to include an aplanatic approach, multiple frequencies, and multiple sources. To illustrate their points, French, Marcoux, and Matzuk used the results from physical models.

After much thought, they had decided that one satisfactory way to study reflected compressional waves was to place a model in a water tank and pass a source and receiver above it, using the water as a propagating medium. Although totally solid systems would have allowed the study of shear and other waves, the inflexibility of such systems ruled them out; the use of air as a propagating medium would have made scaling to real dimensions impossible. French, Marcoux, and Matzuk also recognized that the early experimenters were handicapped because their data were recorded in an analog manner. The aim of the French-Marcoux-Matzuk model system was to produce digital seismic data. The early digitizing system had limitations, and only one sample per trace per shot could be recorded at one time. A trace with 1000 samples, for example, would require 1000 repetitions of the source; thus, experiments were slow. Nevertheless, trace sequential seismic digital data were produced.

As we have noted, the French-Marcoux-Matzuk model system was designed expressly to study the reflection of seismic compressional waves from structures. The source, the receiver, and the model were placed in a water tank that measured three feet on a side. The tank was lined with thick carpet to reduce extraneous reflections from the sides. To reduce spurious reflections as well, the models were supported by fine wires about midway in the tank (Fig. 1.1). In this initial arrangement, the source and receiver were placed at a constant source-receiver (CSR) offset, and they scanned across the model in the y-direction, creating a seismic profile; the x-direction scanner was used only to change the plane of the profile.

As was pointed out by Hall (1956) a reduction in scale to model dimensions required a change in the frequency content of the source. A reasonable reduction in dimensions of 1000:1 (that is, making 1 model foot equivalent to 1000 real feet) would require the working frequencies of reflection seismology (5–100 Hz) to be scaled to 5–100 kHz. This necessitated the use of high-frequency sources. Despite some attempts to use other devices (for example, Clay and McNeil, 1955), sources and receivers making use of the piezoelectric effect are almost universal in seismic modeling. In these transducers, the crystal is pulsed by the data acquisition equipment, and the transmitter generates a source pulse (Fig. 1.2). In the French-Marcoux-Matzuk system, one sample of the return pulse per shot was acquired and stored. A time-sequential digital trace was constructed by the

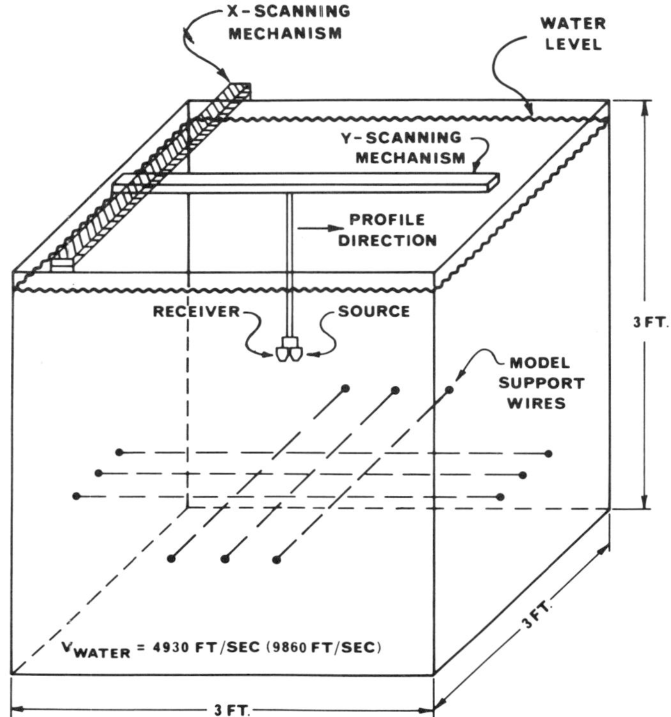

Source: French, 1974.

Figure 1.1. *Sketch of the original physical model tank at Gulf Research.*

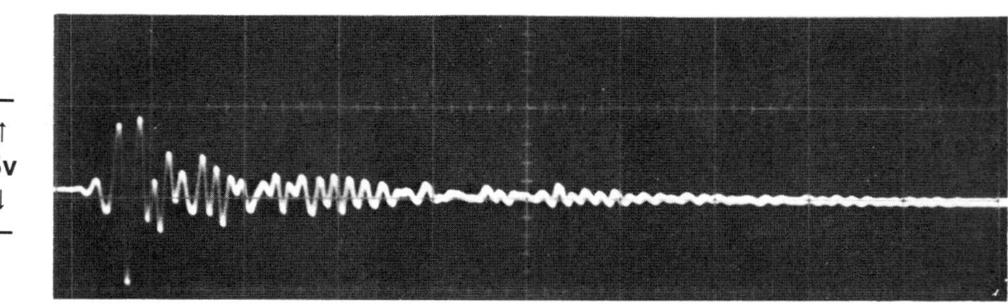

Figure 1.2. *A typical source pulse generated by a piezoelectric transducer in a water tank. In this system, 0.2 μs was equivalent to 1 ms (see Table 1.6).*

Figure 1.3. *The original three-dimensional RTV physical model designed and built at Gulf Research by W.S. French and R.I. Morris.*

computer. In this manner, traces suitable for conventional seismic digital processing were acquired.

The emphasis in physical model research in the mid-1970s was the study of single-interface geologic systems. The energy from the source traveled through water, was reflected back from a single water-model interface to the receiver, and was recorded. The object of the earliest experiments was to image the interface correctly, and therefore such subtleties as the nature of the seismic pulse were disregarded.

Another matter for intense investigation was model-building materials. Many materials were examined for suitabilities and for physical properties; they ranged from brass to paraffin wax to various commercially available plastics. It was eventually found that the most suitable materials for building these single-interface models are room-temperature-vulcanizing (RTV) silicone rubbers. They possess the main seismic property—a high reflectivity to compressional waves when in contact with water. Also, they are easily molded, flowing readily into inaccessible places. Depending on the catalyst used, the vulcanizing (curing) period can be varied from a few minutes to a few hours.

These early experiments at Gulf lead to several important papers, two of which (French, 1974; French, 1975) showed that true three-dimensional seismic data imaging was possible and, moreover, that it provided the correct result in a structural interpretation problem. In Figure 1.3 we show what is now known universally as the French model, constructed by R. I. Morris out of silicone rubber at Gulf Research in 1972. Figure 1.4 shows a series of seismic profiles along line 7 in Figure 1.3. Figure 1.4a shows the unmigrated data; Figure 1.4b presents the two-dimensionally migrated profile, still showing sideswipe from the second dome; and Figure 1.4c is the three-dimensionally migrated profile without sideswipe. The first dome and the fault are correctly positioned, as is shown by the outline.

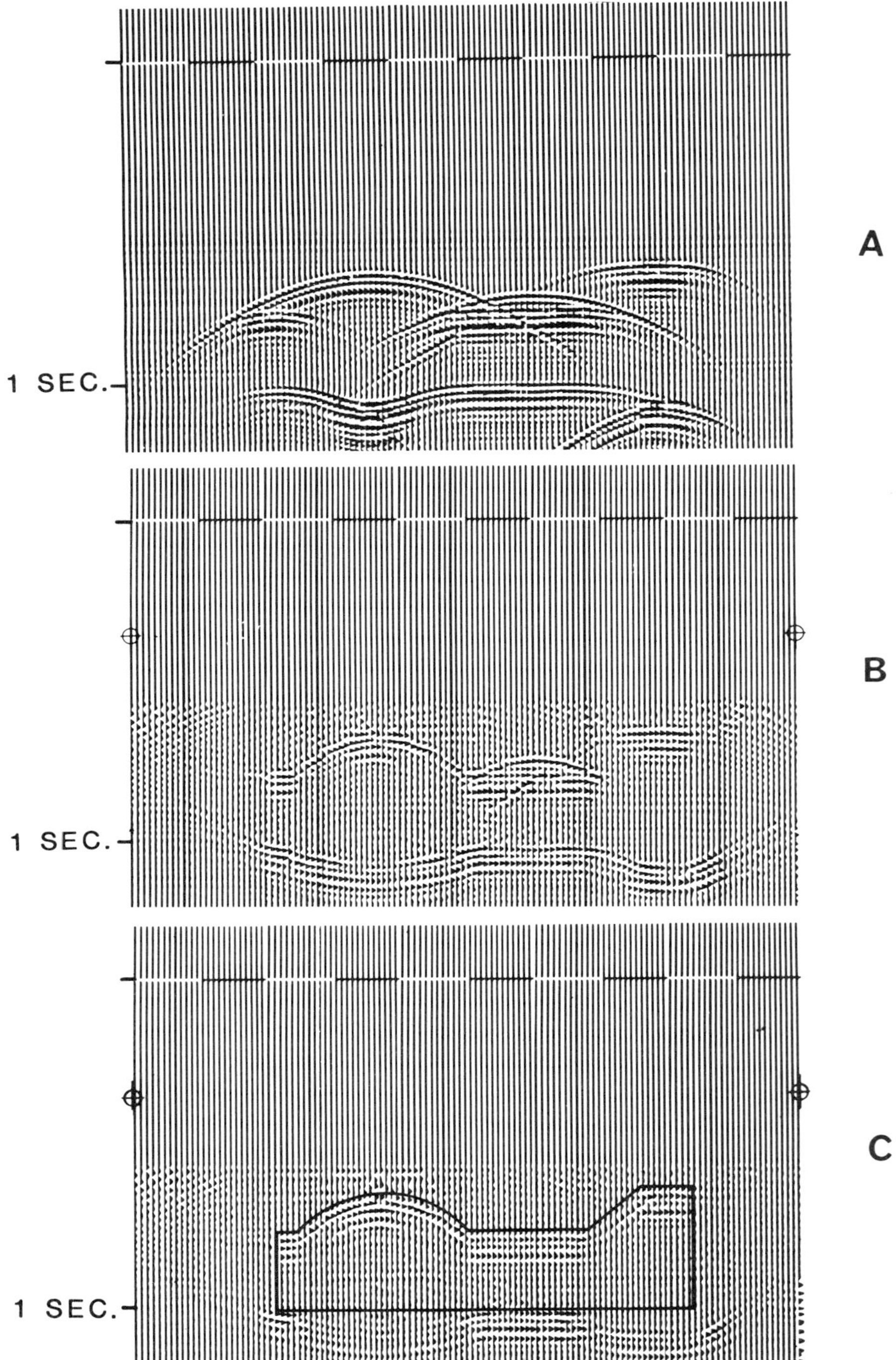

Figure 1.4. *A seismic profile along line 7 in Figure 1.3: (A) the unmigrated data as collected; (B) the data after migration in the plane of the profile; (C) the three-dimensionally migrated data with the shape of profile 7 superimposed.*

Table 1.2. *Directors of the SAL*

Years	Department of Geosciences	Department of Electrical Engineering
1977–1978	F. J. Hilterman	K. Y. Wang
1978–1979	F. J. Hilterman	K. Y. Wang
1979–1980	F. J. Hilterman	G. H. F. Gardner
1980–1981	F. J. Hilterman J. A. McDonald	G. H. F. Gardner
1981–1982	J. A. McDonald	G. H. F. Gardner
1982–1983	J. A. McDonald	G. H. F. Gardner

It was this experiment that showed immediately the value of physical model data. It was possible to collect model data in the same digital formats as field data and to use the model data to develop sophisticated processing techniques. In this experiment, the processing technique was shown to work because the answer was correct. If a similar experiment had been carried out in the field, the data collection costs would have been many times greater and the final result much less clear; the processing technique would probably have remained unproved.

ESTABLISHMENT OF SAL. In 1977 F. J. Hilterman of the Department of Geology at the University of Houston correctly judged that the time was right to make the rest of the industry aware of the benefits of physical modeling. Hilterman had earlier carried out extensive research in numerical modeling (Hilterman, 1970).

With the help of the management of Gulf Research and Development Company, and in collaboration with K. Y. Wang of the Department of Electrical Engineering at the university, Hilterman proposed that about thirteen companies form a consortium. The purpose of the consortium was to support a research project in the use of physical modeling for solving problems in three-dimensional seismic data acquisition, processing, and interpretation. The consortium was proposed in June 1977, and the project began in October. By the end of the first year, 26 companies were supporting the project, at a cost of $10,000 per company.

The story of the early years of the SAL, some of them tumultuous, has been told in detail elsewhere (Taylor, 1982). The management of the laboratory has remained a joint venture between the Departments of Geosciences and Electrical Engineering at the University of Houston. The directorships have undergone three changes (Table 1.2), but the support from industry continues to increase (Figure 1.5).

PHYSICAL FACILITIES AT SAL. The water tank in the system at the SAL is large (10 ft × 8 ft × 5 ft) and contains 12 tons of water. The basic principles are the same as the French-Marcoux-Matzuk design, but there are differences in detail, which will be outlined. Figure 1.6 shows a view of the whole system.

Model Support System. There are two model support systems, depending on the weight of the models to be studied. For light models (that is, up to several tens of pounds), the system is such that the model can be rotated to a preferred direction and tilted to a preferred dip. For larger models (several hundreds of pounds), a three-legged adjustable support rests on the floor of the tank. As the model is only supported at five very small points, extraneous reflections are kept to a minimum.

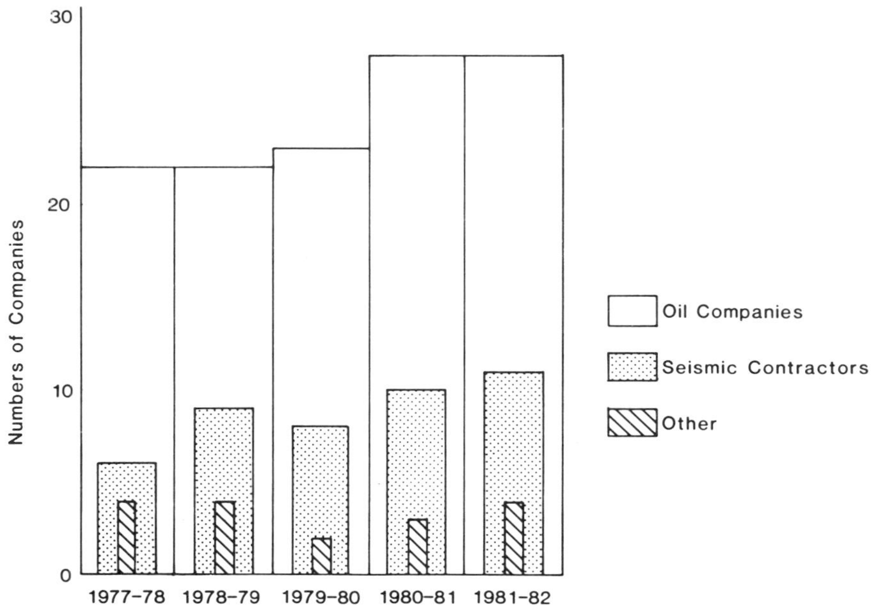

Figure 1.5. *Increasing support for SAL over five years.*

Figure 1.6. *A view of the water tank at the SAL, showing the Wang plotters that carry the source and receiver transducers. The opening in the top of the tank (center) is about 50 in. × 30 in., which scales to 50,000 ft × 30,000 ft (see Table 1.7), but the area actually covered by the transducers is somewhat smaller.*

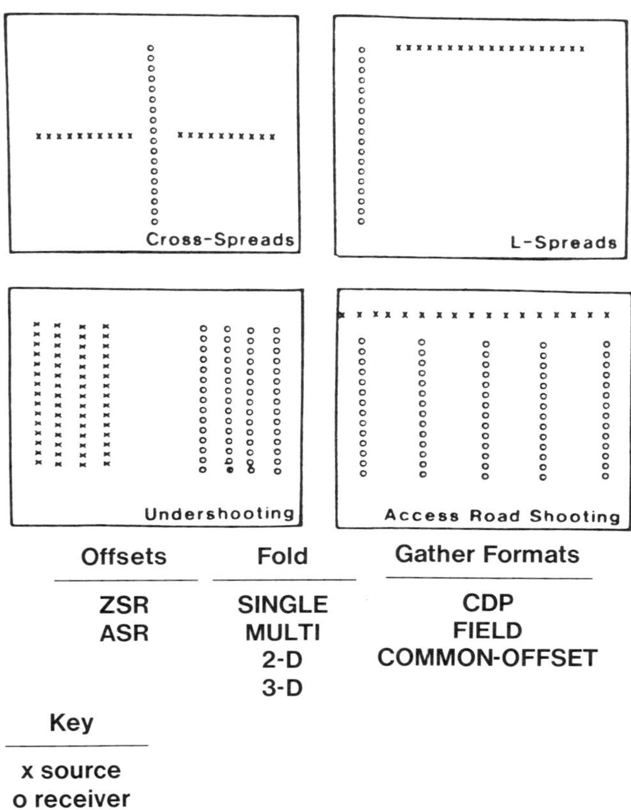

Figure 1.7. *Various source-receiver geometrics may be obtained by computer-controlled movement of the transducers.*

Transducers and Plotter Drives. The system consists of one source and one receiver. In order to simulate normal field geophysical systems, the receiver has to be moved a certain number of times (such as 96) as the source is fired repetitively in the same place. After 96 repetitions, a full 96-trace field record would have been simulated. The source is then moved to a second location and the whole process repeated. In this manner, a towed marine seismic cable is simulated.

Many other source-receiver geometries may be obtained by computer-controlled movement of the transducers (Fig. 1.7). To achieve these movements, the source and receiver are mounted on vertical rods attached to Wang plotters, which move in the horizontal plane at the top of the tank. The vertical rods on which the source and receiver are mounted allow them to be lowered to a suitable plane at a known distance above the model being investigated.

The transducers used as sources throughout these experiments are dependent on the reverse piezoelectric effect; that is, if a suitable crystal is placed in an electric field, it will deform elastically. If one face of the crystal can be kept fixed (with backing), the other will oscillate generating waves in the fluid with which it is in contact. The receivers, of course, depend on the piezoelectric effect—an elastic deformation of the crystal developing an electric field.

The natural frequency f of a freely oscillating crystal depends on its thickness d in the direction of oscillation and is given by

$$f = \frac{1}{2d}\sqrt{\frac{Y}{\rho}},$$

where Y is the Young's modulus of the crystal and ρ is its density. The value of f in this equation is modified by the backing to the crystal.

Equation (1.1) implies that the thinner the crystal, the higher the frequency and hence the smaller the model that can be studied. Larger crystals are needed, however, to increase the energy in the transmitted signal, and compromises usually have to be made. Crystals can be made that have natural frequencies in the megahertz range, but workers at the SAL have found crystals with frequencies in the 100–350 kHz range to be most suitable. These frequencies scale most easily, as will be described later.

Research into suitable transducers is ongoing. Various types of point sources, focused sources, and directional sources have been tried in addition to various types of receivers. The basic object has been to produce a pointlike source that nearly simulates dynamite and to produce a receiver that is not much bigger than a conventional array of geophones. The experiments described in this book generally were unconcerned with source signatures and receiver distortions, and the nature of the source and receiver used in particular experiments often went unrecorded.

The transducers currently in use at the SAL are International Transducer Corporation's ITC-1089, with an operating range of 1 Hz–350 kHz, Celesco's LC-5-2 (1 Hz–600 kHz), and Celesco's LC-10 (0.1 Hz–120 kHz). The last two are usually used as receivers and the first as a source. Despite their careful manufacture, such transducers are often not symmetrical, as can be seen in Figure 1.8. An LC-10 was operated as a source within and above its nominal operating range, with an LC-5-2 as a receiver. One can see that the spectral content of the pulse is a function of azimuth, and subtle variations found in recorded reflections using these transducers may be a function of the transducers, not of the reflecting media.

Data-Acquisition System. The data-acquisition system at the SAL is basically a Petty-Ray ComMand Geophysical Processing System, with front-end additions to interface with the transducers. The acquisition system pulses the source transducer to generate the source pulse. It then amplifies the return signal, providing an adjustable delay time until the start of digitizing. It also provides a variable-gain function, recording up to 4096 samples of the reflected signal, and delivers the samples to the minicomputer on demand. Figure 1.9 shows the data path and control paths.

The pulser/receiver (P/R) has a dual purpose. First, it provides the short-duration, high-voltage pulse for the source transducer. The energy in the pulse is adjustable in four steps. The trigger signal for this pulse is generated by the computer. Second, the P/R contains a preamplifier that provides from 40 to 60 dB of gain for the reflected return signal. This return signal may be acquired by the source transducer, giving a zero-offset condition, or, more normally, by a separate receiver transducer. The P/R has provisions for either method. It also has an adjustable high-pass filter.

The programmable gain amplifier (PGA) provides delay times, variable-gain functions, and final-gain functions, all of which are selectable and are controlled by the computer. Since a limit of 4096 samples may be retained for any trace, the delay time allows the gain function to be moved to the window being recorded.

Provisions are included for six gain functions. Each has a different rate of gain change from the starting gain to the final gain; the rate is determined by the time

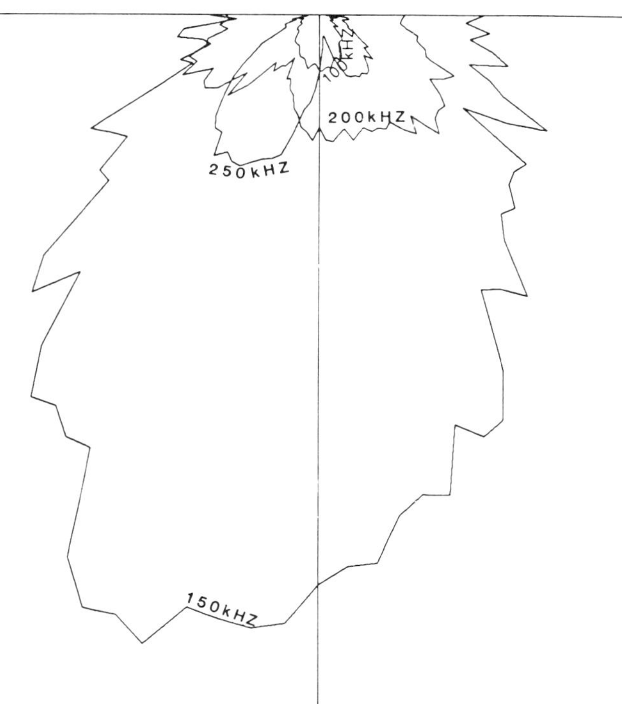

Figure 1.8. *The asymmetric radiation pattern for a pair of piezoelectric transducers acting as source and receiver. In this case the source was operating above its normal operating range.*

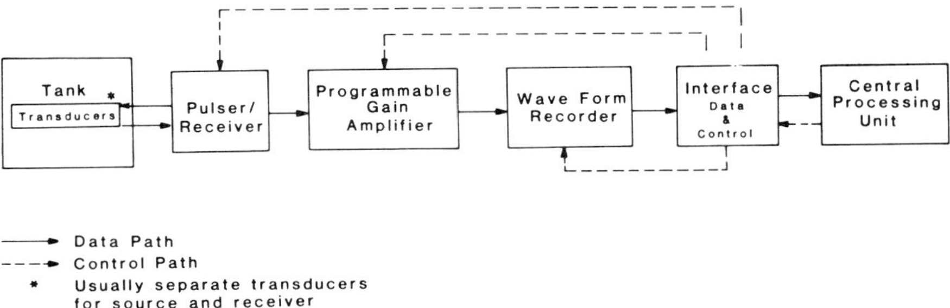

Figure 1.9. *Data & control paths for the SAL physical model acquisition system.*

constant of a resistor and capacitor. Thus, the gain function is merely the increasing exponential time function of the voltage across the capacitor when the series capacitor is charged through a resistor.

The final gain can be selected in 6 dB steps from unity to 36 dB. Resistors are switched in the feedback circuit of a wideband instrumentation amplifier to control the gain. Gain-select signals from the computer are decoded and used to control the relays that actually make or break the resistor circuits.

The waveform recorder (WR) is a very fast 10-bit analog-to-digital converter and a 4096-word memory. The seismic signal can be sampled at an interval as small as

0.1 μs and digitized with a resolution of 1 part in 1024. The trigger signal to begin recording the waveform is generated by the computer. The WR has interval-triggering delay circuits that permit the recorded portion of the waveform to be moved, as a window, up or down the trace.

The recorded waveform may be displayed in analog form on an oscilloscope, using an interval fast-repetition-rate time base, or on an *X-Y* plotter, with the time base adjusted for a very slow rate. At the digital output, the sample word appears in a 10-bit parallel format. It is transferred in a word-serial, asynchronous fashion under commands from the computer.

The interface at the SAL was custom designed by an outside firm for the Petty-Ray ComMand system. The interface provides trigger signals to the P/R, the PGA, and the WR. The PGA also receives final gain and gain function signals from the interface. The WR accepts a command signal from the interface for transfer of the digital word. The WR sends a flag signal and the 10 bits of parallel data to the interface. After these manipulations, traces are recorded on tape on the ComMand system.

Model Materials and Model Building. As we mentioned earlier, many materials were examined for possible use as models. The early physical models, such as those used in the experiments described in this volume, were merely single interfaces. All that was required was that the reflection coefficient at the water/material interface be sufficiently large that a significant signal could be detected by the receiver. The most suitable materials found include room-temperature-vulcanizing (RTV) silicone rubbers, reaction-injection-molding (RIM) epoxy resins, and Plexiglas. These materials are always immersed in water, which is the medium in which the energy is generated and received. The water in the SAL tank is constantly filtered, of course, to prevent buildup of algae and bacteria and is maintained at a constant temperature. The appropriate physical properties of the materials in current use are given in Table 1.3.

The reflection coefficient at an interface between two materials of compressional velocities v_1 and v_2 and densities ρ_1 and ρ_2 is given by

$$\frac{v_2\rho_2 - v_1\rho_1}{v_2\rho_2 + v_1\rho_1}.$$

The reflection coefficients for combinations of the most commonly used materials have been calculated and are presented in Table 1.4.

For a typical simple model, a molding box is used to model the geologic interface in clay (Fig. 1.10). After the clay is set, a plaster of paris cast is poured. Imperfections are removed from this cast, and it is used as the final mold for the silicone rubber. The rubber and its catalyst are mixed under a partial vacuum to prevent entrapment of any bubbles in the fluid. Similarly, bubbles are avoided when the rubber is poured into the mold; such bubbles would act as very effective point scatterers. Much more complex multi-interface models are now being built. Their construction will be described in later papers.

Scaling. The concepts of scaling for physical models have been described in detail by White (1965) and will only be mentioned briefly here. In essence, every dimension in the original field experiment is directly proportional to a corresponding dimension on the model. For example, if the scale factor for distance x is L,

$$x_{\text{original}} = Lx_{\text{model}}.$$

Table 1.3. *Physical properties of typical model materials based on measurements with the modeling system.*

Material	Velocity, V(ft/s)	Scaled Tank Velocity (ft/s)	Density, ρ(lb/cu ft)	Acoustic Impedance, ($\rho V \times 10^3$)	Color	Type
Water at 65°F	4847	11633	62.4	303	Clear	Water
Slygard 170	3197	7673	83.0	265	Black	Silicone rubber
Slygard 184	3410	8184	68.0	232	Clear	Silicone rubber
Slygard 3110	3278	7867	76.0	249	White	Silicone rubber
Slygard 3120	2995	7188	90.0	270	Red	Silicone rubber
Plexiglas	9000	21600	73.0	657	Clear	Plexiglas
Stycast 1266	8094	19426	78.0	631	Clear/Brown	Epoxy resin
Stycast 3180	9157	21977	106.0	971	Black	Epoxy resin
Stycast 2741LV	7727	18545	85.0	657	Black	Epoxy resin
Stycast 1265	5623	13495	69.0	388	Clear/White	Epoxy resin
CPC 19	4218	10123	81.0	342	Pinkish Yellow	Polyurethane
RIM III	4924	11818	62.0	305	Brown	Epoxy resin
RIM I	5167	12401	70.0	362	Clear/Tan	Epoxy resin
RIM IV	5700	13680	57.0	325	Light Brown	Epoxy resin

Table 1.4. *Reflection coefficients between common model materials*

Medium of Incidence \ Reflecting Medium	Water at 65°F	Slygard 170	Slygard 184	Slygard 3110	Slygard 3120	Plexi-glas	Stycast 1266	Stycast 3180	Stycast 2714LV	Stycast 1265	CPC 19	RIM III	RIM I	RIM IV
Water at 65°F														
Slygard 170	0.067													
Slygard 184	0.133	0.066												
Slygard 3110	0.098	0.031	−0.035											
Slygard 3120	0.058	−0.009	−0.076	−0.040										
Plexiglas	−0.369	−0.425	−0.478	−0.450	−0.417									
Stycast 1266	−0.351	−0.408	−0.462	−0.434	−0.401	0.020								
Stycast 3180	−0.524	−0.571	−0.614	−0.592	−0.565	−0.193	−0.212							
Stycast 2741LV	−0.369	−0.425	−0.478	−0.450	−0.417	0.000	−0.020	0.193						
Stycast 1265	−0.123	−0.188	−0.252	−0.218	−0.179	0.257	0.238	0.429	0.257					
CPC 19	−0.060	−0.127	−0.192	−0.157	−0.118	0.315	0.297	0.479	0.315	0.064				
RIM III	−0.003	−0.070	−0.136	−0.101	−0.061	0.348	0.348	0.522	0.366	0.120	0.057			
RIM I	−0.089	−0.155	−0.219	−0.185	−0.146	0.289	0.271	0.457	0.289	0.035	−0.028	−0.085		
RIM IV	−0.035	−0.102	−0.167	−0.132	−0.092	0.338	0.320	0.5	0.338	0.088	0.025	−0.032	0.054	

Figure 1.10. *A collapsible aluminum mold used to cast geologic interfaces in clay. A typical mold would be a 10 in. cube.*

Similarly, for time t, if the scale factor is T,

$$t_{\text{original}} = Tt_{\text{model}}.$$

General scale factors for various quantities are given in Table 1.5. When the elastic moduli of the media have to be considered, a scaling factor for mass m also has to be defined. This is expressed as

$$m_{\text{original}} = Mm_{\text{model}}.$$

One of the strengths of the SAL modeling system is that users are able to conceive their experiments in real dimensions. This is because the proportions of

Table 1.5. General Scale Factor

Quantity	Symbol	General Scale Factor
Distance	x, y, z	L
Time	t	T
Mass	m	M
Velocity	v	LT^{-1}
Stress	σ_{xx}, σ_{yz}, etc	$ML^{-1}T^{-2}$
Elastic moduli	μ, Y, etc	$ML^{-1}T^{-2}$
Poisson's ratio	σ	1
Density	ρ	ML^{-3}

Table 1.6. *Physical model dimensions and scaling factors*

	Dimensionless		
	Model	Original/Oil	Original/Coal
Length	1	12000	3600
Time	1	5000	2500
Velocity	1	2.4	1.44
Frequency	5000	1	2
	Common dimensions		
	Model	Original/Oil	Original/Coal
Length	1 in.	1000 ft	300 ft
Time	0.2 μs	1 ms	0.5 ms
Velocity	5000 ft/s	12000 ft/s	7160 ft/s
Frequency	250 kHz	50 Hz	100 Hz
Smallest x or y movement	1 tank unit	10 ft	3 ft
Smallest z movement	1 tank unit	1 ft	0.3 ft

the wavelengths of the acoustic waves to the dimensions of the geologic structures are retained as they occur in the field. For example, a sedimentary layer with an acoustic velocity of 10,000 ft/s would transmit a 50 Hz pulse with a wavelength of 200 ft. A convenient reduction in dimensions is 1 in. = 1000 ft; that is a scale of 1:12,000. Similarly, a typical piezoelectric transducer operating at 250 kHz would scale by a factor of 5000:1 to 50 Hz. As the dimensions of velocity are LT^{-1}, its scaling factor is 2.4.

Values of dimensions, scales, and scaling factors in use at the SAL are given in Table 1.6.

DEVELOPMENT OF THE ALLIED GEOPHYSICAL LABORATORIES. As the SAL has matured, so have the interests of its researchers, and their horizons have broadened. Two of the newer topics—numerical modeling and three-dimensional seismic data processing—demanded state-of-the-art equipment for adequate investigation. Numerical modeling required a modern computer and, following a donation from the Keck Foundation, a VAX computer was purchased. The technique of three-dimensional or areal seismology produces a volume of information that is inadequately examined in two dimensions. Some method of processing and interpreting the data in three dimensions was required. This led to the purchase of the Adage computer after the Cullen Foundation had also made a donation to the SAL.

The university administration and the SAL researchers felt that exploration geophysics meant more than just physical modeling; therefore, in 1981, the Research Computation Laboratory (RCL) and the Image Processing Laboratory (IPL) were founded. In addition, in order to broaden the scope to real field problems, the Field Research Laboratory (FRL) was established.

The purpose of a university is to educate its students, and, particularly for graduate students, this means involving them in research. With these four laboratories, students can now become involved in many aspects of reflection seismology in exploration geophysics, from data-acquisition problems to those of final interpretaton. So far as we are aware, this is a unique concept in any university. In addition to geophysicists, students of electrical engineering,

Table 1.7. Members of the consortium supporting the SAL

Amoco Production Company	ICI Petroleum Services Limited
Aramco	Japan National
Arco Oil and Gas	Marathon Oil Company
BP International Limited	Mobil R&D
Chevron Oil Field Research	Norsk Hydro
Cities Service Company	Pennzoil Company
Conoco, Incorporated	Petrobras
Control Data Corporation	Petty-Ray Geophysical
Digicon Geophysical Corporation	Phillips Petroleum Company
Exxon Production Research	Prakla-Seismos GmbH
Fairfield Industries	Seiscom Delta, Incorporated
GECO	Seismograph Service Corporation
Geophysical Development Corporation	Shell Development Company
Geophysical Service, Incorporated	Societé National ELF Aquitaine
Geophysical Systems Corporation	Statoil
Geoquest International, Incorporated	Sun Exploration Company
Getty Oil Company	Superior Oil Company
Gulf Research and Development Company	Tenneco Oil Company
Horizon Exploration	Texaco, Incorporated
Hunt Oil Company	Union Oil of California
IBM	Western Geophysical Company

computer science, and mathematics can now be made cognizant of exploration problems.

In 1981, the four laboratories were formally united—under the joint direction of J. A. McDonald and G. H. F. Gardner, representing the Departments of Geosciences and Electrical Engineering, respectively—with the name Allied Geophysical Laboratories (AGL). This organization has rapidly developed into a source of highly trained recruits for industry. Some 18 students have obtained graduate degrees with AGL support, and all are working in energy-related industries. These students include those with degrees in computer science and engineering as well as in geophysics.

ACKNOWLEDGMENTS. Many people have contributed to the success of the SAL and have helped in the formation of the AGL. Principally, Fred Hilterman and Keith Wang had the original inspiration to establish the facility and pursue it through its early development. In recent years, Roice Nelson arrived from Mobil Oil and, with infectious enthusiasm and unbounded energy, solved many problems that others would have found unsurmountable. Nelson played a large part in the foundation of the IPL and is now the general manager of the AGL. Barbara M. Murray was the project administrator during the establishment and development of the new laboratories and has had to oversee the financial functions of all of them.

Outside the laboratories, the support of many must not go unmentioned. Ongoing support from our consortium members has enabled research to continue (members of the SAL consortium are shown in Table 1.7). John C. Butler, chairman of the Department of Geosciences at the University of Houston, has coordinated many fund-raising efforts, and John R. Butler of GeoQuest International, Inc., has helped greatly in acquiring the donations from the Keck and Cullen Foundations.

REFERENCES AND SELECTED BIBLIOGRAPHY

Angona, F. A., 1960, Two-dimensional modeling and its application to seismic problems: Geophysics, v. 25, p. 468–482.

Bennett, A. D., 1962, Study of multiple reflections using a one-dimensional seismic model: Geophysics, v. 27, p. 61–72.

Berckhemer, H., and Ansorge, J., 1963, Wavefront investigations in model seismology: Geophys. Prosp., v. 11, p. 459–470.

Busby, J., and Richardson, E. G., 1957, The absorption of sound in sediments: Geophysics, v. 22, p. 821–828.

Carabelli, E., and Folicaldi R., 1957, Seismic model experiment on thin layers, Geophys. Prosp., v. 5, p. 317–327.

Chowdhury, D. K., and Dehlinger, P., 1963, Elastic wave propagation along layers in two-dimensional models: Bull. Seis. Soc. Am., v. 53, p. 593–618.

Clay, C. S., and McNeil, H., 1955, An amplitude study on a seismic model: Geophysics, v. 20, p. 766–773.

Dampney, C. N. G., Mohanty, B. B., and West, G. F., 1972, A calibrated model seismic system: Geophysics, v. 37, p. 445–455.

Evans, J. F., 1959, Seismic model experiments with shear waves: Geophysics, v. 24, p. 40–48.

Evans, J. F., Hadley, C. F., Eisler, J. D., and Silverman, D., 1954, A three-dimensional seismic wave model with both electrical and visual observation of waves: Geophysics, v. 19, p. 220–236.

Farr, J. B., 1968, Earth holography, a potential new seismic method: Geophysics, v. 33, p. 1046–1047.

French, W. S., 1974, Two-dimensional and three-dimensional migration of model experiment reflection profiles: Geophysics, v. 39, p. 265–277.

———, 1975, Computer migration of oblique reflection profiles: Geophysics, v. 40, p. 961–980.

French, W. S., Marcoux, M. O., and Matzuk, T., 1973, Technical limitations of seismic holography: Geophysics, v. 39, p. 1199.

Gardner, G. H. F., French, W. S., and Matzuk, T., 1974, Elements of migration and velocity analysis: Geophysics, v. 39, p. 811–825.

Goodman, R. E., and Appuhn, R. A., 1966, Model experiments on the earthquake response of soil-filled basins: Bull. Geol. Soc. Am., v. 77, p. 1315–1326.

Gregson, V., 1967, A model study of elastic waves in a layered sphere: Bull. Seis. Soc. Am., v. 57, p. 959–981.

Hall, S. H., 1956, Scale model seismic experiments: Geophys. Prosp. v. 4, p. 348–364.

Harper, D. R., 1965, Observed reflection and diffraction wavelet complexes in two-dimensional seismic model studies of simple faults: Geophysics, v. 30, p. 72–86.

Healy, J. H., and Press, F., 1960, Two-dimensional seismic models with continuously variable velocity depth and density functions: Geophysics, v. 25, p. 987–997.

Hilterman, F. J., 1970, Three-dimensional seismic modeling: Geophysics, v. 35, p. 1020–1037.

Hilterman, F. J., Nelson, H. R., Jr., and Gardner, G. H. F., 1981, Physical modeling: an aid for production geophysicists, Proc. O.T.C., v. 2, p. 139–147.

Howes, E. T., Tejada-Flores, L. H., and Randolph, L., 1953, Seismic model study: J. Acoust. Soc. Am., v. 35, p. 915–921.

Ivakin, B. N., 1960, Methods for controlling the density and elasticity of a medium during the two-dimensional modeling of seismic waves: Bull. Acad. Sci. USSR, Geophys. Ser. (English transl.), No. 8, p. 761–771.

Ivakin, B. N., and Vasilev, Y. V., 1963, The wave properties of perforated plates for seismic modeling: Bull. Acad. Sci. USSR, Geophys. Ser. (English transl.), No. 2, p. 149–156.

Kaufman, S., and Roever, W. L., 1951, Laboratory studies of transient elastic waves: Proc. Third World Petroleum Congress, The Hague, Sec. I, p. 537–545.

Knopoff, L., 1955, Small three-dimensional seismic models: Trans. Am. Geophys. Union, v. 36, p. 1029–1034.

Koefoed, O., van Ewyk, J. G., and Bakker, W. T., 1958, Seismic model experiments concerning reflected refractions: Geophys. Prosp., v. 6, p. 382–393.

Kuo, J. T., and Thompson, G. A., 1963, Model studies on the effect of a sloping interface on Rayleigh waves: J. Geophys. Res. v. 68, p. 6187–6197.

Lavergne, M., 1961, Étude sur modèle and ultrasonique du problème des couches minces en sismique refraction. Geophys. Prosp., v. 9, p. 60–73.

Levin, F. K., and Hibbard, H. C., 1955, Three-dimensional seismic model studies: Geophysics, v. 20, p. 19–32.

McDonald, J. A., Gardner, G. H. F., and Kotcher, J. S., 1981, Areal seismic methods for determining the extent of acoustic discontinuities: Geophysics, v. 46, p. 2–16.

Newman, P., 1980, 3-D Acoustic model experiments: Geophysics, v. 46, p. 416.
Northwood, T. D., and Anderson, D. V., 1953, Model seismology: Bull. Seis. Soc. Am., v. 43, p. 239–245.
O'Brien, P. N. S., 1955, Model Seismology—the critical refraction of elastic waves: Geophysics, v. 20, p. 227–242.
Oliver, J., Press, F., and Ewing, M., 1954, Two-dimensional model seismology: Geophysics, v. 19, p. 202–219.
Press, F., 1957, A seismic model study of the phase velocity method of exploration: Geophysics, v. 22, p. 275–285.
Press, F., Oliver, J., and Ewing, M., 1954, Seismic model study of refractions from a layer of finite thickness: Geophysics, v. 19, p. 388–401.
Rieber, F., 1937, Complex reflection patterns and their geological sources: Geophysics, v. 2, p. 132–160.
Riznichenko, Y. V., 1957, The development of ultrasonic methods in seismology: Bull. Acad. Sci. USSR, Geophys. Ser. (English transl.), v. 11, p. 31–37.
Roever, W. L., Vining, T. F., and Strick, E., 1959, Propagation of elastic wave motion from an impulsive source along a fluid/solid interface. I Experimental pressure response. II Theoretical pressure response. III The pseudo-Rayleigh wave: Phil. Trans. Roy. Soc. London, Ser. A, v. 251, p. 455–523.
Rykunov, L. N., and Feofilaktov, V. D., 1961, A piezoelectric emitter of single-stroke ultrasonic pulses for modeling seismic waves: Bull. Acad. Sci. USSR, Geophys. Ser. (English Trans.), No. 2, p. 131–136.
Sarrafian, G. P., 1956, A marine seismic model: Geophysics, v. 21, p. 320–336.
Silverman, D., 1969, Mapping the Earth with elastic wave holography: IEEE Trans. Geoscience. Electron., v. GE-7, p. 190–199.
Siskind, D. E., and Howell, B. F., 1967, Scale-model study of refraction arrivals in a three-layered structure: Bull. Seis. Soc. Am., v. 57, p. 437–442.
Taylor, G., 1982, Allied Geophysical Labs' story reveals joint success of University, industry: AAPG Explorer, v. 3, p. 1–8.
Teng, T. L., and Wu, F. T., 1968, A two-dimensional ultrasonic model study of compressional and shear-wave diffraction patterns produced by a circular cavity: Bull. Seis. Soc. Am., v. 58, p. 171–178.
Terada, T., and Tsuboi, C., 1927, Experimental studies on elastic waves (Part 1): Bull. Earth. Res. Inst., Tokyo Imperial Univ., v. 3, p. 55–63.
Tsuboi, C., 1927, Experimental studies on elastic waves (Part 2): Bull. Earth. Res. Inst., Tokyo Imperial Univ., v. 4, p. 9–20.
Troitskiy P., Husebye, E. S., and Nikolaev, A., 1981, Lithospheric studies based on holographic principles: Nature, v. 294, p. 618–623.
White, J. E., 1965, Seismic waves: radiation, transmission, and attenuation: New York, McGraw-Hill.
Woods, J. P., 1956, The composition of reflections: Geophysics, v. 21, p. 261–276.
———, 1975, A seismic model using sound waves in air: Geophysics, v. 40, p. 593–607.

2. LIMITATIONS IN THE RECONSTRUCTION OF THREE-DIMENSIONAL SUBSURFACE IMAGES

Thomas A. Smith and Fred J. Hilterman

INTRODUCTION. Seismic reflection data by common-depth-point (CDP) shooting are collected by recording signals where as many as 24 different source and receiver combinations are symmetrically disposed about a point on the earth's surface known as the midpoint. The pairs of sources and receivers are usually laid out along a line and organized so that there is an ordered range of offsets between them. The reflections from subsurface surfaces of acoustic impedance discontinuities give valuable information about the shapes of these surfaces and about the acoustic properties of the materials between the discontinuities.

The migration process in the analysis of these data attempts to construct the geometric arrangement of the reflecting surfaces. This process usually takes place after velocities in the ground have been estimated. The velocity distribution in the earth is used to convert the recorded arbitrary source and receiver (ASR) position data to coincident source and receiver (CSR) position data. The common solution involves the assumption of hyperboloid time moveout surfaces that are dependent on velocity and independent of source-receiver separation. When the reflecting surface dips, a dip-dependent velocity function may be used if the angle between the recording line and the strike of the plane is known everywhere (Levin, 1971).

Since this much information is seldom available a priori, imaging the offset data directly may be preferable. If it may be assumed that the raypaths to and from a point diffractor are straight, the Kirchhoff summation migration may be used to moveout-correct the downgoing and upcoming raypaths separately. Imaging in this manner, which is dip-independent, has been presented elsewhere (Gardner and Kotcher, 1977).

This chapter will discuss the diffraction effects that give rise to waveform distortion within the CDP gather. Experimental results will show amplitude changes within the gather, and synthetic models will show phase changes as well. Gathers of data will be stacked using velocities appropriate for flat and dipping planes. Where recorded reflection data from these planes interfere, we demonstrate that the estimate of a coincident source-receiver wavefield may be questionable.

HALF-PLANE MODEL. Diffraction wave theory using boundary integration was compared numerically with the solid angle approach (Trorey, 1970; Hilterman, 1970, 1975). Theoretically and numerically, they are consistent. Figure 2.1 shows the diffraction response of a half-plane using the boundary integration method. The half-plane is parallel to the observation plane and 2500 ft below it. The edge of the plane is 5000 ft from the origin. The large arrow points toward the time trace where a coincident source-receiver is immediately above the edge. The amplitude of this event is half as large as the reflection from the plane. Positive and negative diffraction arrivals are seen to the right and left of the arrow.

To test boundary diffraction wave computations (SAL Progress Review, 1979) against earlier results, impulse-response traces were computed for the model. These data, convolved with a 48 ms symmetrical wavelet (bandpass 10–40 Hz), are shown in Figure 2.2. A spherical divergence gain correction has been made in both Figures 2.1 and 2.2.

This chapter was previously published in *Acoustical Imaging*, Volume 9 (New York: Plenum Publishing Corporation, 1980), pp. 699–736.

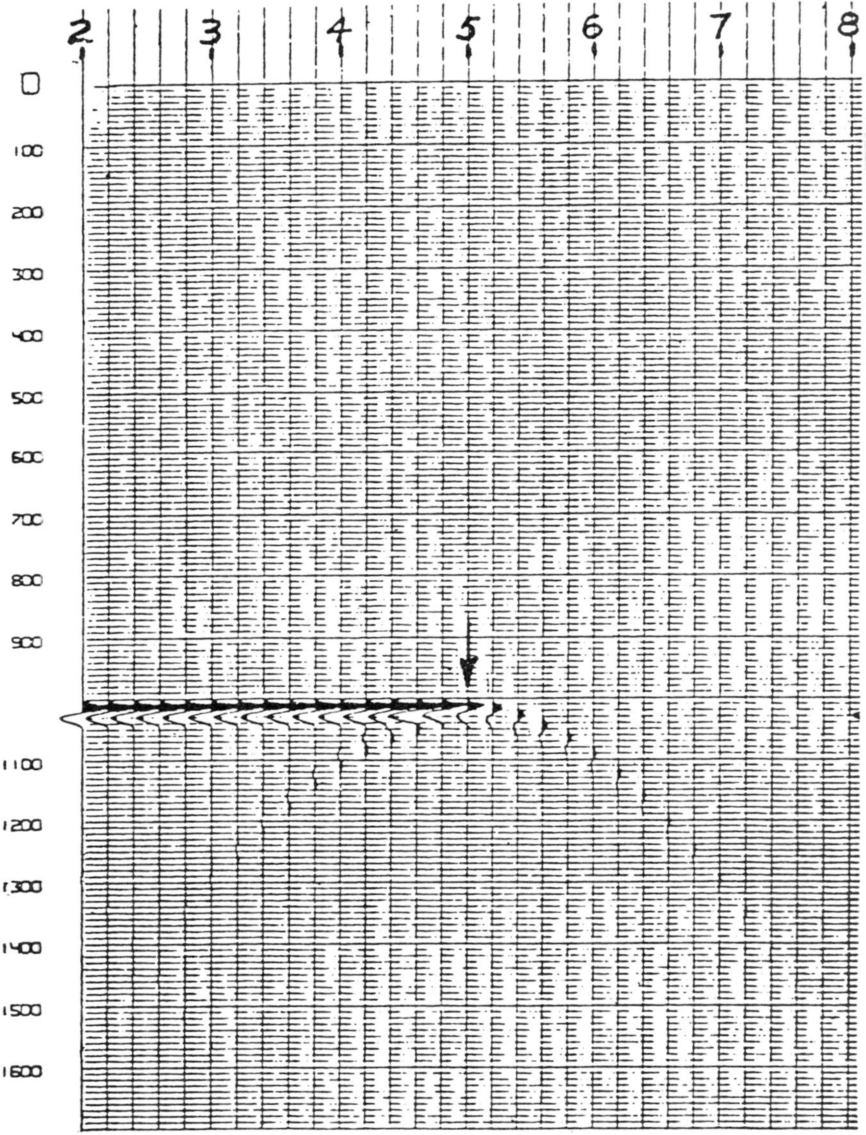

Source: After Trorey, 1970.

Figure 2.1. *Coincident source-receiver diffraction response over a half-plane, using a wide-band wavelet.*

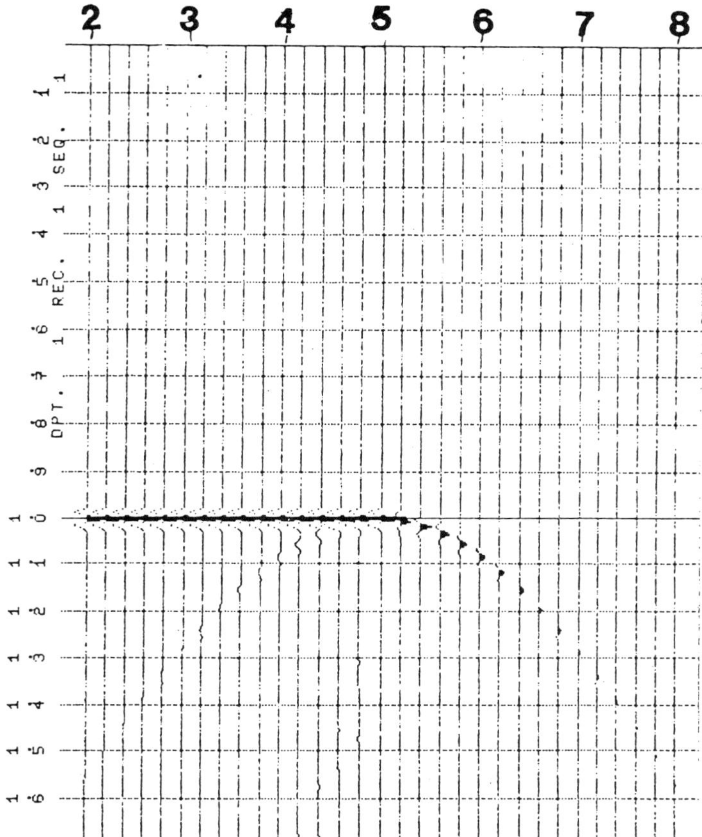

Figure 2.2. *Coincident source-receiver diffraction response, using the Maggi-Rubinowicz line integral and a 10–40 Hz wavelet.*

In Figure 2.3, the source and receiver have been separated by 1250 feet by advancing the receiver location 625 ft and retarding the source location 625 ft in relation to the midpoint coordinate. The specular reflections are delayed slightly because of the extra travelpath length when the source and receiver are separated. The positive and negative diffraction events fall on a curve that is slightly flatter than before and is no longer hyperbolic.

MONOCLINE MODEL. In order to understand the diffractions recorded when source and receiver are separated, a simple two-dimensional monocline model has been tested. The model consists of a flat reflecting plane at a depth of 4000 ft on the left and another flat reflecting plane at a depth of 3650 ft on the right (see Fig. 2.4). The plane connecting the two half-planes has a dip of slightly more than 30°. The velocity of the medium between the observation plane and the model is 10,000 ft/s. The dotted lines on Figure 2.4 encompass the surface location of the specular reflections from the dipping plane.

The CSR diffraction responses are displayed in Figure 2.5. Once again, the impulse response traces have been convolved with the 48 ms symmetric wavelet. The trace spacing is 150 ft, and the synthetic data for this model are plotted at a

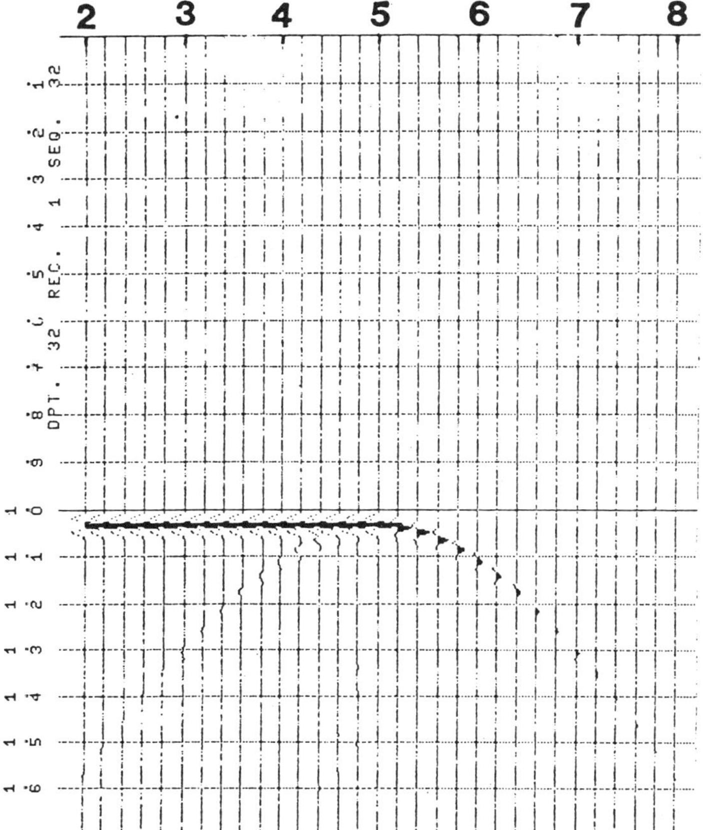

Figure 2.3. *Diffraction responses where the source-receiver separation is 1250 ft.*

fixed gain. The section is strongly controlled by the diffraction responses of the left and right half-planes. There is a noticeable distortion of waveform due to the interaction of these diffractions with the diffractions from the dipping strip. Because of the narrow bandwidth of the wavelet, resolving these superpositional effects is difficult.

These interfering diffraction effects are more easily understood by studying the model impulse response data rather than the filtered data. Although the impulse responses could never be recorded by a physical experiment, they do reveal the mechanism by which real waveforms are distorted. They might also show what data would look like if the wavefield could be perfectly recorded.

The model impulse responses in Figure 2.6 demonstrate that the specular reflections from the dipping plane fall on the three traces centered about the −2000 ft coordinate. For traces with coordinates less than −2500 ft, the positive diffraction arrivals from the dipping strip and the negative diffraction arrivals from the lower flat half-plane may be seen. Traces with coordinates between −1500 and 0 ft have two diffractions (lower pointer) and one reflection. The positive diffraction associated with the upper flat half-plane (upper pointer) is stronger than its negative diffraction counterpart because of the extra positive diffraction contribution from the dipping strip.

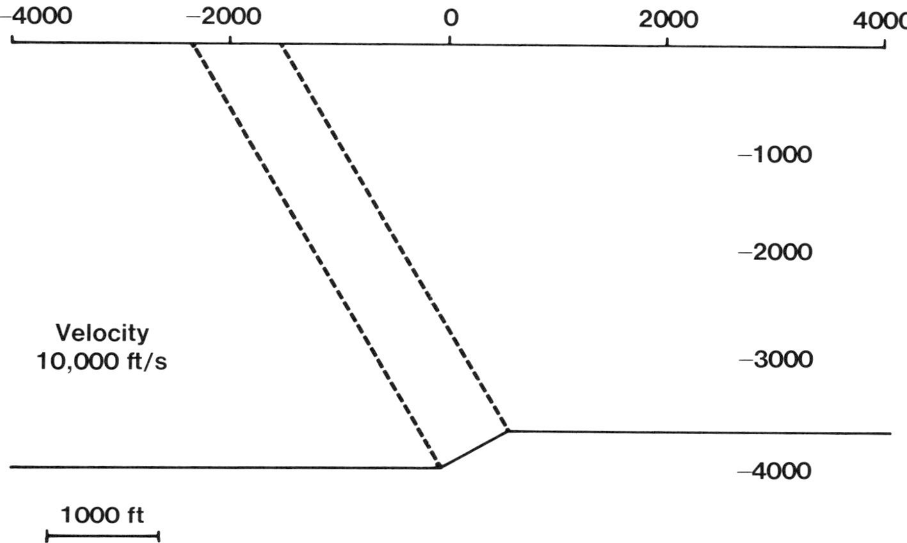

Figure 2.4. *Two-dimensional depth model of a step with a displacement of 350 ft. The dotted lines project the limits of specular reflections to the surface.*

CDP gathers were computed for midpoint coordinates of −1800, 0, and 1800 ft. The source-receiver separation ranged from 0 to 4000 ft. The filtered and impulse response gathers are presented in Figures 2.7 and 2.8. For the most part, the waveforms in Figure 2.7 are uniform in shape and do not appear to vary with offset.

The impulse responses in Figure 2.8 reveal a different story. Although the filtered gather at coordinate −1800 ft looks remarkably uniform, the impulse response diffractions for the lower event in Figure 2.8 (pointer) reveal a significant variation with offset. What appeared to be a specular reflection from the dipping plane is actually a closely spaced interference of the specular reflection, with both a positive diffraction from the upper flat plane edge and a negative diffraction from the lower flat plane edge.

RECTANGULAR PLATE MODEL. A rectangular plate model has been used to investigate the diffraction and specular reflection patterns from a simple three-dimensional body. A horizontal plate with side lengths of 2000 and 4000 ft was positioned 1000 ft below the observation plane. A velocity of 5000 ft/s was used.

In the first set of calculations, four lines of fixed source-receiver separation were chosen. The separation distance was 1250 ft, and the sources and receivers were aligned with the direction of recording. The midpoint spacing was 250 ft, and the line spacing was 1000 ft. The line locations are plotted in Figure 2.9. In Figure 2.10, the impulse responses have been convolved with the symmetric wavelet described previously. A spherical divergence correction has been applied.

On line 1, the only detected diffractions are those caused by the left edge, parallel to the line direction. Line 2 passes over the left edge of the model, and so the reflections from the edge are half-amplitude. The flat events in lines 3 and 4 are reflections from the plate. A hint of a side diffraction from the left edge is visible on line 3 (pointer).

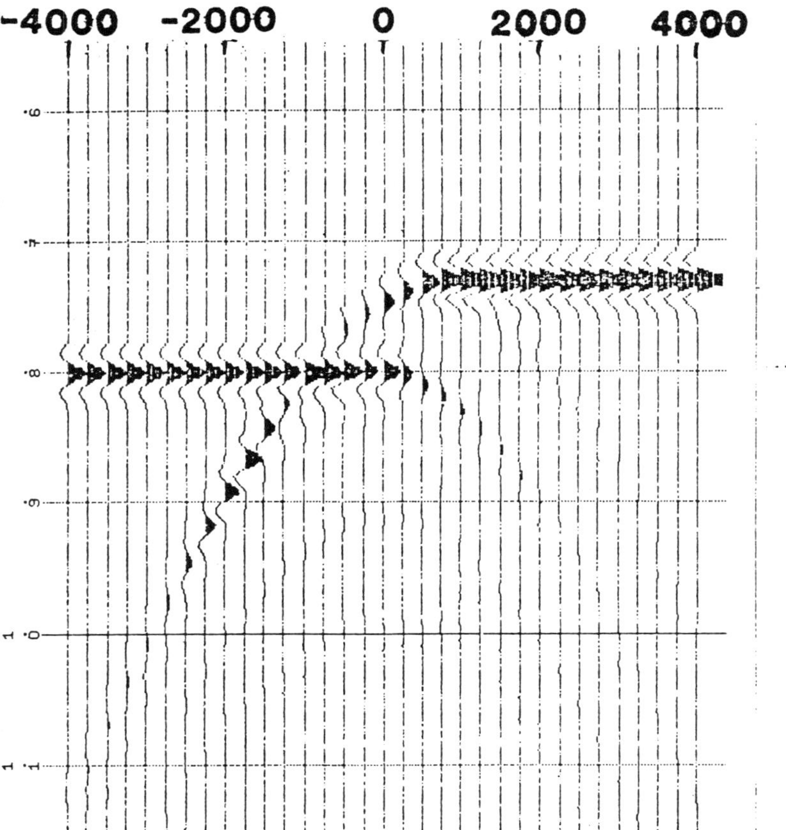

Figure 2.5. *Coincident source-receiver diffraction responses over the monocline model, using a 10–40 Hz wavelet.*

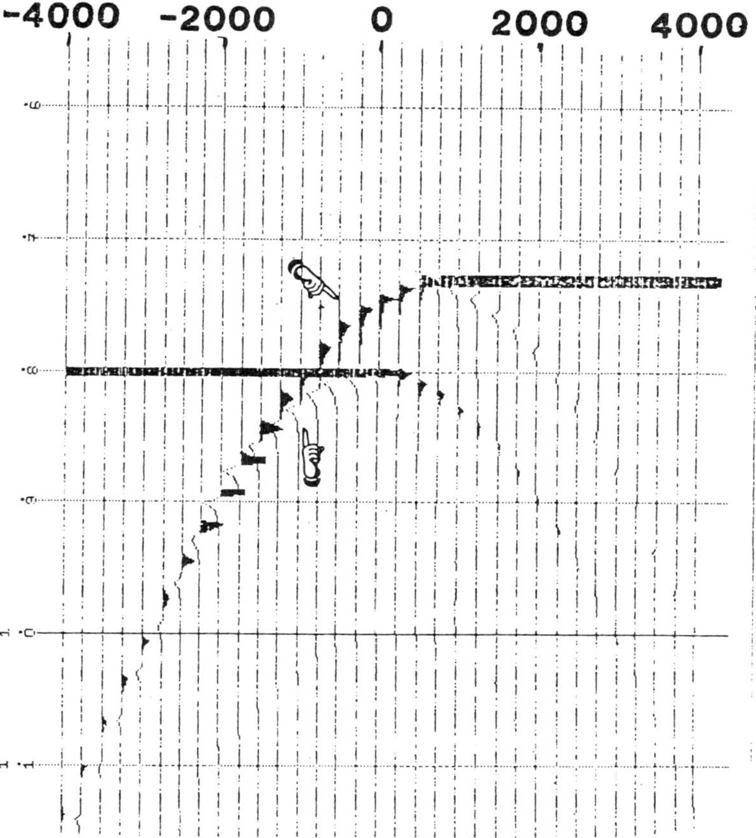

Figure 2.6. *Diffraction responses when source and receiver are coincident and not band-limited by a source wavelet.*

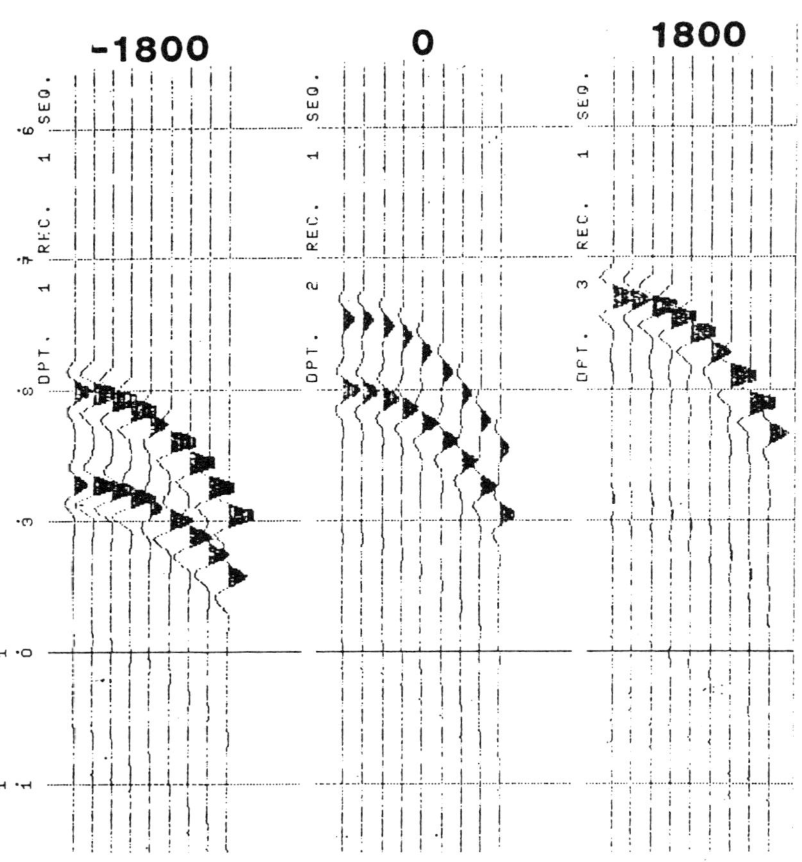

***Figure 2.7.** Three CDP gathers over the monocline model, using a 10–40 Hz wavelet. Midpoint coordinates are indicated. Source-receiver separation ranges from 0 to 4000 ft.*

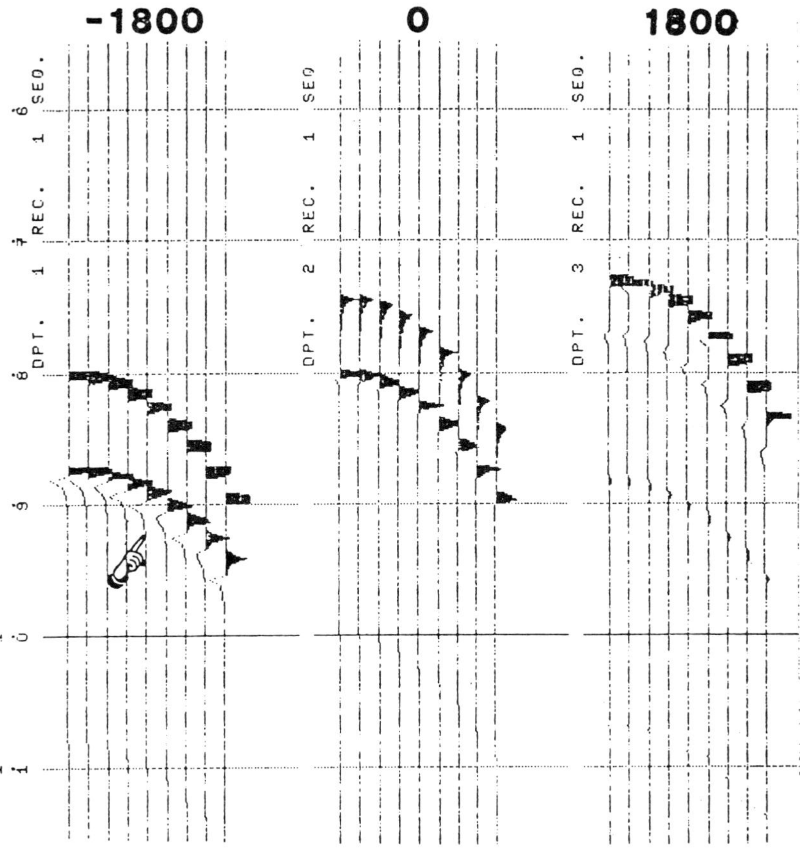

Figure 2.8. *Three CDP gathers without the bandlimiting wavelet. Note the change of the negative (left) diffraction event with increasing source-receiver separation.*

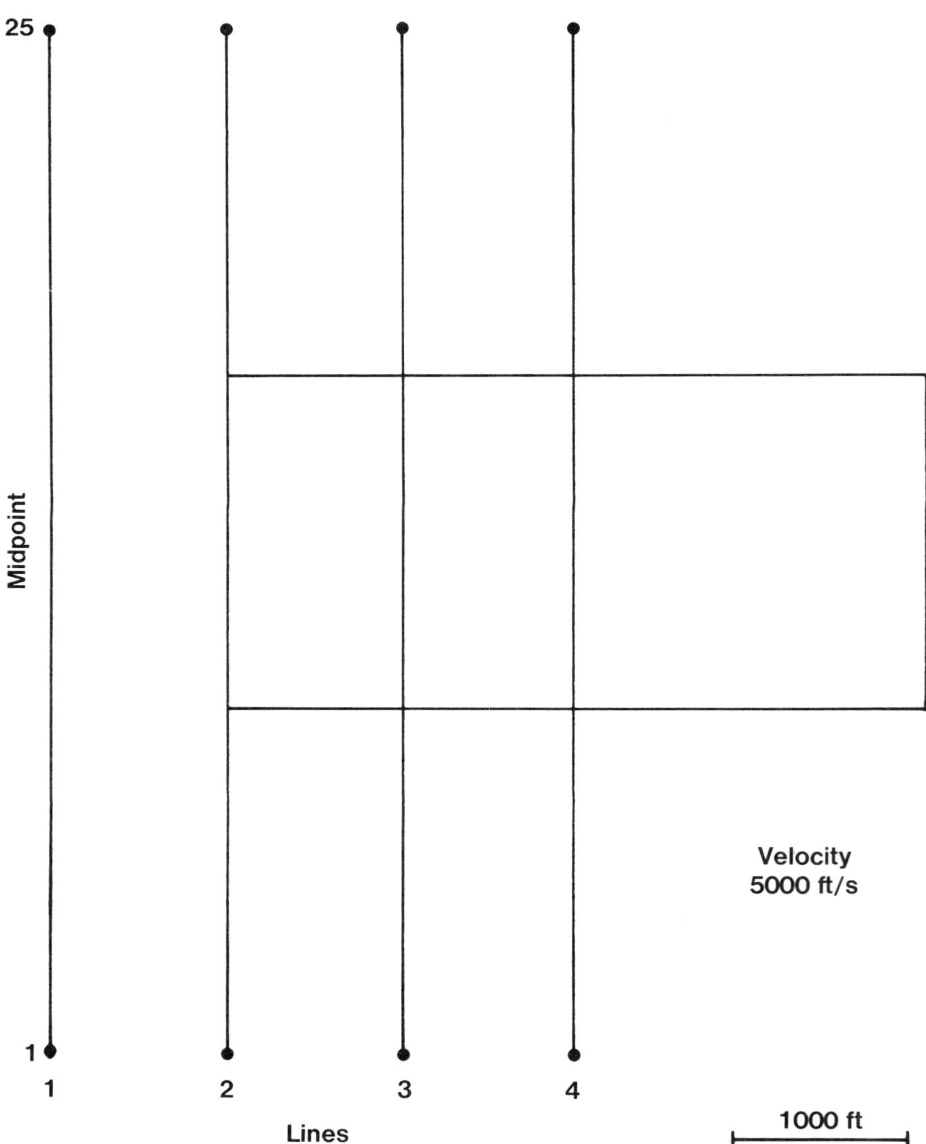

Figure 2.9. Map view of a rectangular plate model at a depth of 1000 ft, indicating the location of four lines of data.

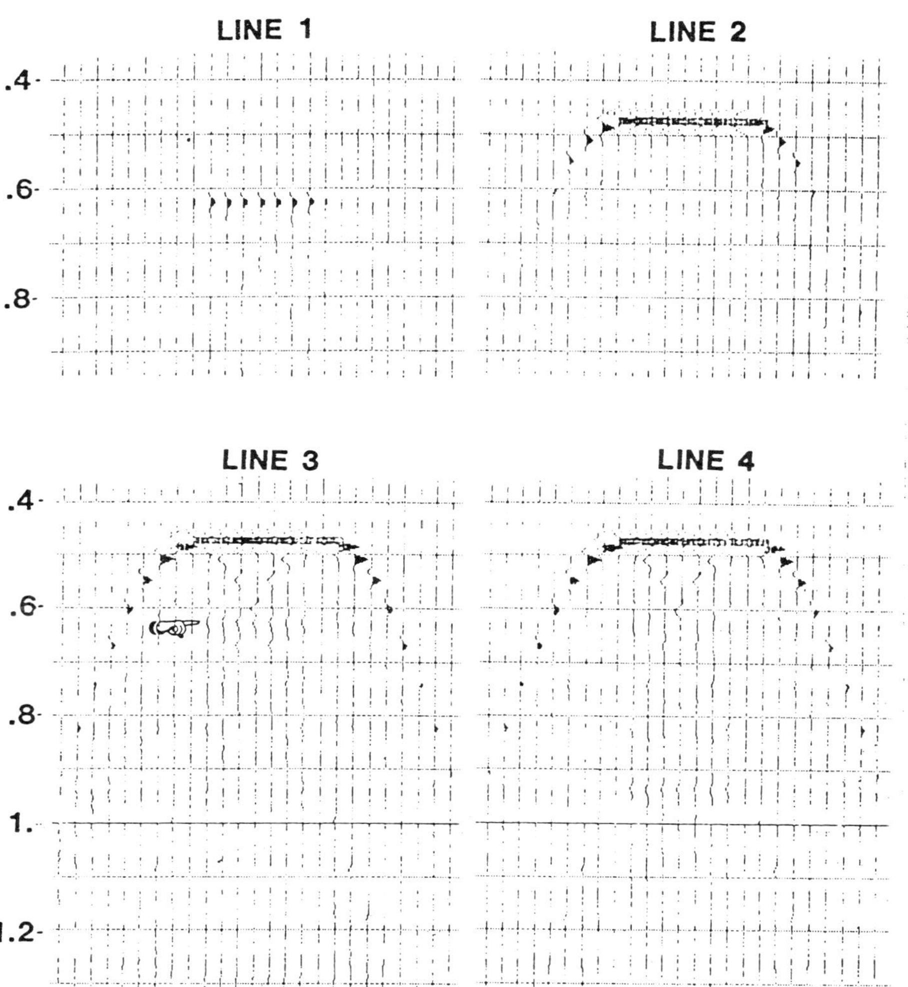

Figure 2.10. *Bandlimited diffraction responses for four lines over the rectangular plate. The source-receiver separation is 1250 ft.*

The model impulse responses are presented in Figure 2.11. Now, on line 3, the left-edge diffraction is easily recognized (upper pointer), and the right-side diffraction (lower pointer) is also identifiable. On line 4, these two side diffractions add constructively (right pointer).

A synthetic shot gather has also been computed for the model. The source was placed above the center of the left edge, and a line of 41 receivers was placed along the long axis of the model. The receiver locations are identified in Figure 2.12. Figure 2.13 shows the filtered responses for these receivers. The data are plotted with a spherical divergence correction. The strongest events are caused by the diffraction from the left edge and from the specular reflections over the plate. It is obvious that there is a complicated interaction of diffraction effects, but, because of the narrow bandwidth of the wavelet, their identification is difficult.

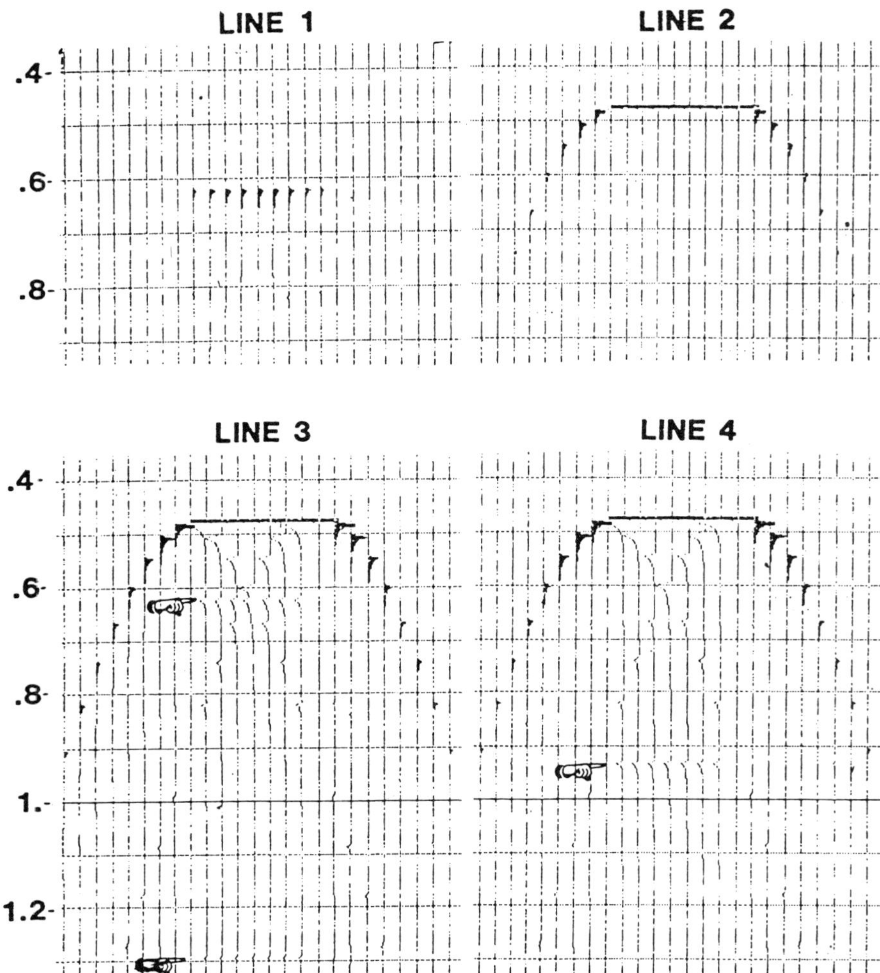

Figure 2.11. *Diffraction responses for the four lines without the bandlimited wavelet. The sideswipe diffractions on line 3 add constructively on line 4.*

In Figure 2.14, the impulse responses for the shot gather reveal a complicated superposition of diffraction events. The diffractions from the right edge fall on a symmetric hyperbola whose apex falls on trace 21 (center pointer). Its polarity is negative until trace 37 (right pointer). Beyond this, its polarity is positive. Between traces 5 and 37, the first arrival is a positive specular reflection. The left-edge diffraction has negative polarity and arrives at the same time as the right-edge diffraction on trace 21. Beyond trace 21, the left-edge diffraction has the latest arrival time. The side diffractions become visible on trace 6 (left pointer) and maintain negative polarity on all traces. At trace 21, the side diffractions arrive before the left- and right-edge diffractions. Near trace 37, they arrive after the right-edge diffraction.

WRENCH FAULT MODEL. The boundary diffraction wave theory was compared to experimental data collected in our laboratory. The map diagram in Figure 2.15

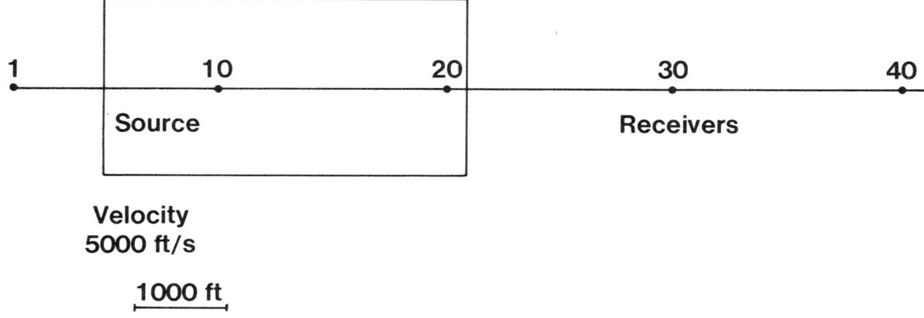

Figure 2.12. *Location map of 41 receiver positions over the rectangular plate, using a fixed source position over the left edge.*

shows the orientation of a room-temperature-vulcanized (RTV) silicone rubber model as it rested in the water tank. The model represents a stylized version of a wrench fault, in which a monocline is terminated against a fault edge. In the figure, the lower quarter-plane is in the upper right corner of the diagram, and the upper three-quarter plane is on the left and bottom. The difference in elevation between the two planes is 1000 ft. The observation plane is 5300 ft above the upper surface.

A set of eight parallel CDP lines was run over the model. Each line comprised 121 midpoints spaced 100 ft apart; at each midpoint, a set of six source-receiver positions was used, ranging in offset from 2000 ft to 7000 ft. The line spacing was 1000 ft. Lines 3 and 5 were selected, and their locations are indicated by the pointers in Figure 2.15. The results in this section are all plotted with fixed gain.

Near- and far-trace common-offset gathers for line 3 are shown in Figure 2.16. The upper and lower flat planes are represented by the flat reflections on the left and right, respectively. Diffraction from the right edge of the lower plane is just detectable. Note the amplitude thinning at the break between the upper flat plane and the dipping plane.

The same types of gathers on line 5 are shown in Figure 2.17. There is now a new event above the reflection from the lower edge. These are side diffractions from the fault edge along the upper plane. There is also a noticeable weakening of the reflection from the lower plane, since the line is near the fault.

The boundary diffraction wave theory was used to simulate the experimental data. The near- and far-trace offset gathers are presented in Figure 2.18. The filter operator chosen for these data was taken from the first trace of line 3 of the experimental data.

The amplitude dimming associated with the joint between the upper flat plane and the dipping plane is similar to that run on the experimental data. Note that the interference of edge diffractions with the specular reflection causes a leggy appearance in one portion of the data (upper pointer). The dead zone (lower pointer) is caused by destructive interference. These phenomena are observed in both the synthetic and the experimental data.

The numerical modeling of line 5 is presented in Figure 2.19. The event above the lower plane reflection (pointer) is caused by the side diffraction from the upper plane edge. There is remarkable agreement with the experimental data (Fig. 2.17). Note that, in both cases, the side diffraction causes an arrival time waviness and an amplitude fluctuation.

In Figure 2.20, lines 3 and 5 were rerun, assuming that the source and receiver were coincident. Except for the small time delay, the features of these data are

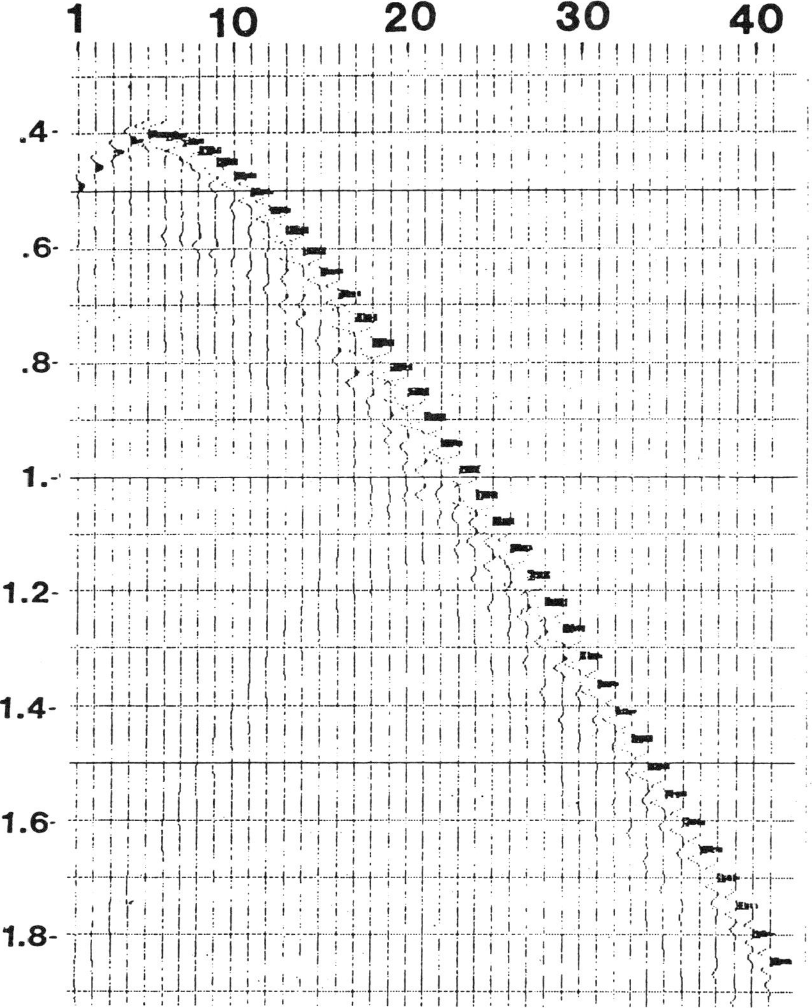

Figure 2.13. *Diffraction responses over the rectangular plate, using a 10–40 Hz wavelet. The source and receiver are coincident on the fifth trace from the left.*

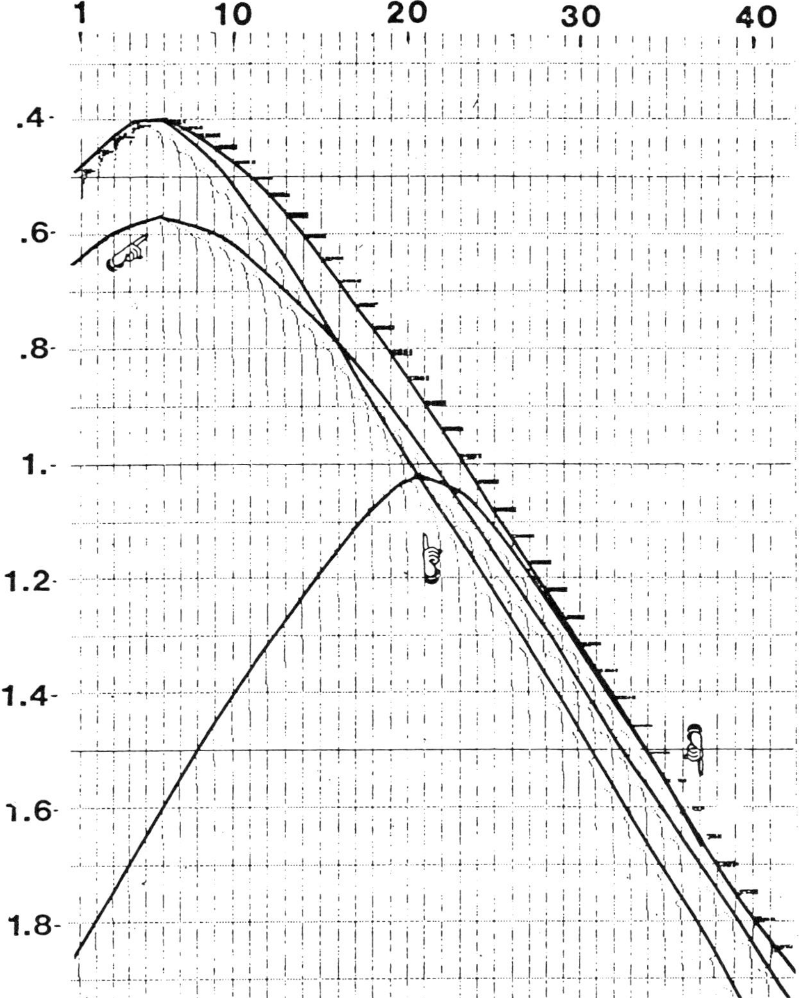

Figure 2.14. Diffraction responses without filtering.

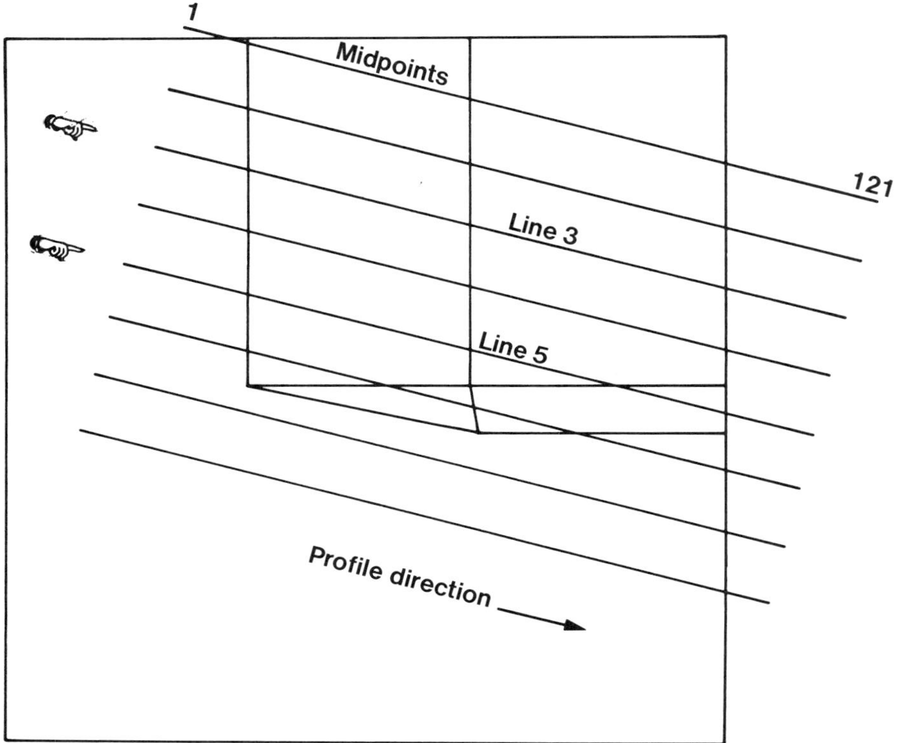

Figure 2.15. *Map view of a wrench fault model and the location of lines 3 and 5. The lower plane of the monocline below the right portion of line 3 abuts a wrench fault below midpoint 100 on line 5.*

very similar to the experimental data (Figs. 2.16a and 2.17a). These wavefields would be appropriate to use with an imaging method that assumes the CSR configuration.

THE MIDPOINT AND BISECTOR IMAGE POINTS. Numerical computations using this diffraction theory can be time-consuming, and so some effort should be given to finding acceptable approximations for the line integral. Recently (Berryhill, 1977; Trorey, 1977), it has been suggested that a CSR image point could be used to approximate the diffraction response for separated source-receivers. This method places the image point on a line that connects the minimum time point on the edge to the midpoint of the line joining source and receiver. The image point on this line would also be chosen so that the traveltime would be the same as the traveltime for separated source and receiver (see Fig. 2.21).

In another choice, a line is chosen that bisects the angle between the lines drawn from source and receiver to the edge (Gardner and Kotcher, 1977; Ilukewitsch, 1977). As in the midpoint approximation, the image point is chosen to fall on the line so that the traveltime is equivalent to the traveltime with source and receiver separated.

Using the bisector approximation, the near- and far-trace gathers for line 3 were computed; they are shown in Figure 2.22. They are very similar to the earlier results (Fig. 2.18).

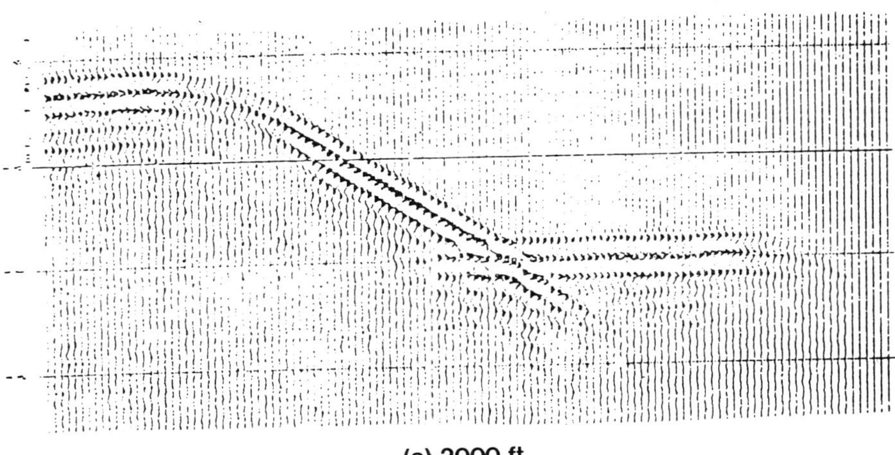

(a) 2000 ft

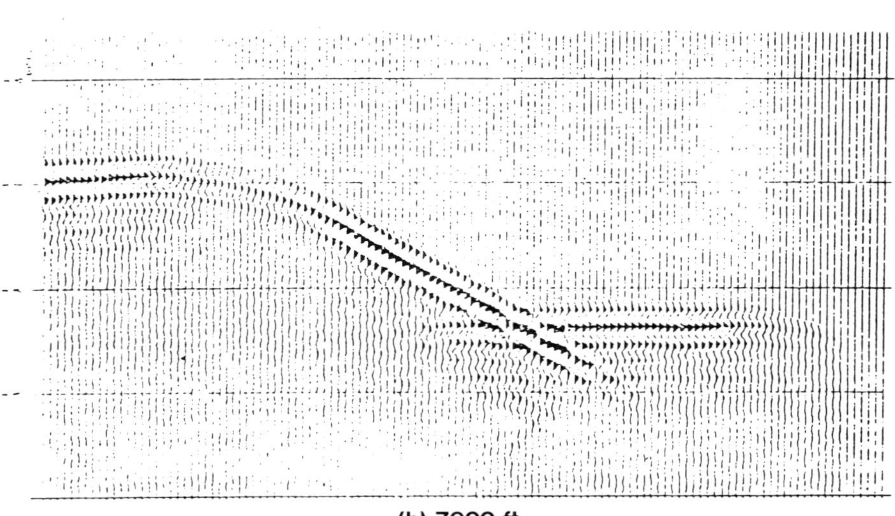

(b) 7000 ft

Figure 2.16. *Gathers of traces with 2000 ft and 7000 ft source-receiver separation along line 3, from the tank experiment.*

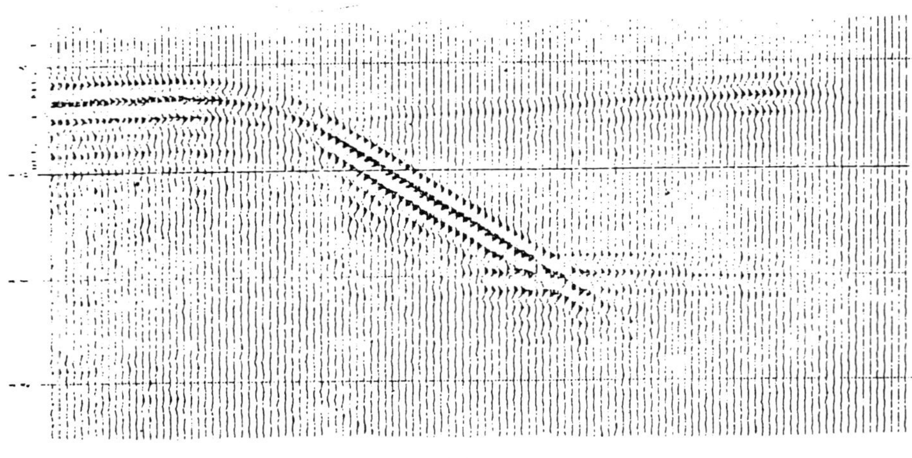

(a) 2000 ft

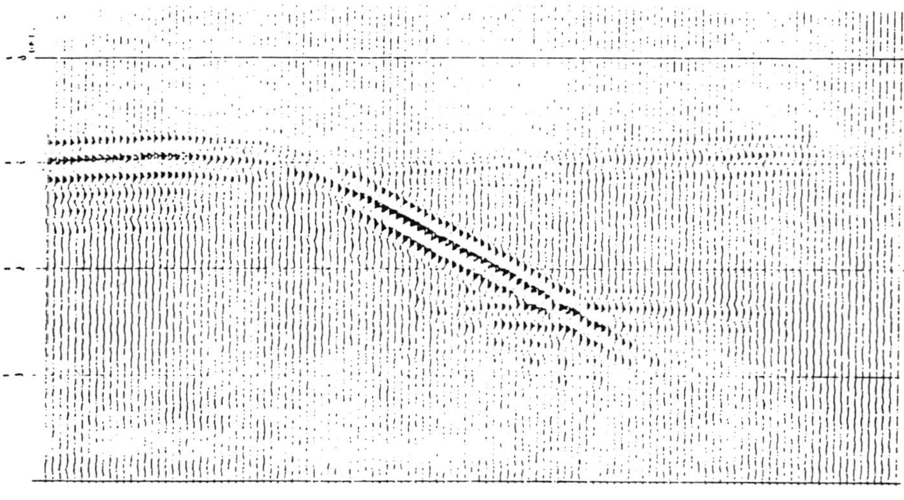

(b) 7000 ft

Figure 2.17. *Gathers of traces with 2000 ft and 7000 ft source-receiver separation along line 5, from the tank experiment.*

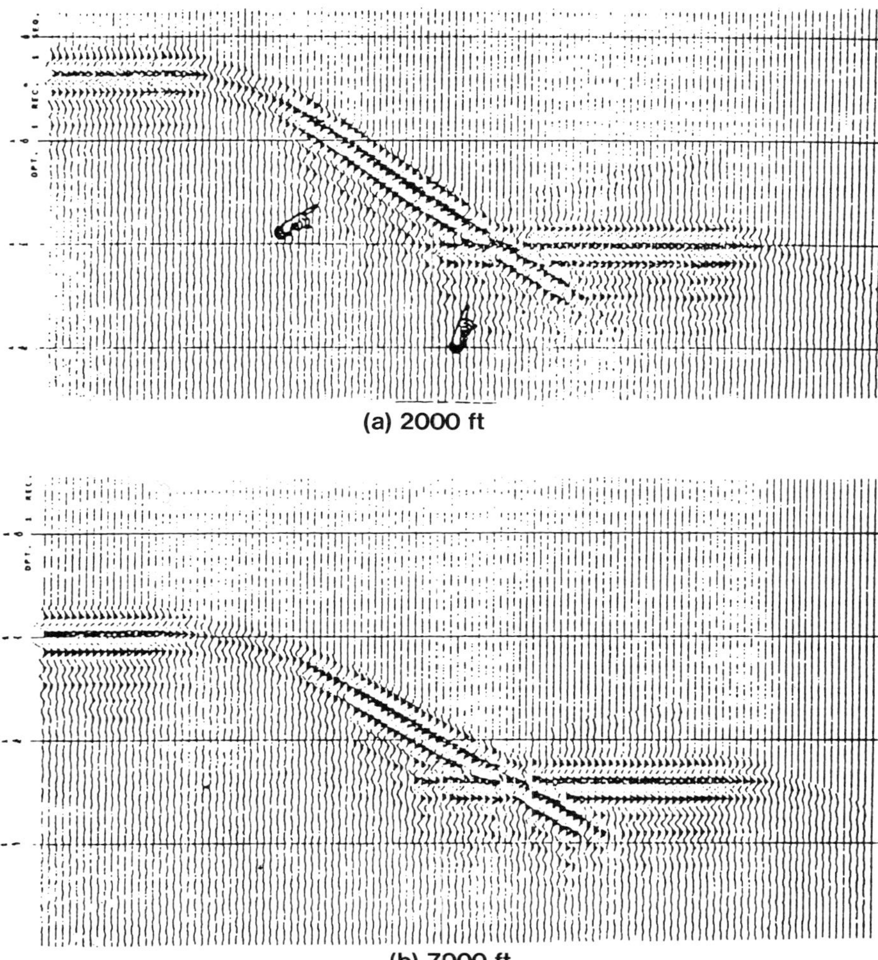

Figure 2.18. *Numerical simulation of the experimental data on line 3. The diffraction responses have been filtered with the reflection event on the upper left trace in Figure 2.16.*

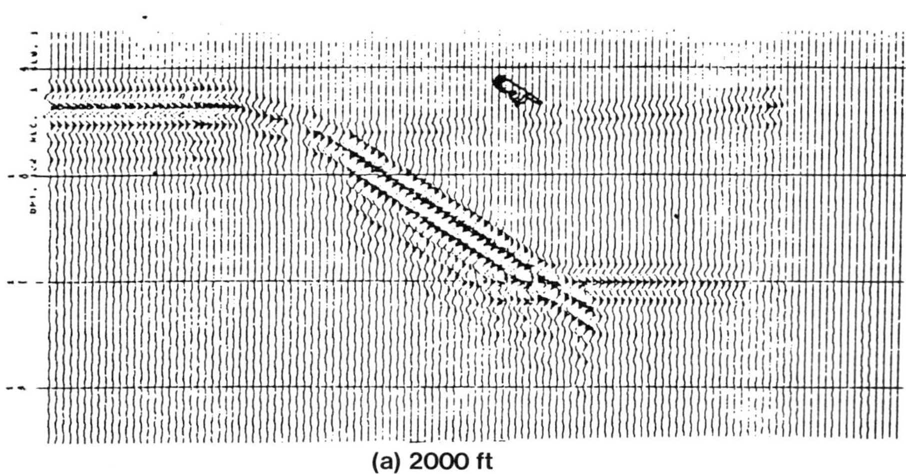

(a) 2000 ft

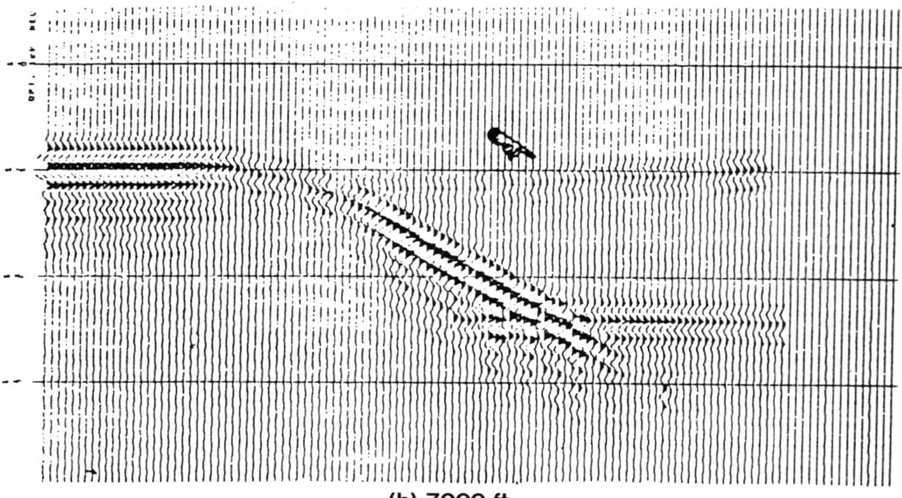

(b) 7000 ft

Figure 2.19. *Numerical simulation of the synthetic data on line 5. Compare with Figure 2.17.*

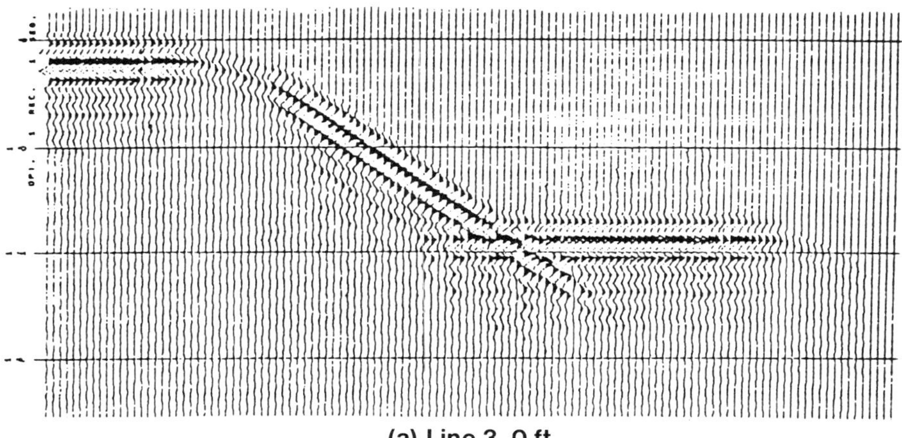

(a) Line 3, 0 ft

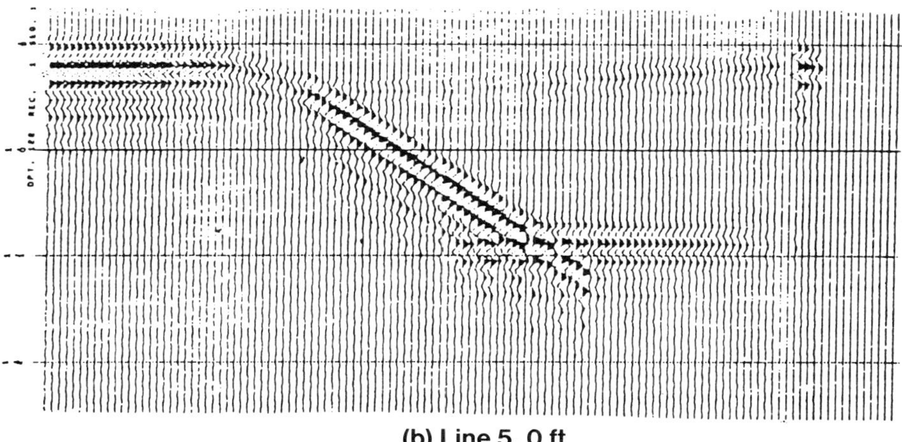

(b) Line 5, 0 ft

Figure 2.20. *Numerical simulation of CSR data along lines 3 and 5.*

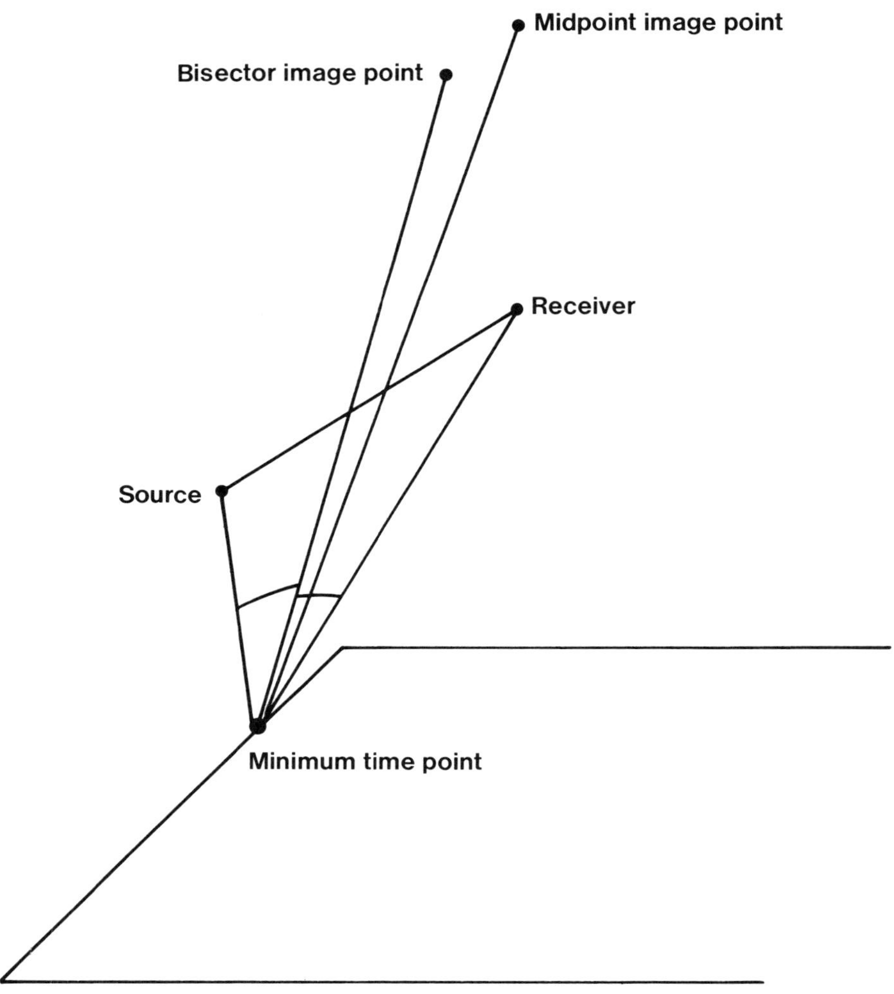

Figure 2.21. *Bisector and midpoint positions for approximating the diffraction response of an edge. The responses with source and receiver separated may be approximated by placing both at one of these positions (and simplifying computations).*

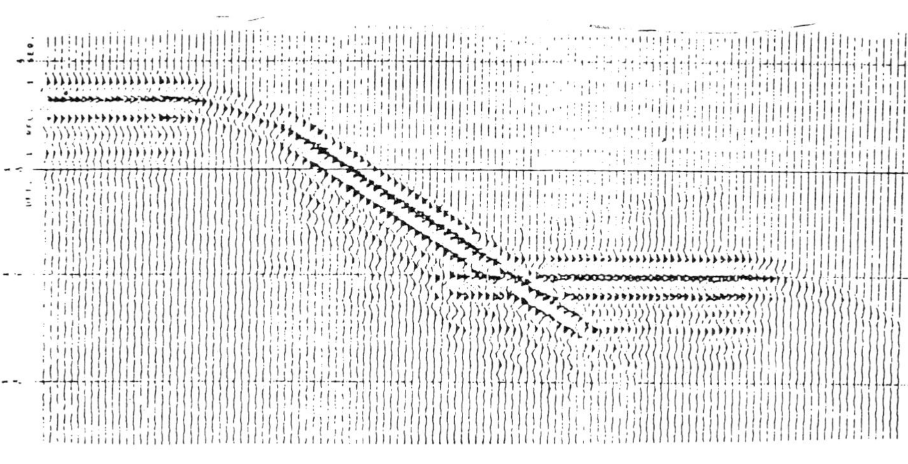

(a) 2000 ft

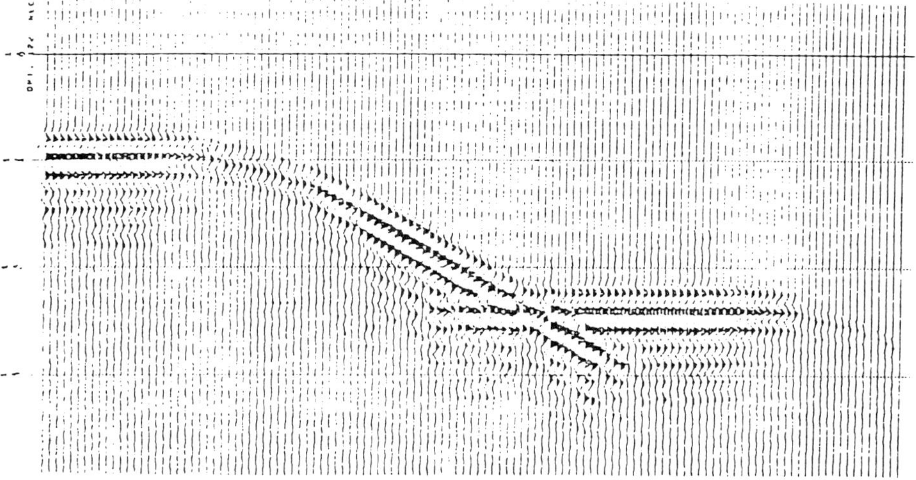

(b) 7000 ft

Figure 2.22. *Numerical simulation using the bisector approximation along line 3. Compare with Figure 2.18.*

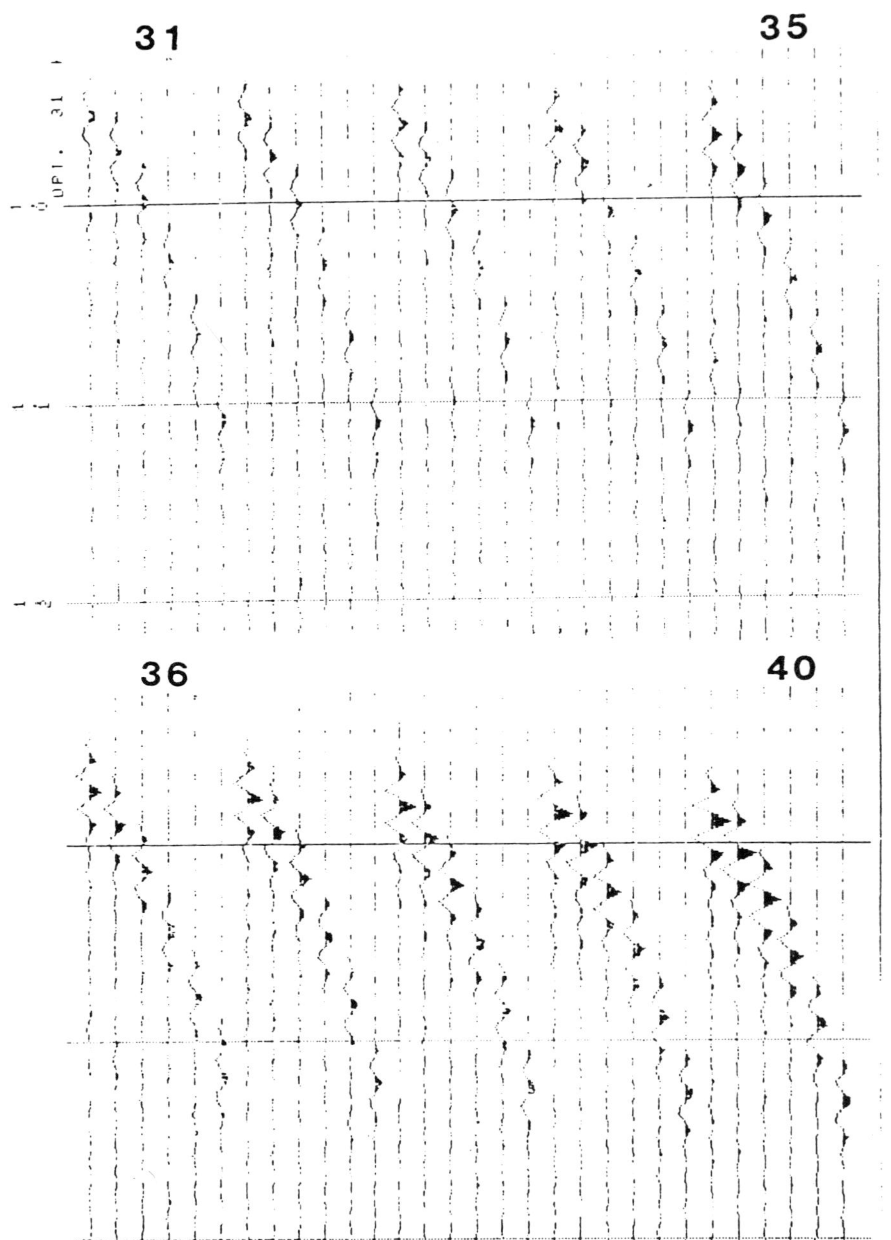

Figure 2.23. *CDP gathers from midpoints 31 to 40 of line 3, from the tank experiment. Source-receiver offsets ranged from 2000 ft to 7000 ft (scaled).*

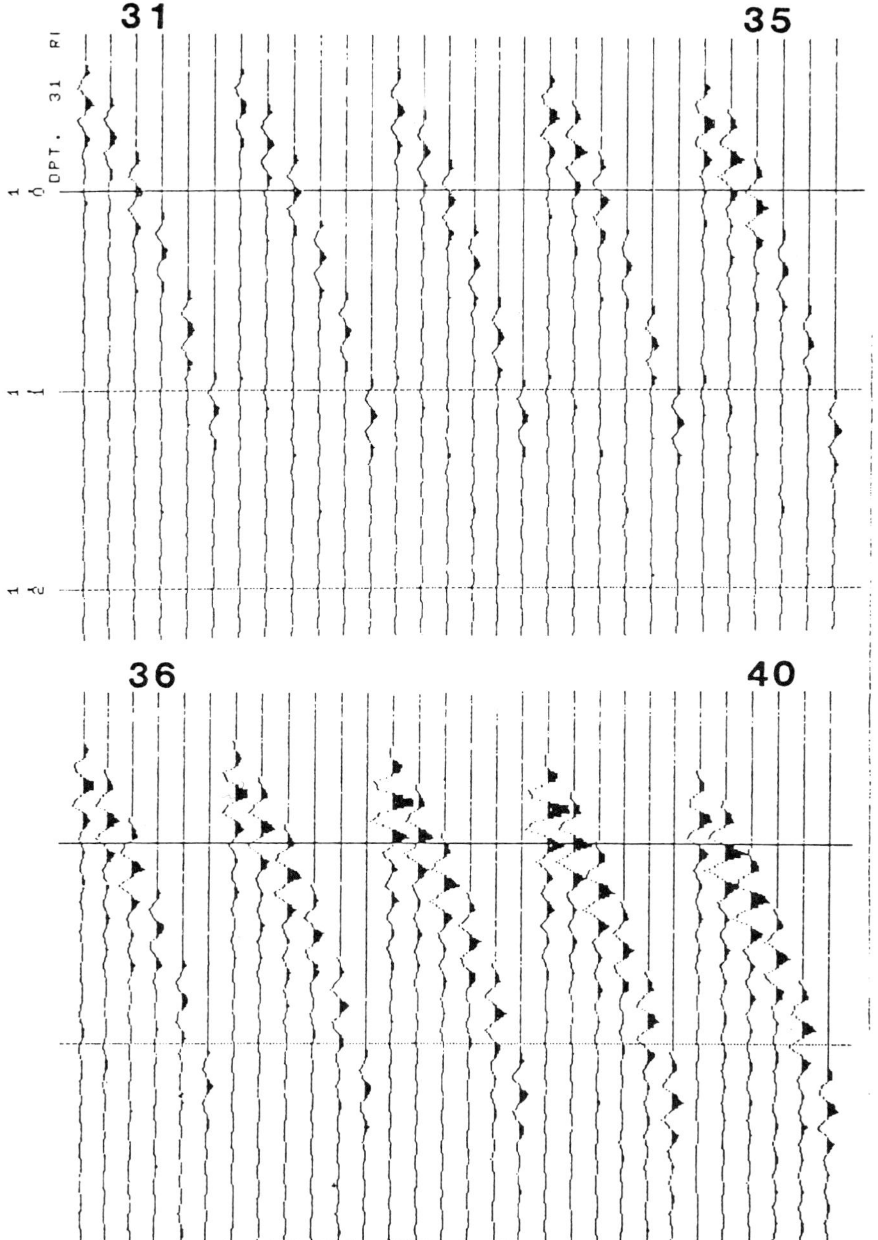

Figure 2.24. *CDP gathers from midpoints 31 to 40 of line 3, synthetic data, using the line integral.*

The experimental CDP gathers from midpoints 31 to 40 on line 3 are shown in Figure 2.23. This portion of the line represents the transition zone from a positive diffraction to a specular reflection from the left portion of the dipping plane. Within a gather, the amplitude falls off with offset because of the extra travel distance. From gathers 31 to 40, the amplitude increases as the specular reflection point is reached.

In Figure 2.24, the numerical simulations exhibit an amplitude balance and waveform shape that is very similar to those of the experimental data.

In Figure 2.25, the ten near-offset and far-offset traces from each set of data have been replotted. Notice that there is an amplitude buildup on both the experimental and synthetic near-trace gathers. The amplitude buildup is not present in the two far-offset gathers in the second row.

It is clear that, within this transition zone, some mechanism is causing certain traces in a CDP gather to change and not others. For this waveform bandwidth, the observed effect is mainly an amplitude variation.

To understand the change in waveform within a CDP gather caused by a diffraction effect, consider a gather above the diffracting edge of a half-plane, as in the upper portion of Figure 2.26. The raypath for each of the source-receiver pairs is a diffraction raypath involving the edge. If the half-plane is rotated and shifted updip, the inside traces will record two arrivals—one a diffraction from the edge and the other a specular reflection from the half-plane (pointer). For a far-offset trace, the specular reflection path does not exist.

A situation could also arise in which the far-offset traces have a specular raypath and the near-offset traces do not. This case would arise when the half-plane extends updip, and the position of the edge will determine how many far-offset traces will have a specular reflection. In Figure 2.27, the unfiltered synthetic model impulse responses show that, for sufficient source excitation bandwidth, there is not only an amplitude change within the gather but also a change in phase. On the near-offset traces, the impulse responses grade from a positive diffraction near CDP 31 to a combination of specular reflection and negative diffraction near CDP 40 (upper pointer). On the far-offset traces, the impulse responses are all positive diffractions (lower pointer).

STACKED WAVEFIELDS. The tank CDP gathers in line 3 were stacked assuming a constant-scaled water velocity (Fig. 2.28a). The specular reflections from the upper and lower flat portions of the model stack in-phase and are very similar to those seen in the synthetic estimate of the CSR wavefield (Fig. 2.28a). These data were also stacked with a velocity that accounts for the known dip (15.3°) of the model (Fig. 2.28b).

The specular reflections from the dipping plane may be seen to stack in-phase (Fig. 2.28b). The reflections from the flat portions of the model are not so strong as before because of destructive interference. In addition to the amplitude dimming, there is also an apparent polarity reversal between the two sections. The stack section with the higher stacking velocity has the appearance of delaying the flat-lying reflections by approximately half a wavelength, and so gives the appearance of deepening the event by approximately 400 ft. This mispositioning of the reflector is difficult to detect on the section because the reflection grades smoothly into a diffraction (Fig. 2.28, pointer).

The sometimes difficult selection of proper stacking velocity functions is well known. It is made less ambiguous by increasing the number of traces in the gather. In addition to the signal-to-noise ratio improvement, there is the benefit that finer spatial-offset sampling lessens the risk of misidentification of the primary arrival in the presence of multiples and aliasing effects. The stack multiplicity cannot resolve the problem of interference where the velocity function is truly multivalued, however. In Figure 2.29, the two stacked sections between CDP 50 and 90 have been replotted on a larger scale. Within this CDP

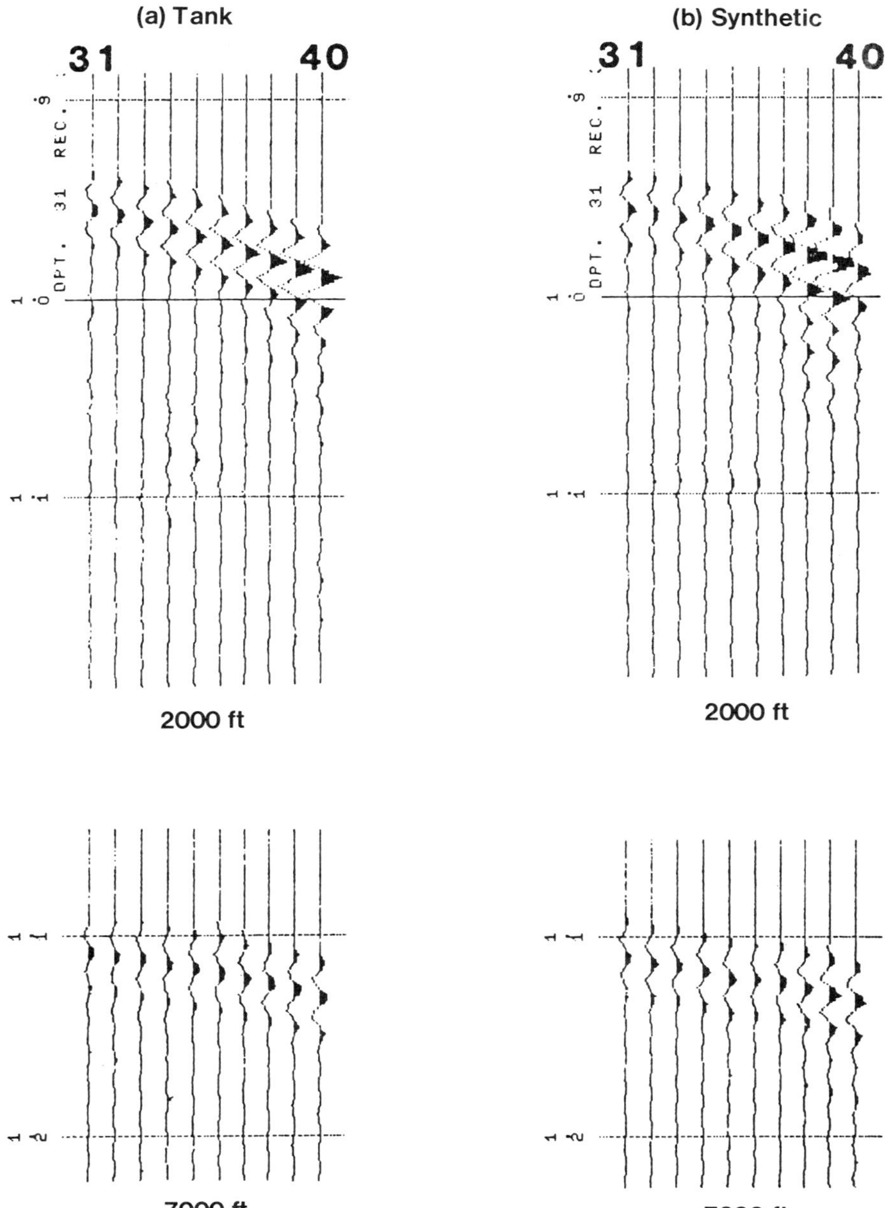

Figure 2.25. *Near- and far-offset traces of data in Figures 2.23 and 2.24.*

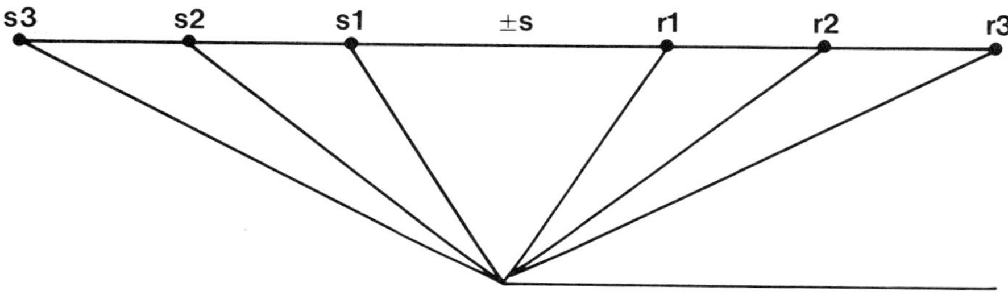

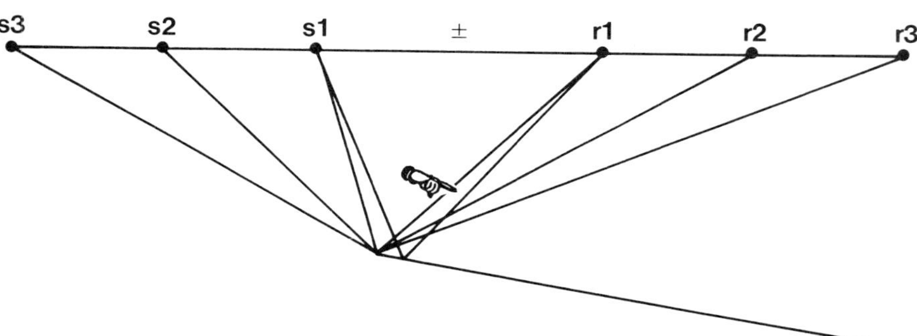

Figure 2.26. *The upper diagram depicts how the edge diffraction is recorded on all traces in the CDP gather over the edge. The lower diagram shows that, by rotating and shifting the half-plane, the near-offset receivers record a diffraction and specular reflection, while the far-offset receivers record only a diffraction.*

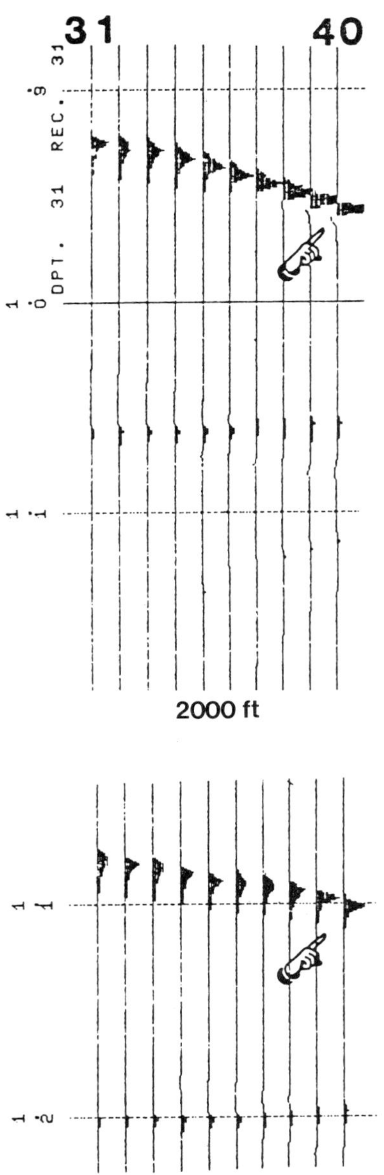

Figure 2.27. *The numerical diffraction responses from Figure 2.25b, without a bandlimiting wavelet.*

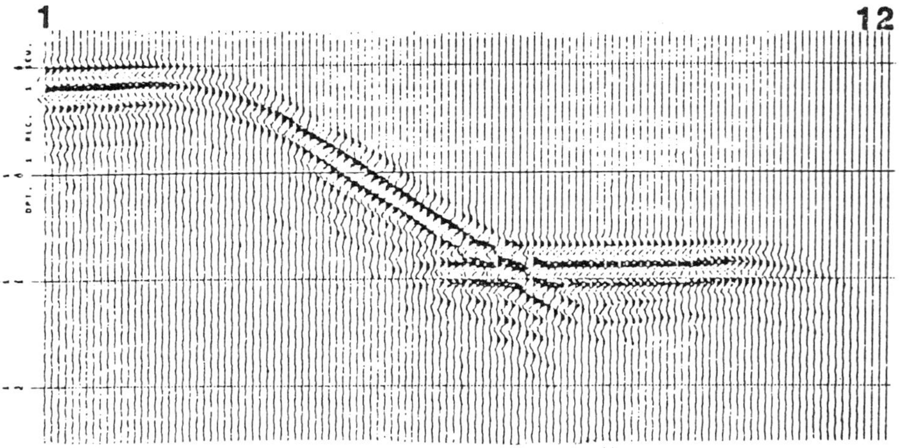

(a) VSTACK 11928 ft/s

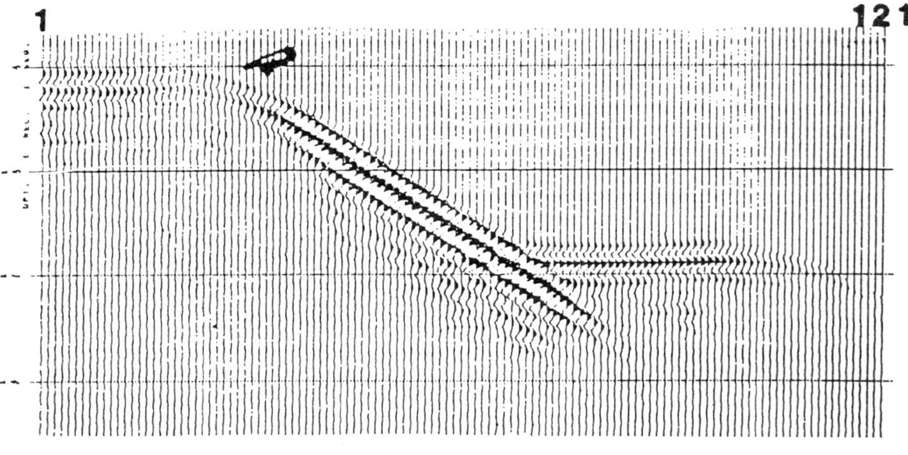

(b) VSTACK 12366 ft/s

Figure 2.28. *Estimation of a CSR wavefield by stacking the six traces per midpoint, using a scaled wave velocity (a) and a velocity adjusted for the dipping plane (b). Compare to the CSR wavefield in Figure 2.20.*

range, the data from the dipping plane and the flat plane actually should be decomposed and stacked independently. As can be seen in the figure, choosing one velocity function properly stacks one wavefield at the expense of the other.

CONCLUSIONS. The boundary wave theory calculations have been shown to predict accurately the experimental data collected with our tank model. The temporal bandwidth of the source waveform appears to play a critical role in the resolution of interfering diffraction phenomena. It has been shown that there can be significant waveform variations within a CDP gather, so that, even if the arrivals within the gather could be time-aligned, the estimation of a CSR trace

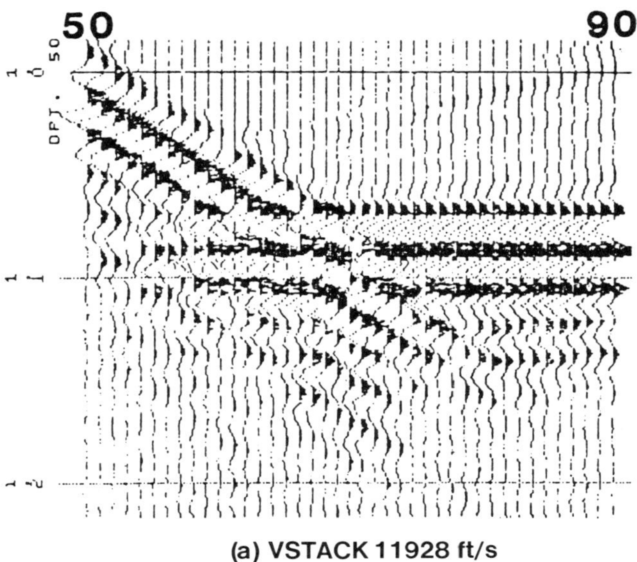

(a) VSTACK 11928 ft/s

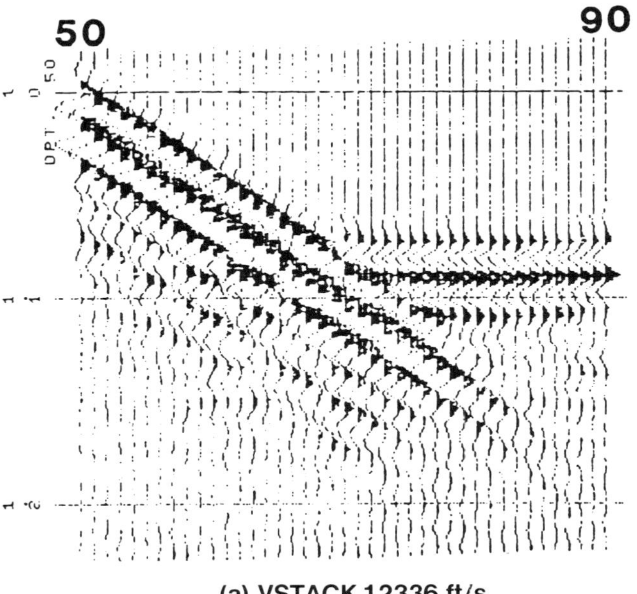

(a) VSTACK 12336 ft/s

Figure 2.29. *An enlargement of a portion of Figure 2.28, revealing the differences in the estimate of the CSR wavefield.*

could be in error. The more well-known problem of interference of conflicting dip data is also demonstrated.

REFERENCES

Berryhill, J. R., 1977, Diffraction response from nonzero separation of source and receiver: Geophysics, v. 42, p. 1158–1176.

Gardner, G. H. F., and Kotcher, J. S., 1977. Improvements in dip-analysis, velocity analysis by the use of Kirchhoff summation, *in* Modeling and migration, a symposium: Dallas, Dallas Geophysical Society.

Hilterman, Fred J., 1970, Three-dimensional seismic modeling: Geophysics, v. 35, p. 1020–1037.

______ 1975, Amplitude of seismic waves—A quick look: Geophysics, v. 40, p. 745–762.

Ilukewitsch, A. G., 1977, Three-dimensional seismic modeling with source-receiver offset: M.S. thesis, Univ. of Houston.

Levin, F. K., 1971, Apparent velocity from dipping interface reflections: Geophysics, v. 36, p. 510–516.

Seismic Acoustics Laboratory, Semi-Annual Progress Review, Boundary diffraction wave: 1979, v. 3, p. 373–377.

Trorey, A. W., 1970, A simple theory for seismic diffractions: Geophysics,v. 35, p. 762–784.

______ 1977, Diffractions for arbitrary source-receiver locations: Geophysics, v. 42, p. 1177–1182.

3. *SEISMIC COAL SEAM MODELING CONSTRAINED BY DEPOSITIONAL ENVIRONMENT*

Robert E. Duffy, Jr.

INTRODUCTION. Coal often has appeared on a conventional seismic section as an extremely high amplitude event. Many a seismic bright spot has been drilled in quest of movable hydrocarbons, only to discover a small coal deposit. The high amplitude of seismic reflections is because of the low acoustic impedance of coal. Acoustic impedance is the product of velocity and density of the reflecting medium and is often plotted in a fashion similiar to Figure 3.1. The reflectivity of the coal depends on the acoustic impedance contrast across its contacts. Typically, this contrast is large between coal and its adjacent sedimentary units—hence, the high energy return. Only loosely compacted shales or porous gas-filled sandstones can have equal or lower acoustic impedance.

Seismic stratigraphers utilize the presence of a discontinuous, high-amplitude coal reflector on conventional seismic sections to infer qualitatively a nonmarine depositional environments are modeled to analyze the seismic response. The Widmier, 1977). With the advent of high-resolution seismic techniques, this burst of energy can now be measured quantitatively. In shallow, high-resolution coal exploration, the recovered seismic signal commonly possesses a central frequency of 100 Hz. In some areas, this frequency range has been reported to reach 250 Hz (Coon, Reed, and Dunster, 1978).

The increased frequency range of the seismic signal permits three-dimensional (3-D) resolution of the coal body. To aid in this resolution, the seismic response of various coal geologic settings are characterized in this chapter. Two separate despositional environments are modeled to analyze the seismic response. The effects of a meandering sand-filled channel, both dry and gas-saturated, that disturbs a coal seam and localized deltaic coals that overlie a thick coal are discussed. Also, the subtle variations imposed on the frequency and amplitude of the seismic response by transitional contacts and vertically and horizontally thinning coals are investigated. Finally, a method is developed for estimating the total volume of these small coal bodies.

GEOLOGIC BACKGROUND. Coal is metamorphosed peat that accumulates in both shallow, undrained depressions and plains, provided that the environments are favorable to luxuriant plant growth (Thiessen, 1947). This environment for vast accumulation of peat requires a delicate balance between geologic and biological processes. As an example, Spackman, Reigel, and Dolsen (1969) reported that the biological phenomenon of plant succession, which is responsible for development of new types of sedimentation on certain sites within the southern Florida marshes, is accomplished without the aid of geologic activity. On the other hand, such geologic processes as differential sediment compaction, differential bedrock solution, or differential subsidence can reverse the biological processes and return the sites to their previous mode of sedimentation. These biological processes are not a major concern here but are assumed to be correct for peat accumulation when the geologic process is in a favorable state.

Vertical Sequence Geometry. Among the geologic requirements for peat accumulation are shallow water and a region free from appreciable detrital influx (Ferm, 1970). One geologic environment that meets this requirement is a delta, in which a sedimentary wedge forming at a river mouth has an upper surface that

This chapter was also presented as a paper at the SEG convention in Houston, November 1980.

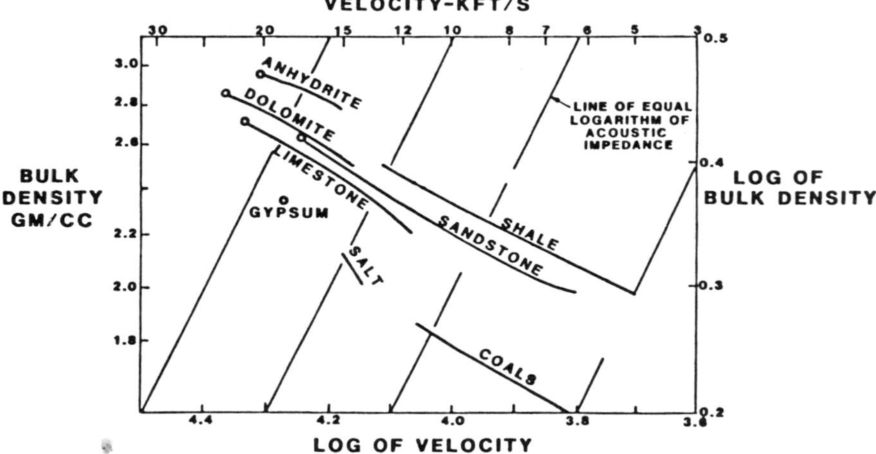

Source: Modified from Meckel and Nath, 1977.

Figure 3.1. *Acoustic impedance graph.*

typically reaches sea level. The emergent portion of this prograding wedge, the deltaic plain, provides several environments that are suitable for peat accumulation. These range from minor accumulations adjacent to levees in the active lower delta to vast accumulations over an abandoned deltaic lobe.

The vertical sequence on an abandoned deltaic lobe is controlled by continual subsidence within the mobile deltaic environment, which then can allow for a very thick accumulation of sediments in overlapping deltaic lobes. As an example of this, an 1160 ft vertical section of Pennsylvanian rocks in southeastern Illinois contains 14 repetitions of inorganic sequences (Fig. 3.2). The first inorganic sequence is a delta front or prodeltaic sediments, which are composed basically of gray, silty shales; the other inorganic sequence is deltaic interbedded sandstones, siltstones, and shales (Wanless et al., 1970). These sequences comprise 72 percent of the total footage. The other 28 percent of the 1160 ft section is composed of 14 subareal soil layers or underclays, 11 coal measures that accumulated in deltaic marshes, and 14 layers of marine limestones or calcareous shales. The balance between geologic and biological processes can be inferred in this 1160 ft section, where the selective accumulation of either underclay, coal, or calcareous sediments depends primarily on the height of sea level with respect to the deltaic plain. The geologic and biological constraints for accumulation of coal and limestone are similiar, in that both are biochemical sediments requiring a restricted depositional environment.

The upper and lower contacts of the coal measure in a delta have different gradational patterns. The contact between a lower coal measure and an upper calcareous sedimentary unit is normally very sharp. A rapid rise in sea level halts the peat accumulation and starts calcareous deposition. The base of the coal measure is normally transitional, in that its inorganic deltaic sediments grade upward into a soil profile that is composed of organic, root-filled underclays capped by a coal seam that is composed primarily of all organic material.

In addition to deciding whether a coal or a limestone accumulates, relative sea level changes also affect the vertical succession of the deltaic coal. As an example, after a rapid regression during the Pennsylvanian, the Higginsville limestone, a marine transgressive limestone of the eastern midcontinent, provided a platform for coal accumulation. The Lexington, Mystic, and Mulky Coals overlie this

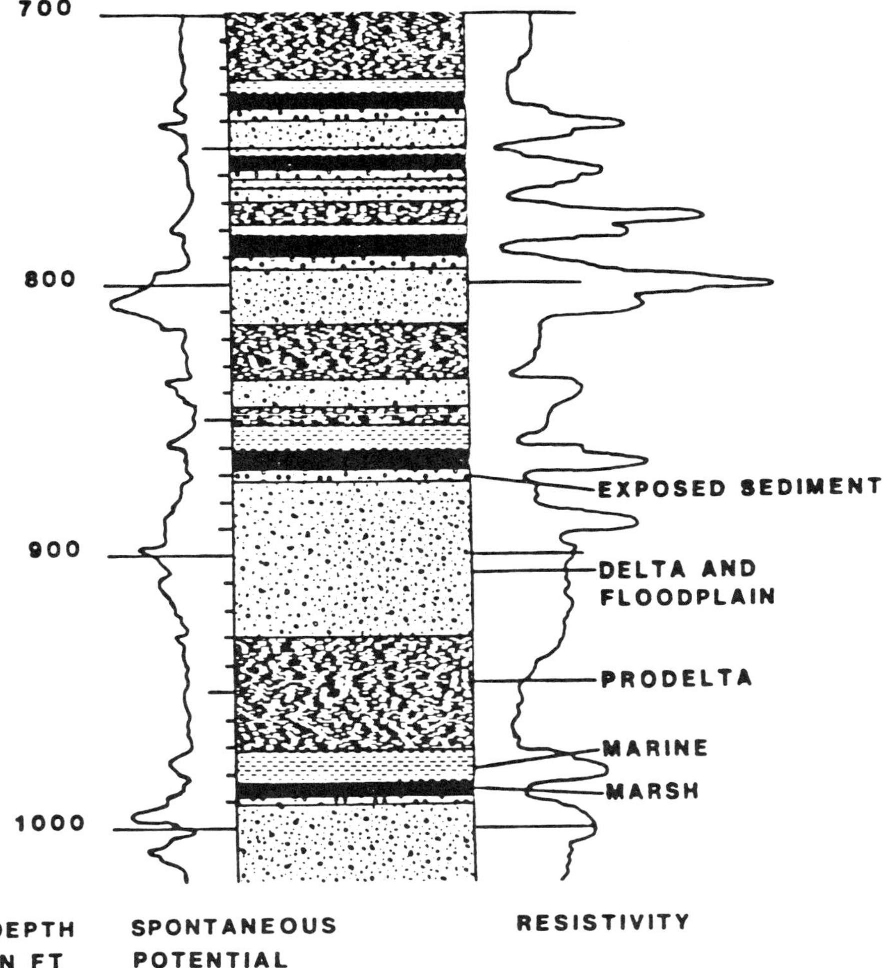

Source: Modified from Wanless et al., 1970.

Figure 3.2. *The interpreted electric log for a 300 ft portion of an 1160 ft sequence of Pennsylvanian deltaic rocks.*

limestone and are strip-mined in Iowa and Missouri. The thickness of these coals ranges up to 4 ft (Wanless, Baroffio, and Trescott, 1969). Thus, a coal measure with a sharp lower contact can be deposited on a limestone bed.

Three-Dimensional Sequence Geometry. A second depositional environment of coal is the coastal marsh, or back-barrier lagoon. Although these coals are not so volumetrically significant as deltaic coals, they can be economically important because of locally thick coal deposits. One type of back-barrier lagoonal coal is the Pennsylvanian Clarion Coal. Initially, the Clarion Sandstone accumulated as a series of barrier islands at the marineward extent of a small delta. After a slight marine regression, the Clarion Coal was deposited within the partially drained lagoon and delta complex. Medium-thick coals, 4 to 5 ft, are mined from this paleolagoon and delta (Wanless, Baroffio, and Trescott, 1969).

A second back-barrier coal is the Beckley Coal of West Virginia. The coal bodies trend parallel to a paleobarrier environment and are segregated into separate pods by penecontemporaneous tidal channels. The thickness of this coal varies, with the thickest beds over pre-existing topographic lows and the thinnest coals over highs.

Horne et al. (1976) summarized the expected 3-D shape of coals deposited in active deltas and barrrier-bar environments from their experiences with the Carboniferous of the Appalachian region. The thickness and lateral continuity of a coal deposit depend primarily on the depositional environment of the peat, with the preceding and succeeding environments modifying the coal deposit. The pre-existing topography can regulate the thickness of the coal, and postdepositional erosion and channeling can disrupt its lateral continuity. On a regional scale, a knowledge of the depositional environment of a coal enables one to predict its trend.

The back-barrier coals are elongated parallel to depositional strike and are coincident with the trend of the barrier-bar system. The coals are laterally discontinuous and form pod-shaped bodies. Continuous peat deposition is prevented by tidal channels along the barrier-bar system.

The lower delta coals, accumulating along the narrow levees of the distributary channels, are extremely continuous along depositional dip but discontinuous along strike, forming a trend orthogonal to the trend of the back-barrier coals. In a river-dominated coastal environment, the channels tend to be straight, prograding rapidly in the direction of depositional dip. Lower delta coals are normally thin, with numerous clay splits from crevasse splays.

Upper delta plain-fluvial coals are elongated parallel to depositional dip. These pod-shaped coals accumulate on flood plains adjacent to coexisting meandering channels. They display abrupt variations in thickness and are often disturbed by postdepositional channeling.

Coals deposited within the transition zone between the lower and upper delta are laterally extensive and slightly elongated parallel to depositional strike. Volumetrically, the transitional zone coals comprise the largest percentage of coals on an active delta, accumulating on the broad platform of a sediment-filled interdistributary bay. The coals are often mud-split, and contain washouts, where postdepositional channels locally erode the coal seam.

The Harrisburg No. 5 Coal is an extensively mined upper delta plain-fluvial coal (Fig. 3.3). The Pennsylvanian coal from the Carbondale Formation of the Illinois Basin accumulated in the alluvial valley of a meandering stream. Preserved as a sand-filled river cut, the penecontemporaneous stream deposits do not extend stratigraphically higher than the coal. The Harrisburg Coal is thickest immediately adjacent to the meandering stream deposit and lacks extensive mud-splits, as the river very seldom topped its levees during a flood (Wanless et al., 1970).

These deltaic coal bodies, regardless of deltaic position, normally lack a sharp contact with the adjacent sedimentary units. The coals bounded by penecontemporaneous stream deposits with flood-resistant levees or postdepositional washouts are exceptions. Typically, a coal seam interfingers with the neighboring sedimentary deposit. Laterally, the coal seam diminishes as the original environment becomes too dry or the water table becomes too dry, too low, or too saline for continued peat accumulation. Vertically, the coal seam pinches out. The seam may thin as the unit transgresses from a topographic low to a high or may split in the direction of the source of the inorganic sediments. The split is filled by a thickening wedge of the adjacent sands or shales.

PHYSICAL MODEL CONSTRUCTION. Meandering Channel Models. This sequence of models simulates a washout—that is, a channel cut in the coal that is

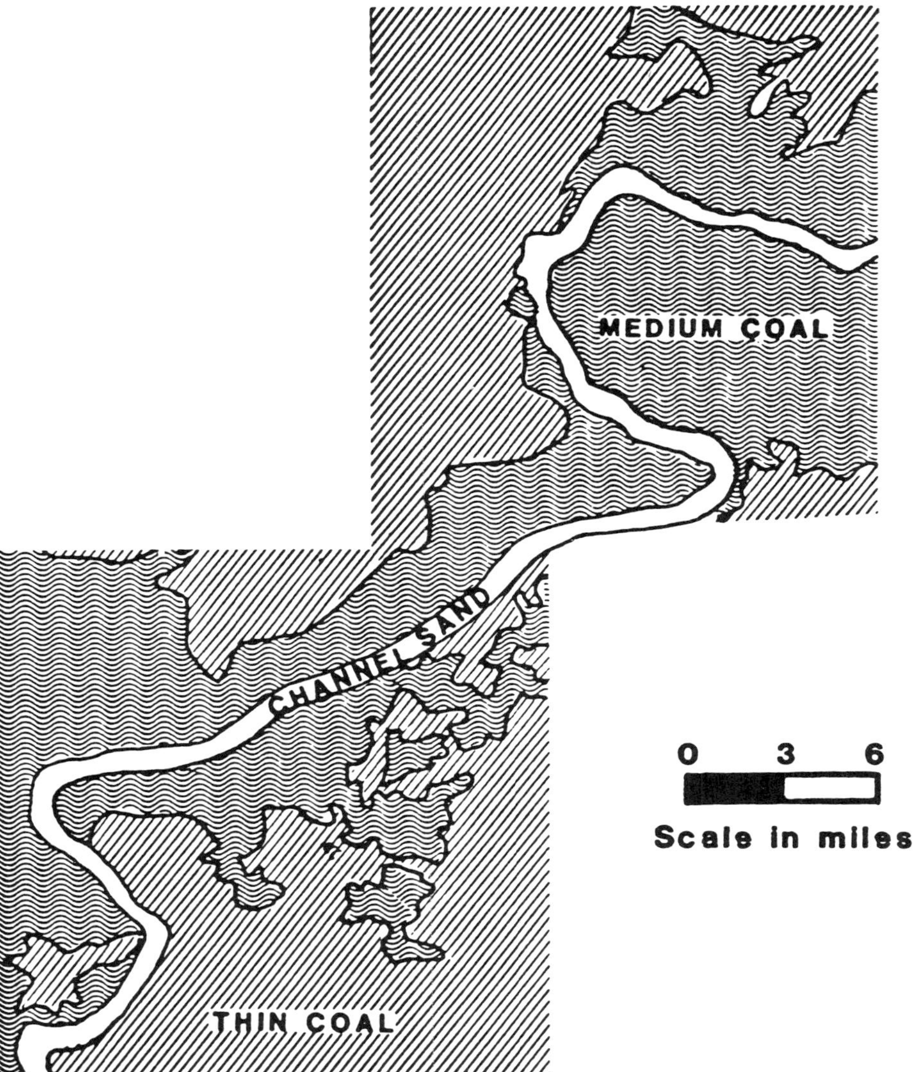

Source: Modified from Wanless et al., 1970.

Figure 3.3. *The Harrisburg No. 5 Coal, a Pennsylvanian upper delta plain-fluvial coal disturbed by a meandering sand-filled channel.*

filled and overlaid with either sand or shale. To help interpret seismic sections across the model, a theoretical model was designed with identical physical characteristics. A program was written to contour and plot the theoretical sand channel with the proper dimensions for the physical model, as illustrated in Figure 3.4.

The computer plot of the theoretical sand channel was traced onto the base of the model construction box. This box, constructed of half-inch aluminum plate, has inside dimensions of 16 in. × 16 in. × 8 in.

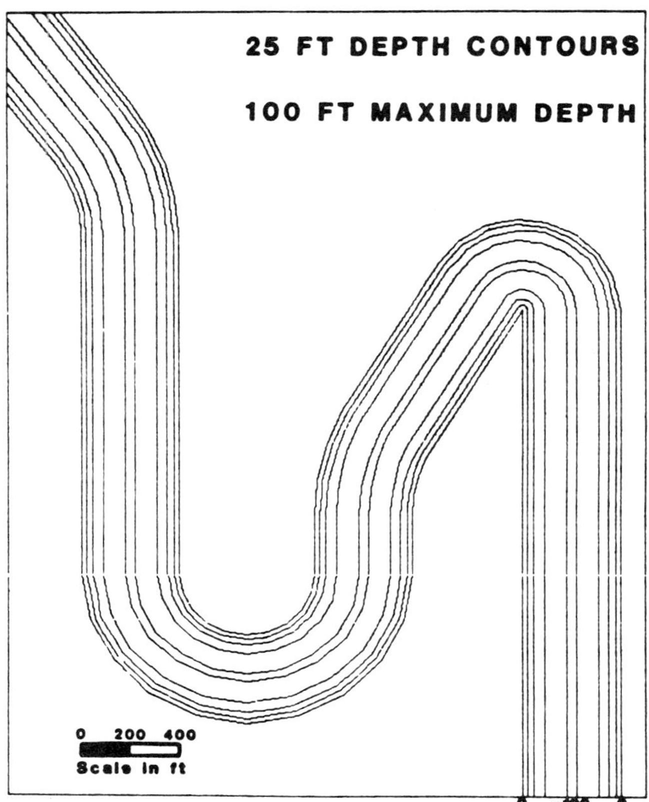

Figure 3.4. *Theoretical sand channel contoured at 25 ft depth intervals.*

The optimum areal dimensions of the physical model were determined to be 3600 × 3600 ft, with a minimum of 600 ft between the edge of the physical model and most of the channel. This clearance ensures separation of the primary reflections from those events caused by the edges of the model. (These events are defined and illustrated later.)

The channel fill was molded from artist's modeling clay. A plaster of paris cast of the clay channel fill was poured and allowed to set overnight. After removing the cast from the box, the clay was peeled out to reveal the preserved channel structure. The surface of the plaster cast was smoothed, and the holes were filled with a thinned plaster mixture. Meticulous smoothing of the cast was necessary to eliminate scattering of seismic energy by a rough surface on the final silicone model. After returning the plaster cast to the construction box with the channel concave upwards, the cast was leveled with putty placed below its four corners.

RTV silicone rubber was selected for the physical model. This rubber has a density of 1.29 g/cm^3 and a compressional wave velocity of 3110 ft/s. Before being allowed to set, the mixture was evacuated in a vacuum chamber and then poured carefully onto the cast to minimize trapped air. Extreme care had to be exercised at this point, because any bubbles contained within the cured RTV would be point scatterers, which would create noise on the seismic section. After curing overnight, the RTV model was trimmed with a razor. The final thickness of the

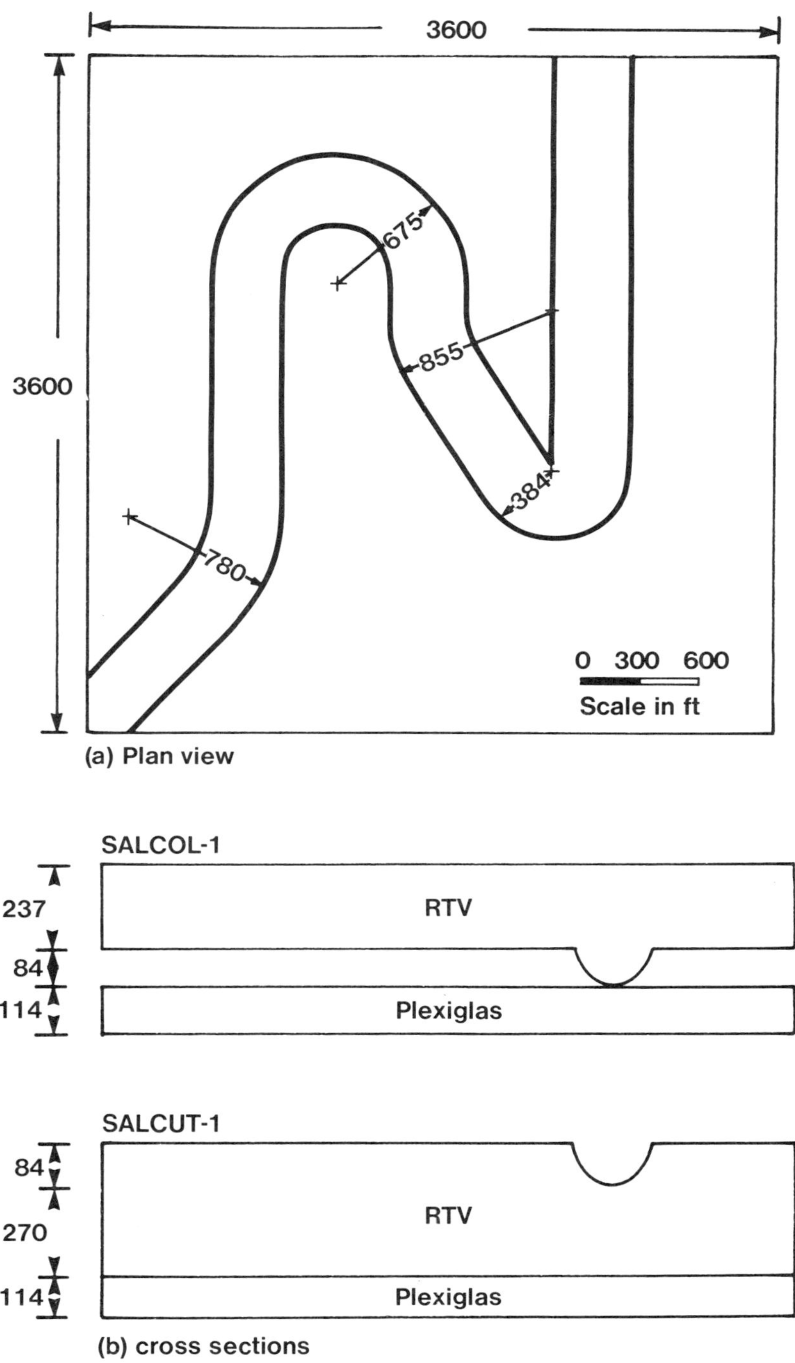

Figure 3.5. *Plan view and cross sections of the meandering stream models.*

model, after leakage of some of the RTV prior to curing, was 243 ft at the edge and 337 ft at the center of the channel.

To help understand wave propagation phenomena in seismic sections taken over this model, an inverse of the meandering channel model was made. After lightly greasing the original plaster cast, an extra-thick mixture of plaster of paris was poured onto the cast. Since drying plaster of paris is an exothermic process, an extra-thick mixture was used to shorten the cooling time and reduce the chance that the new plaster cast would adhere to the original one. After separating the two casts, the new plaster cast was smoothed, and a second physical model of RTV was generated. In plan view, the new physical model is identical to the original one; in cross section, however, the location and structure of the channel changes (Fig. 3.5). The different channel structure permits simulation of varied geologic relations that have similar geometries.

Prograding Delta Models. These models, which simulate a series of prograding delta and back-barrier coals, consist of vertical and horizontal sequences of Plexiglas suspended in water. For ease in construction and greater velocity variations, a sequence of Plexiglas shapes was selected, rather than a multi-layered silicone model. The densities of the two materials are similar, but the compressional wave velocity of Plexiglas is 9000 ft/s compared to 4970 ft/s for water.

The wavelength of a 100 Hz seismic signal in Plexiglas is 130 ft. A 600 ft thick Plexiglas block, 3600 ft on a side, serves as the base of the model, ensuring complete time separation of wavelets reflecting from the top and base of the block. The upper layers were made of 37 ft thick Plexiglas. This is approximately the tuning thickness, where constructive interference of the wavelets from the top and base increases the amplitude of the observed wavelet.

The lateral dimensions of the thin Plexiglas bodies, which were the upper layers and corresponded to the coal measures, were determined by the model ratios and realistic prototype depositional environments. The Plexiglas shapes were cut with an electric jigsaw and smoothed.

The thin Plexiglas shapes were fashioned into tiers, with approximately 100 ft vertical separation between boundaries, and were supported by thin, straightened paper clips that fitted snugly into 10 ft diameter holes. This wire was thin enough to avoid significant reflection of seismic energy, yet the construction was durable enough to resist accidental disassembly.

SCALING FACTORS. For the coal prototype, the scaling factors of length and time are 3600:1 and 2500:1, respectively. This means that 1 in. on the physical model is equivalent to 300 ft in the field. Also, the physical model sampling rate of 0.2 μs and central frequency of 250 kHz corresponds to a sampling rate of 0.5 ms and a central frequency of 100 Hz for prototype field data. All dimensions subsequently listed in both the text and the figures are scaled to these dimensions unless otherwise noted. The scaling factors are given in Chapter 1 (Table 1.6).

MEANDERING STREAM MODELS. The SALCOL[1] and SALCUT series of physical models, shown in Figures 3.5 and 3.6, simulate a coal seam cut by a meandering, sand-filled channel cut. These three physical models with two different geometries resemble the geometry of the Harrisburg No. 5 Coal, which was deposited adjacent to a meandering stream. The seismic time sections from profiles across the washout in various geologic settings have been interpreted and classified according to the geometries of the physical models. For these three models, the 2-D and 3-D seismic characteristics enable the interpreter to discern undisturbed

[1] The acronym SALCOL comes from Seismic Acoustics Laboratory coal. Similarly, other model names have SAL inhouse identification codes.

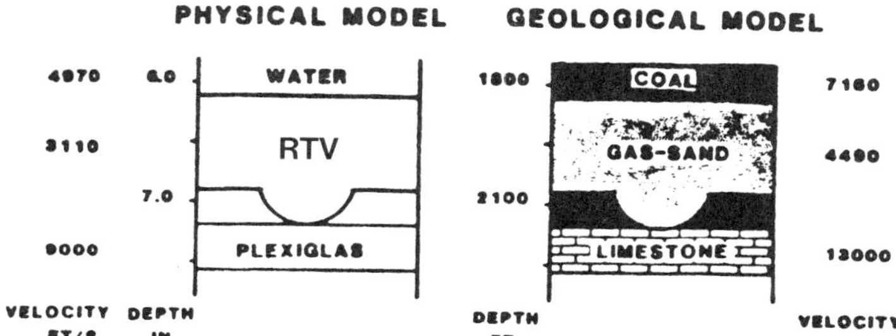

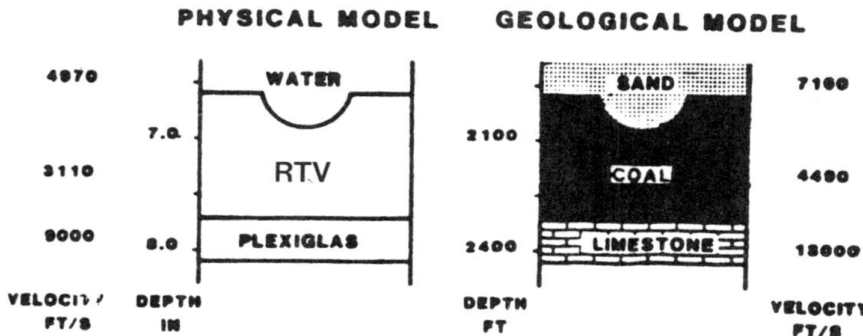

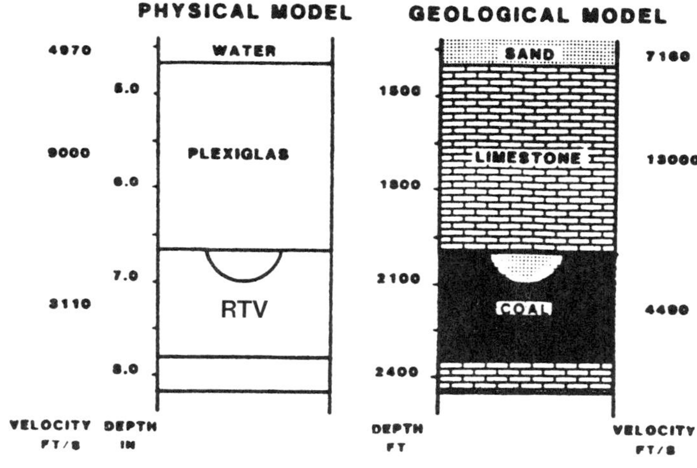

Figure 3.6. *Physical model and its geologic counterpart for the meandering stream models.*

coal from channel sand and to chart the location of the meandering channel cut. When mining coal, it is very important to have an accurate map of the channel to allow sufficient lead time for designing new coal faces.

In the meandering stream models, a high-velocity limestone underlies the objective coal (see Fig. 3.5), providing a prominent reflector to mark the base of the coal. A sand-filled channel completely cuts the thin coal in the SALCOL-1 model, but the coal seam is thicker in the SALCUT models, and the channel cuts only its upper portion.

The channel sand in the SALCOL-1 model is part of a thick sand deposit overlying the objective coal. This sand unit, overlaid by a second coal, is a gas-filled, low-velocity sand. The channel sand in the SALCUT models is gas-free, with a higher velocity than the adjacent coal. In the SALCUT-1 model, this channel sand is again part of a thick overlying sand unit. In the SALCUT-2 model, however, a thick limestone lies between the channel sand and the thick sand unit.

The course of the meandering channel approximates a portion of the sand channel disturbing the Harrisburg No. 5 Coal (see Fig. 3.5). The channel doubles back on itself twice, with a gentle meander bend of 180° and a tight meander bend of 150°. Also, two gentle bends change the stream course by approximately 30°.

Thirty-seven seismic lines were gathered over each physical model at 90 ft line spacing, with a 0.5 ms sampling rate and filter limits of 40–160 Hz. The lines contained 141 shotpoints (SP), with 30 ft spacing. A common-offset source-receiver configuration was employed, with 540 ft between the source and the receiver. This approximated a zero source-receiver (ZSR) within the limitations of the modeling tank.

Theoretical Sand Channel. To help interpret time sections observed from the physical model, sections were computed across the theoretical sand channel. The sand channel model consists of only one interface, which corresponds to the sand-coal contact of the SALCUT-1 physical model. In the model, 710 triangular tiles describe the interface (Fig. 3.7). Only one interface was coded, because the computer modeling program assumes a constant-velocity medium.

Six theoretical time sections, with 151 ZSR shotpoints each, were generated for a 0.5 s time window. These theoretical time sections were bandlimited with a 100 Hz zero-phase Ricker wavelet. From these six time sections, three sections were selected that contained the characteristic seismic features of a meandering sand channel.

Line A. (Figs. 3.8 and 3.9). Of the three seismic responses shown in Figure 3.9, the uppermost one is line A, whose shotpoint map is given in Figure 3.8. The flat reflector at 0.16 s is the sand-coal contact. Note that a negative reflection coefficient appears as a trough on the time section. Therefore, the onset of this reflection is measured at the central trough of the zero-phase Ricker wavelet. The event under SP 36 on the left-hand side of this time section is the seismic response of the sand channel traversed perpendicular to its trend. Two sets of diffraction events originate at the upper edge of the channel. The first is a strong event, with the same polarity as the reflection, and the second is a weaker event of opposite polarity. This diffraction emanates from the steep channel sides and is similar to the diffraction response of a vertical fault (Trorey, 1970).

Below these diffraction events is a strong apparent anticlinal feature—the buried focus (herein referred to as the 2-D focus). The physical surface at the bottom of the syncline has a radius of curvature that is focused beneath the simulated surface of the ground (see Fig. 3.5). To describe the physical surface in the syncline completely, each point or small area can be described as a portion of an elliptical paraboloid that requires two radii of curvature. The planes in which the radii of curvature are swept are orthogonal to each other. For each radius of

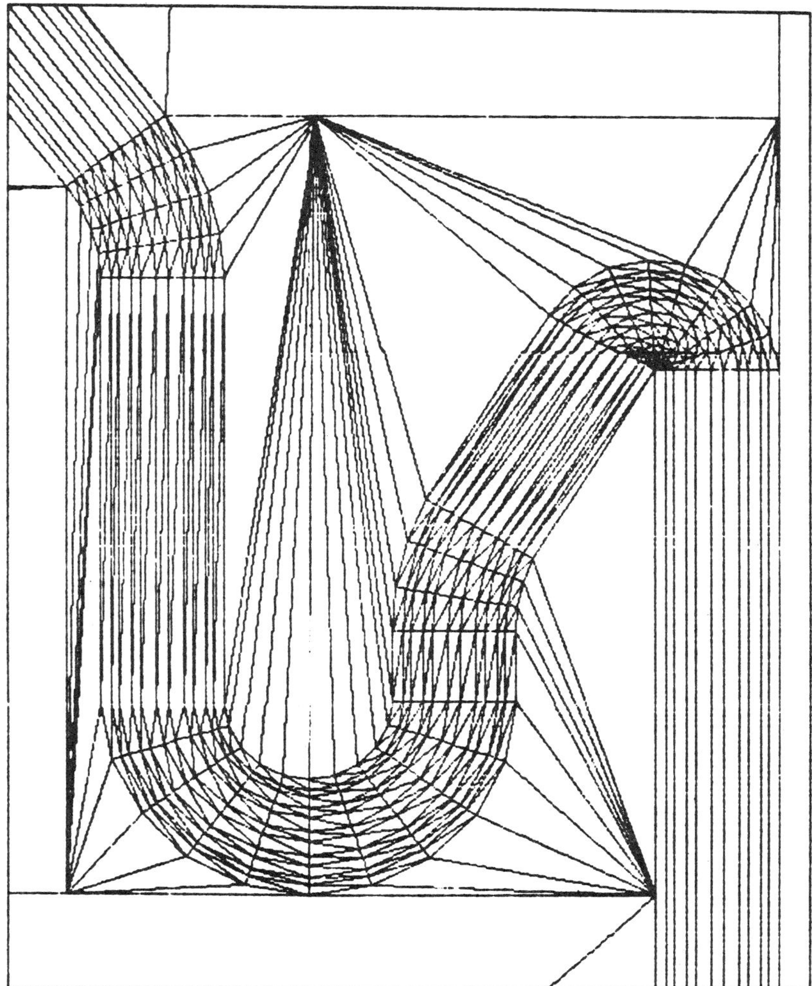

Figure 3.7. *The tile representation of the theoretical sand channel, with the 710 triangular tiles plotted.*

curvature that has its focus beneath the ground surface, a 90° phase change is observed. Consequently, the wavelet of the 2-D focus is 90° out of phase with the reflected wavelet (Dix, 1952, p. 365; Hilterman, 1970). Since the reflected wavelet is zero-phase, the onset of the 2-D focus is at the zero-crossing of the wavelet, which is at time 0.2 s on SP 36.

The seismic line traverses the edge of the tight channel bend. There is a second apparent anticlinal event with the same polarity as the 2-D focus. The apex of the event is at 0.22 s, and the event is broader than the 2-D focus, demonstrating that the seismic line fails to cross the center of the channel. Immediately above, around 0.16 s, a gentle time sag in the reflector implies false continuity of the sand-coal contact. This earlier event is an alignment of the strong diffractions from the near corner of the channel bend. The later event is the reflection

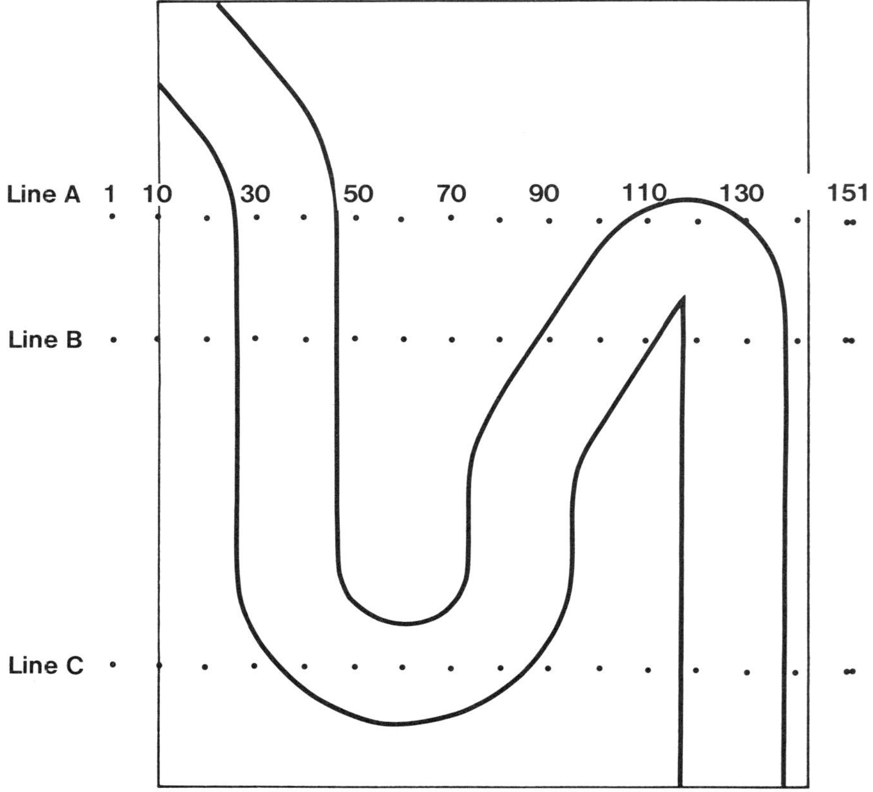

Figure 3.8. *Shotpoint map for the theoretical sand channel, lines A, B, and C, each with 151 ZSR shotpoints at 20 ft spacing.*

(diffraction)[2] from the far channel wall of the bend. Once again, the event is 90° out of phase with the reflected wavelet.

Line B (Figs. 3.8 and 3.9). Line B intersects the meandering channel three times. There are three 2-D foci, with the center one wider than the others. The extra width results from traversing the channel obliquely rather than perpendicularly. A fourth event at 0.27 s below SP 115 represents an event that is 90° out of phase with the 2-D focus and 180° out of phase with the reflected event (herein referred to as the 3-D focus). This 3-D focus is a reflection off the far wall of the tighter meander bend, which the profile line passes inside rather than directly over.

Line C (Figs. 3.8 and 3.9). This time section exhibits the problem of trying to classify events to a rigid structural chart. The right-hand event under SP 126 is definitely a 2-D focus by our standards, but the event under SP 60 meets the standards of neither a 2-D nor a 3-D focus but falls somewhere in between.

The gentle bend of the channel is traversed just slightly off center, generating the broader anticlinal event. This event arrives only slightly later than the 2-D

[2]Because the apparent anticlinal event is reflected not from a point but from a surface area, there is some confusion regarding whether to call this seismic event a reflection or a diffraction. In this chapter, the term *diffraction* is reserved for events generated at sharp corners, unless otherwise noted.

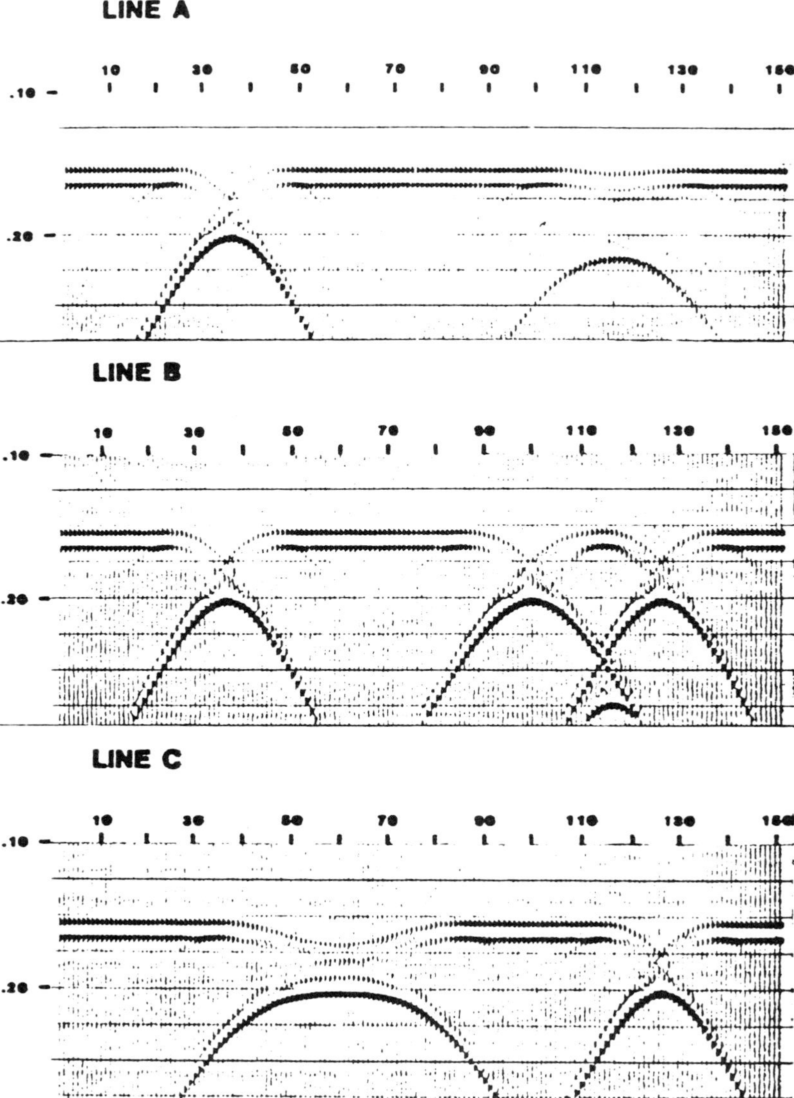

Figure 3.9. *The seismic response of lines A, B, and C of the theoretical sand channel. Note the apparent anticlinal features of the 2-D foci (apexes at 0.20 s on line A, SP 36; line B, SP 36, SP 100, and SP 125; line C, SP 125); the 3-D focus (apex at 0.275 s on line B, SP 115); and the meander bend sideswipes (apexes at 0.215 s on line A, SP 117, and at 0.20 s on line C, SP 55–65).*

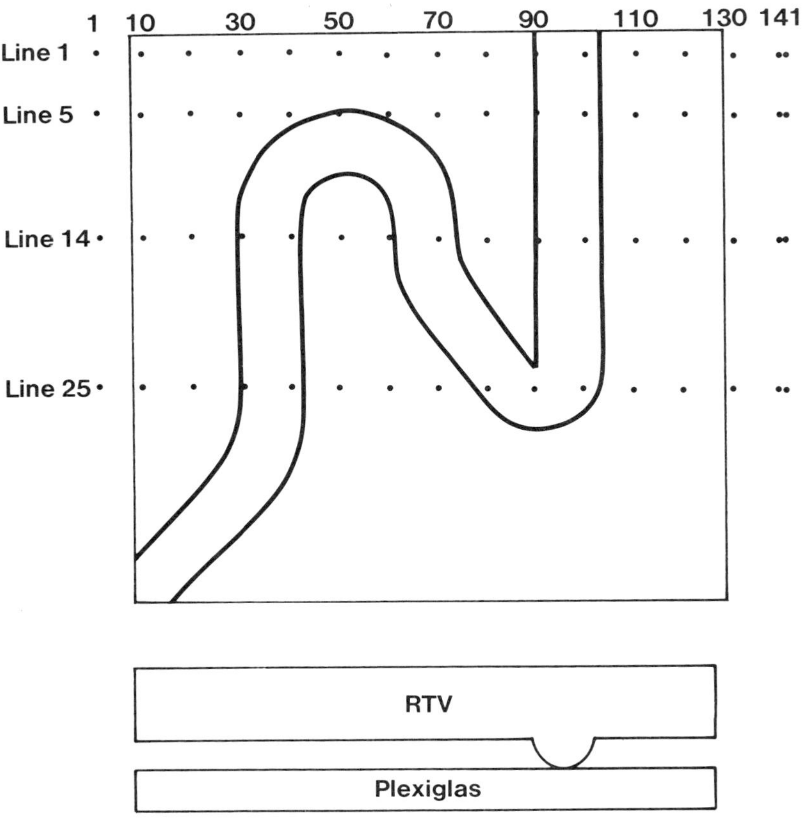

Figure 3.10. *Shotpoint map for SALCOL-1, lines 1, 5, 14, and 25, each with 141 common-offset (540 ft) shotpoints at 30 ft spacing.*

focus; consequently, the time sag in the upper reflector is greater than the example on line A.

SALCOL Physical Model. The SALCOL-1 physical model, as shown in Figure 3.5, features the original meandering stream model overturned, resting on a 112 ft Plexiglas sheet. Several small cubes of RTV were inserted near the edges, between the slab and the Plexiglas, for additional support. The center of curvature of the channel lies within the slab.

Four characteristic lines are presented with the polarity reversed, so that a trough marks the top of the coal.

SALCOL-1, Line 1 (Figs. 3.10 and 3.11). Four flat-lying reflectors are visible on the upper portion of this section (Fig. 3.11). The series of zero-phase peaks at 0.625 s is the reflection from the top of the gas-filled sand. Troughs at 0.735 s and 0.760 s and peaks at 0.780 s correspond to the top of the undisturbed portion of the coal seam, its base, and the base of the limestone, respectively.

The channel can be located on this time section by several methods. First, the 2-D focus associated with the channel is prominent on the right-hand side of the section. This apparent anticlinal event is broadened by the refractive properties of the upper gas-sand contact and interferes with the limestone reflectors, so that it

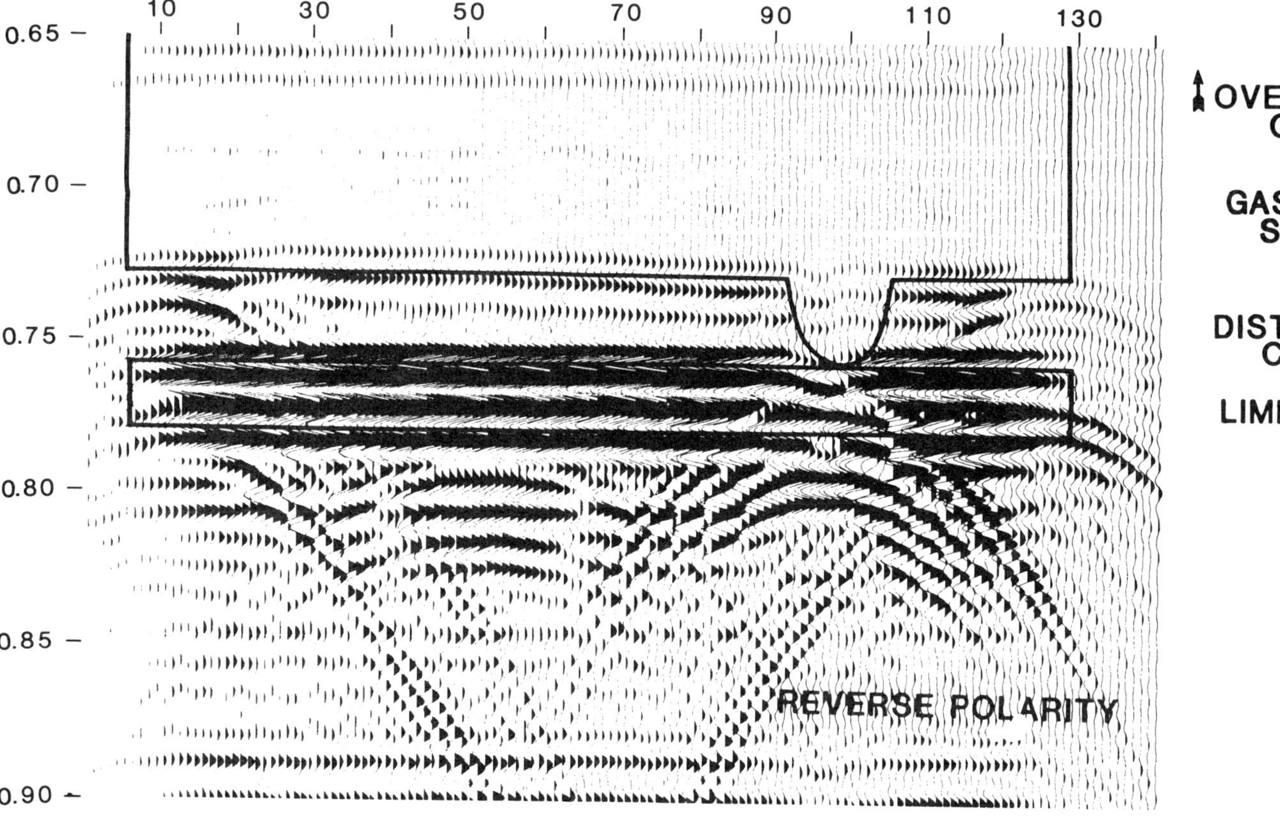

Figure 3.11. *The seismic response of SALCOL-1, line 1. Note the 2-D focus (apex at 0.76 s, SP 96) and the velocity sag in the limestone reflector (SP 92 to SP 104).*

is hard to pick the 2-D focus event. Also, the upper coal reflector loses continuity over the channel.

Indirectly, the channel can be located by velocity effects. The low-velocity channel produces a time sag in the underlying limestone reflector. As the coal seam yields laterally to slower gas-sand, the traveltime increases for any raypath that traverses the channel. The true time sag, 13 ms, exceeds the 4 ms sag that appears on the time section. The reflection coefficient at the top of the limestone increases where the slower gas-sand rests directly on the limestone. This increase in reflection amplitude, coupled with the 13 ms delay yields a false pick under the channel. Rather than picking the center trough under the channel, the earlier side-lobe trough appears to line up with the undisturbed portion of the reflector, and this appears, at first, to be the continuation of the reflector. Thus the event is picked 9 ms too early under the channel.

SALCOL-1, Line 5 (Figs. 3.10 and 3.12). The actual waveform of the 2-D focus is clearer on this time section than on the time section for line 1. Spurious events from the edge of the model masked the character of this event on line 1. The reflected waveform from the flat-lying lower gas-sand contact has an odd number of high-amplitude legs. If the apparent anticlinal event represents a diffraction rather than a 2-D focus, it should have the same number of legs as the reflected wavelet. This waveform contains an even number of legs and is therefore the derivative waveform, 90° out of phase with the reflected event.

Line 5 grazes the outer edge of the channel meander bend. Between SP 40 and SP 65, the amplitude of the reflection from the top of the coal seam is reduced, because only half the area is available for reflection. Also, in this same shotpoint zone, there is an apparent anticlinal event that is deeper and broader than the 2-D focus on the right-hand side. The broad reflection event emanates from the far channel wall and interferes with the base of the limestone unit, thus creating a reeflike feature, with an associated velocity sag.

SALCOL-1, Line 14 (Figs. 3.10 and 3.13) Line 14 crosses the channel three times, each time creating a 2-D focus and time sag on the underlying limestone reflector. There are two sideswipe events on this time section. A 3-D focus lies between SP 47 and SP 61, which is below the intersecting arms of the 2-D foci from the left two channels. The apex of the 3-D focus at 0.84 s comes from the inside of the meander bend near the center of curvature of the meander. A similar 3-D focus below the other intersecting 2-D foci between SP 70 and SP 90 is not evident, because the line is too far from the center of curvature of this tight meander bend.

The second 3-D event appears where the line crosses the middle channel, just above a gentle bend in the channel, and is shown between SP 80 and SP 95, dipping to the right. The bend focuses energy on the right arm of the center 2-D focus, but the energy arrives slightly later than the 2-D focus event. The amplitude of this arm is increased and appears to lengthen the waveform.

SALCOL-1, Line 25 (Figs. 3.10 and 3.14). Line 25 crosses the tight bend and, within the bend, intersects the center of the channel twice. The resulting broad focus resembles the one on the left-hand side of the time section for line 5, except that it arrives earlier, interfering with the upper surface of the limestone rather than the lower surface. Thus, even though a broad focus implies channel sideswipe, the timing of the event confirms that the center of the channel is crossed. This energy is reflected from the side of the bend and implies false continuity of the base of the low-velocity gas-sand.

SALCUT Physical Models. The SALCUT physical models, like the SALCOL model, were made from the same type of dark blue RTV. The SALCUT models are also meandering stream models, with the RTV resting upright on a 112 ft Plexiglas sheet (see Fig. 3.5). In these models, the RTV represents the coal seam, with the channel disturbing only the upper portion. Unlike the SALCOL model, this channel has a higher velocity than the coal. In the SALCUT-1 model, there is only

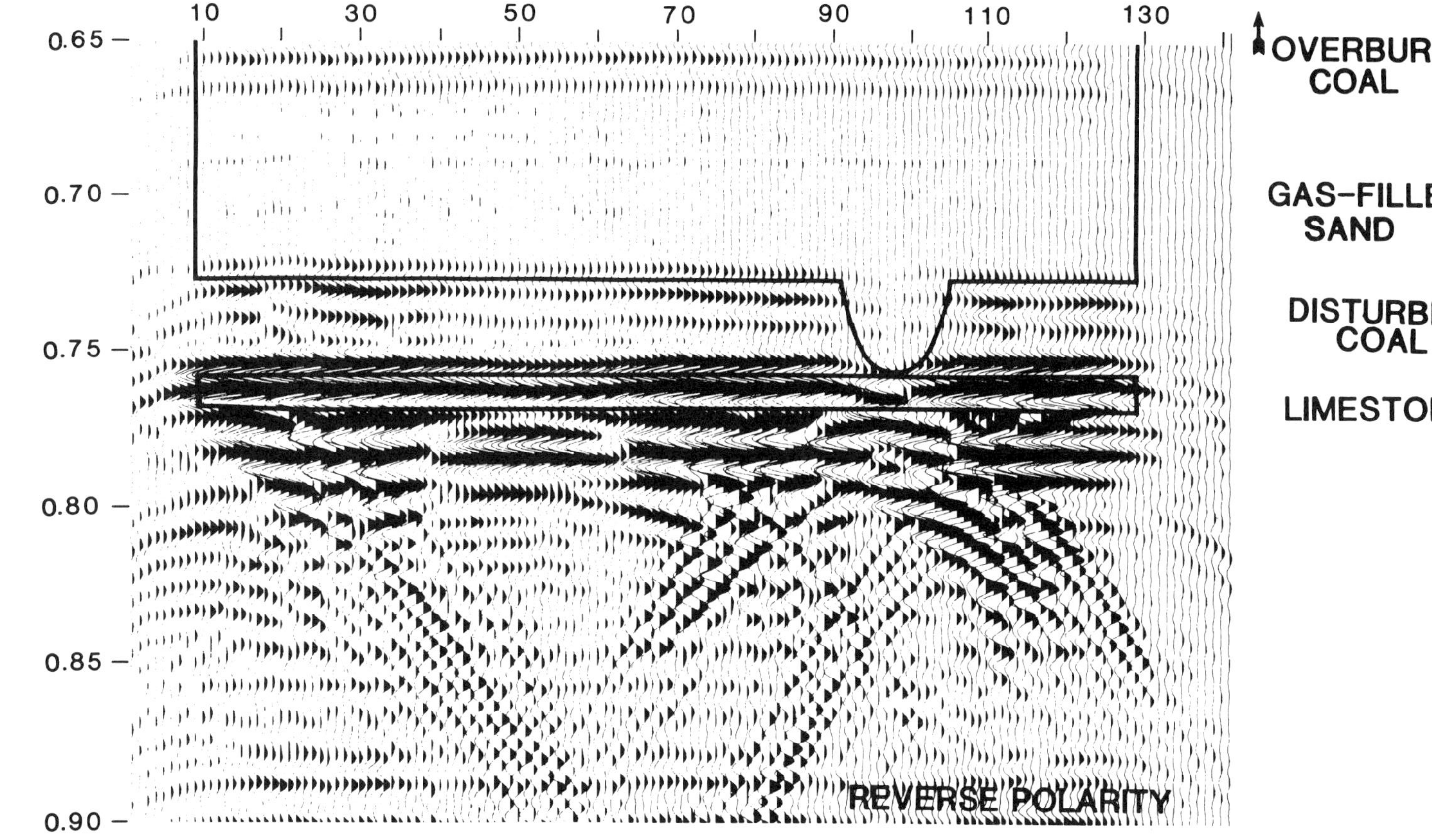

Figure 3.12. *The seismic response of SALCOL-1, line 5. Note the 2-D focus (apex at 0.75 s, SP 96) and the reeflike feature (0.78 s, SP 38 to SP 64), with velocity sag of the reflector directly below.*

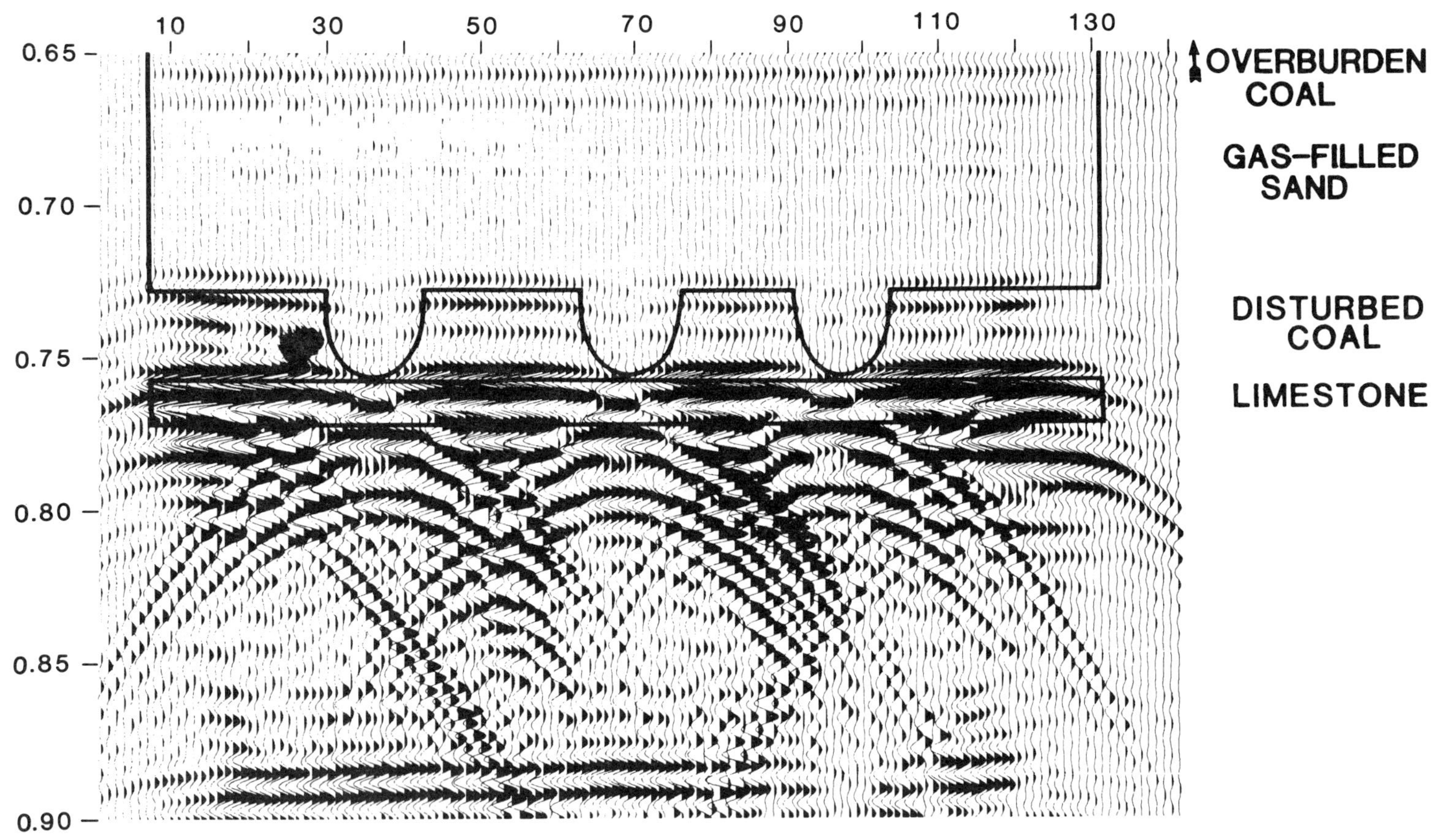

Figure 3.13. *The seismic response of SALCOL-1, line 14. Note the three 2-D foci (apexes at 0.76 s, SP 35, SP 68, and SP 96) and the 3-D focus (apex at 0.84 s, SP 53).*

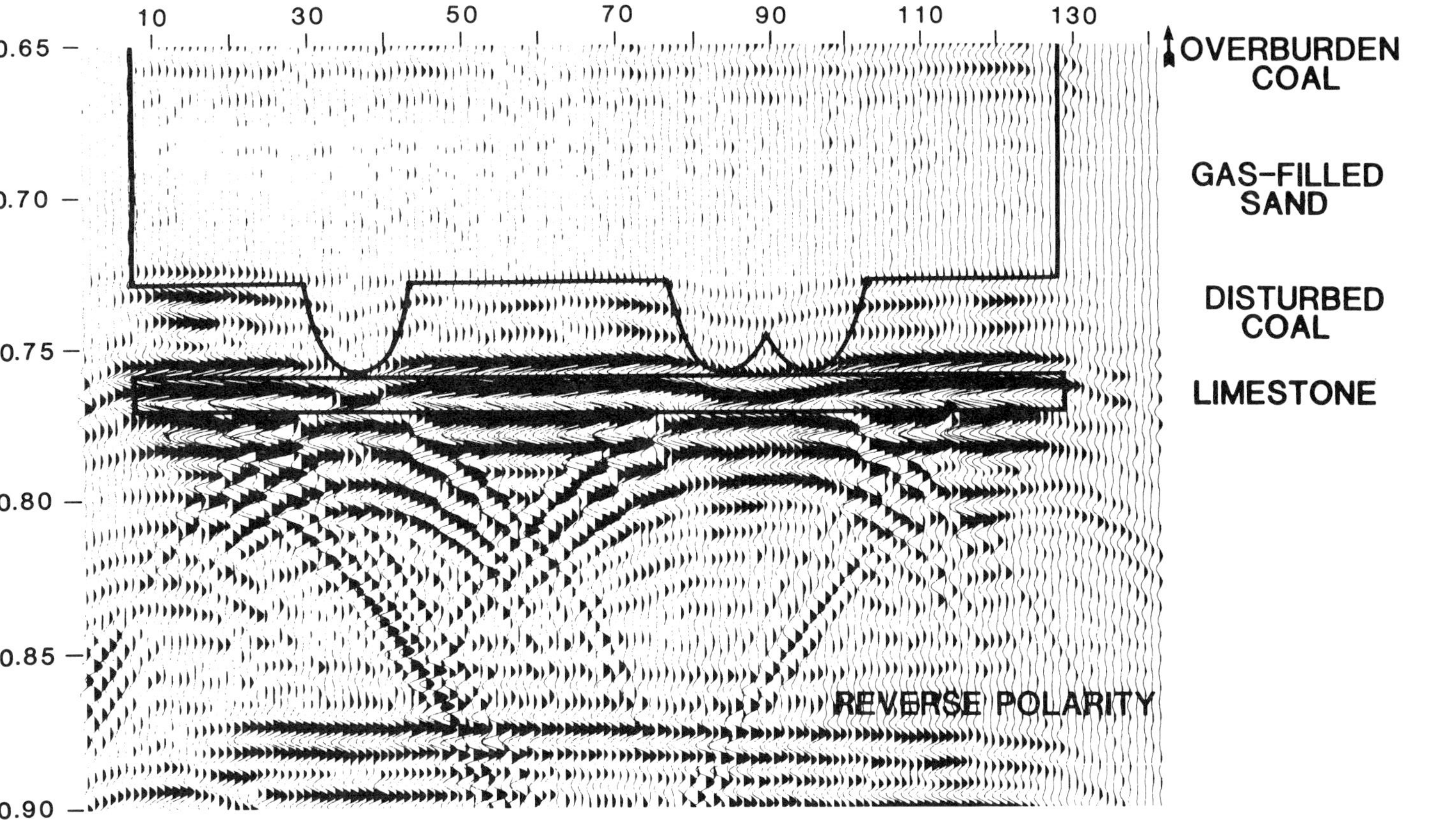

Figure 3.14. *The seismic response of SALCOL-1, line 25. Note the 2-D focus (apex at 0.76 s, SP 35) and the channel sideswipe (apex at 0.76 s, SP 78 to SP 96).*

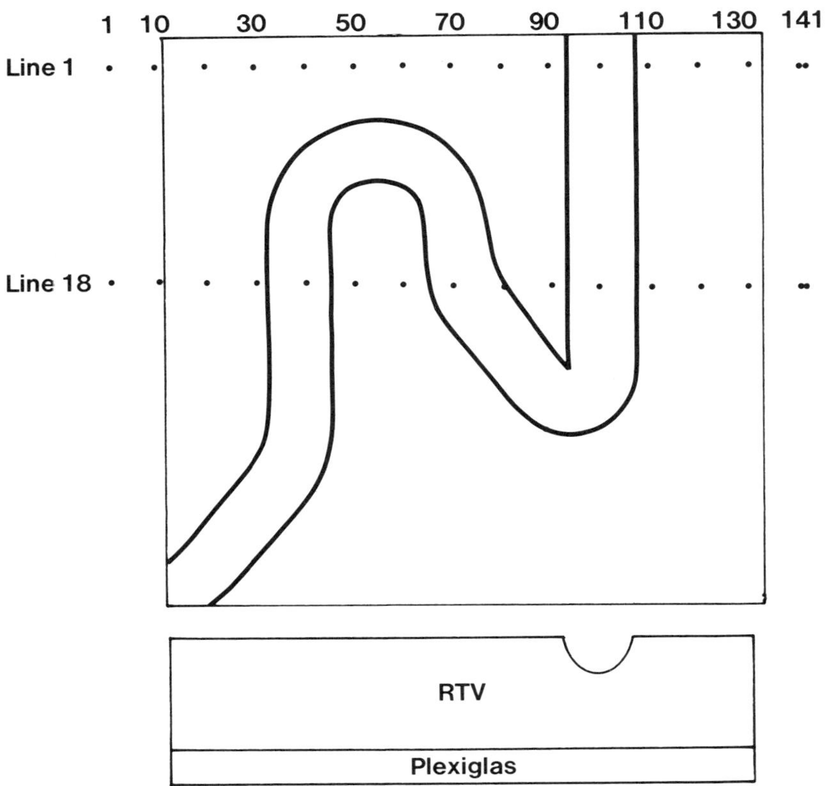

Figure 3.15. *Shotpoint map for SALCUT-1, lines 1 and 18, each with 141 common-offset (540 ft) shotpoints at 30 ft spacing.*

water between the RTV and the source and receiver. In the SALCUT-2 model, a square block Plexiglas rests on top of the RTV. This provides an interface between the channel wall and the center of curvature of the channel.

Three characteristic lines, two from SALCUT-1 and one from SALCUT-2, are presented, with the polarity reversed, so that a peak marks the top of the coal.

SALCUT-1, Line 1 (Figs. 3.15 and 3.16). The series of dipping peaks at 0.66 s is the seismic response of the top of the undisturbed portion of the coal seam. Troughs at 0.815 s and peaks at 0.835 s correspond to the top and base of the limestone, respectively. These events are all inclined, because the model dips slightly to the lower shotpoints.

The absence of another layer between the channel and the receiver permits unaltered development of the 2-D focus. Also, the thicker coal seam restricts interference with the deeper limestone, thus allowing observation of more subtle features. The 2-D focus is fully evolved, containing both sets of diffraction events that precede the 2-D focus.

The channel can be located in this model by velocity effects. The velocity pull-up of the underlying limestone below the channel is pronounced. The limestone reflection arrives earlier under the channel, because the higher velocity of the channel material significantly reduces the traveltime of the deeper reflection.

Two subtle sideswipe features are also highlighted here. A weak apparent anticlinal event appears on the left-hand side of this time section, with its apex at

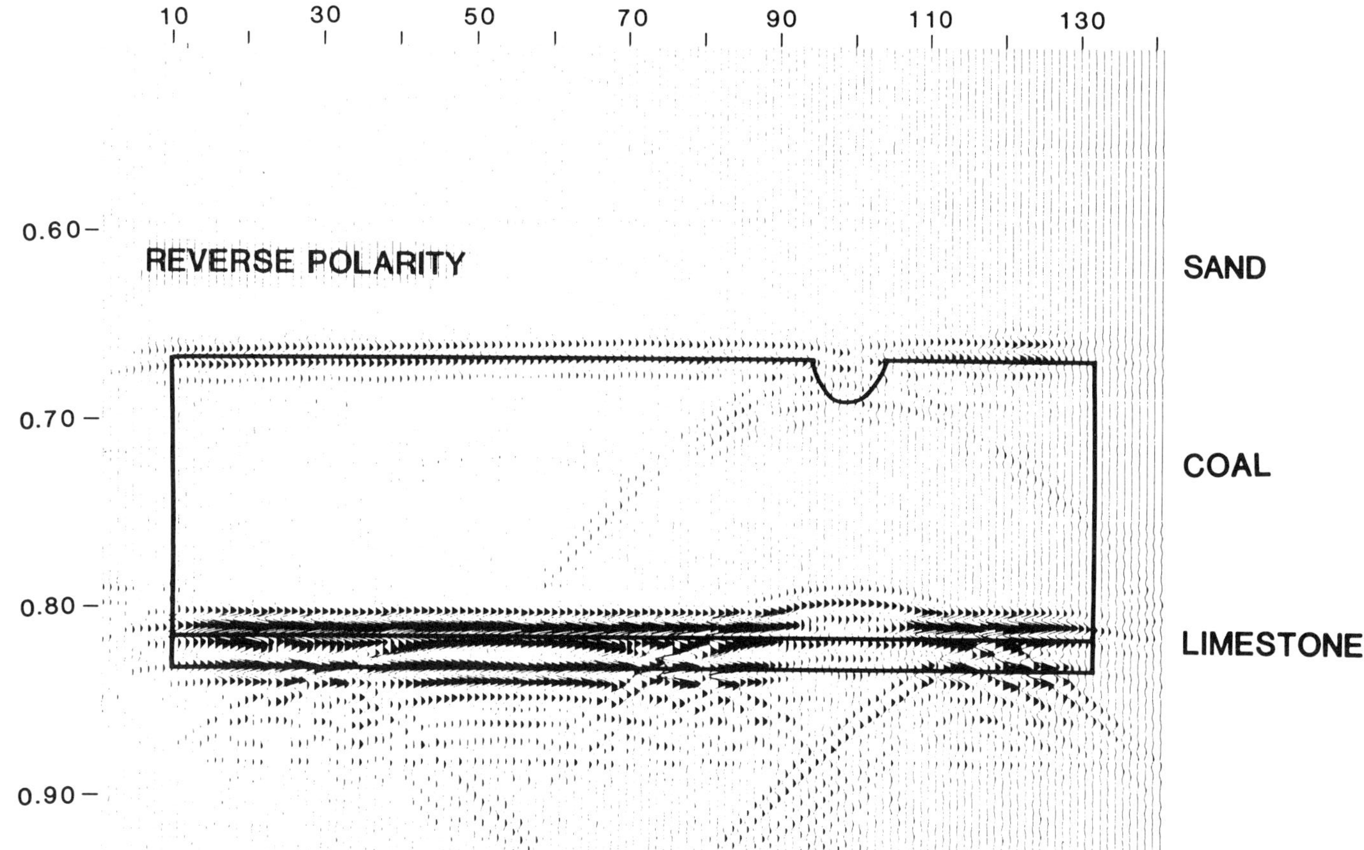

Figure 3.16. *The seismic response of SALCUT-1, line 1. Note the 2-D focus (apex at 0.685 s, SP 98), with an underlying diffractionlike velocity pull-up of the limestone reflector, and the channel sideswipe (apex at 0.72 s, SP 52), with a similar underlying diffractionlike feature.*

0.725 s at SP 53, originating at the far inside channel wall of the meander bend, well removed from the line. As successive seismic lines approach this meander bend, the event arrives at an increasingly earlier time, of course.

There is a second, broader and deeper apparent anticlinal event associated with the meander bend. This event interferes with the basal limestone reflection. The strong peak from this contact breaks into a doublet between SP 50 and SP 70. It would be easy to conclude that phase changes in this event represent local changes in lithology. The event arrives earlier as each successive seismic line gets closer to the meander bend. When the line passes directly over the bend, the event is the primary limestone reflection.

The origin of the sideswipe explains the diffractionlike appearance of the velocity pull-up associated with the channel. The deeper apparent anticlinal event and the diffractionlike tails of the velocity pull-up are very unusual and only occur if the channel fill has a higher velocity than the coal. In addition to the direct vertical raypath through the undisturbed coal, a second raypath to the limestone exists, with normal incidence at the limestone, which is refracted by the near channel wall and deflected toward the normal upon entering the slower coal medium. The traveltime of this second reflection increases directly with distance to the channel.

SALCUT-1, Line 18 (Figs. 3.15 and 3.17). The time section is very complex in the coal zone for such a simple model. The section for line 18 displays three 2-D foci and at least two 3-D foci. All three 2-D foci have the same arrival time, after correcting for regional dip. The middle 2-D focus is wider, as the channel is crossed at an obtuse angle to its trend. The velocity pull-up of the limestone reflector is also wider under the center channel.

The arrival time of the 3-D foci reflects the distance from the shotpoint to the outer channel wall of each meander bend. The 180° phase change of the reflected event expected with a 3-D focus is prominent. The limestone refracted reflection present on line 1 is absent below these 3-D foci. This may be a result of the noisy appearance of this section below the coal seam.

SALCUT-2, Line 14 (Figs. 3.18 and 3.19). The seismic reflection from the thick limestone layer is the series of troughs at 0.505 s, correlating to its top, and the series of peaks at 0.595 s, corresponding to its base. The discontinuous series of troughs and peaks at 0.745 s and 0.760 s represent, respectively, the top and base of the thin limestone.

Line 14 traverses a similar section of the model as SALCUT-1, line 18 (see Figs. 3.15 and 3.18), intersecting the meandering sand channel three times. On the time section of line 14, all direct evidence of the presence of this sand channel is absent. Apparently, the base of the thick limestone positioned between the channel wall and its center of curvature has diffused the energy of the 2-D and 3-D foci. At a greatly enhanced gain, the lower limestone reflectors become continuous, displaying the diffractionlike velocity pull-up seen previously. However, there is still no evidence of either a 2-D or a 3-D focus.

Amplitude can be an indirect method of determining the channel location. The amplitude of the reflection from the base of the limestone should be greater where it rests on undisturbed coal. The magnitude of the reflection coefficient (−0.383) is larger for a limestone-coal interface than it is for a limestone-sand interface (−0.286). Amplitude differences among the series of peaks at 0.595 s are not obvious.

Because data for each time section were recorded digitally, the weight of each sample can be printed individually. A computer program was written to measure the peak-to-trough amplitude of this event. Its magnitude is plotted for each shotpoint on Figure 3.20. The undisturbed coal between SP 42 and SP 62 is defined by a broad platform of high amplitudes. However, the amplitude anomaly associated with the coal between SP 74 and SP 89 is less pronounced.

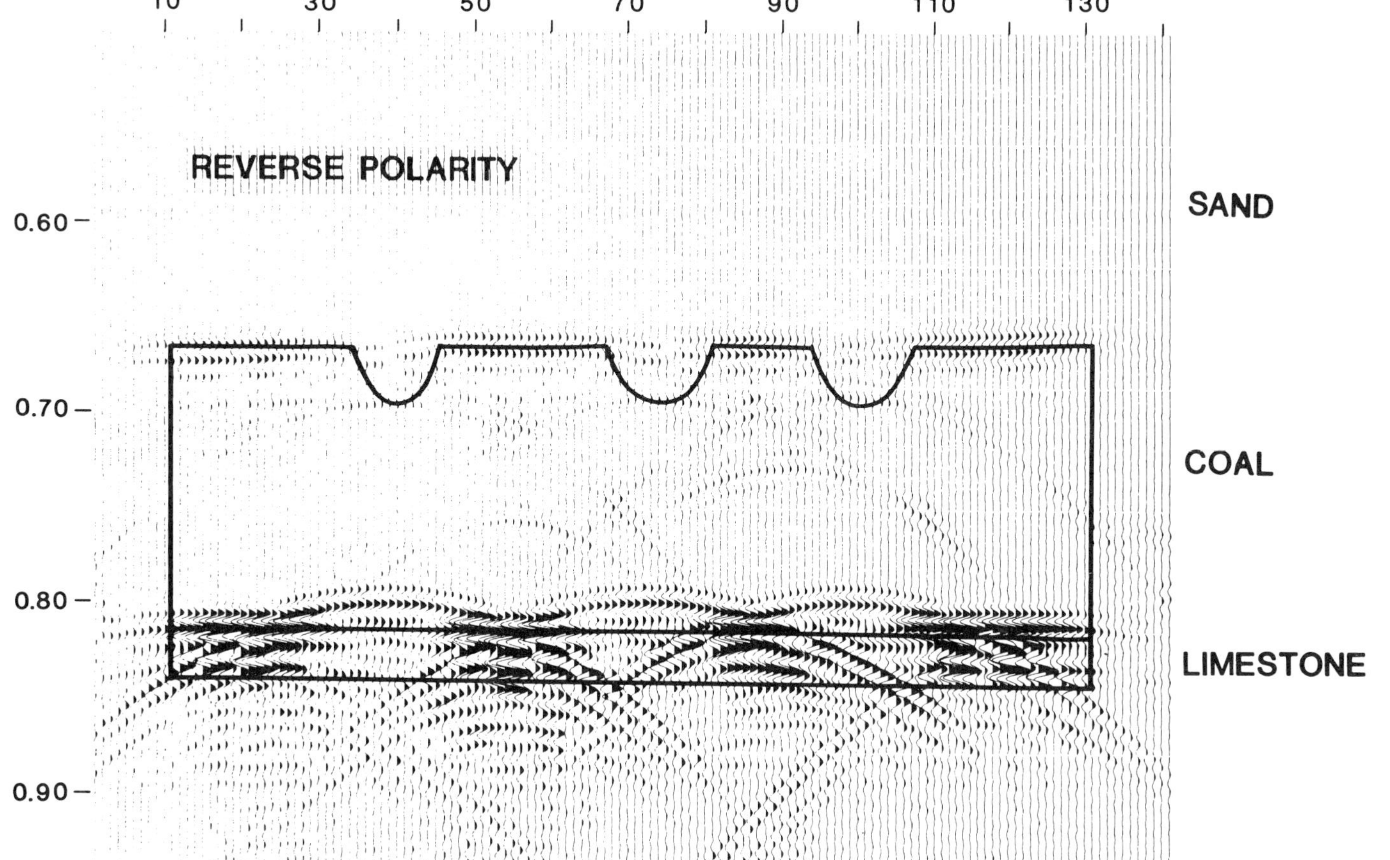

Figure 3.17. *The seismic response of SALCUT-1, line 18. Note the complex appearance of this seismic section because of the three 2-D foci (apexes at 0.69 s, SP 39, SP 73, and SP 98), with underlying diffractionlike velocity pull-up of the limestone reflector, and the two 3-D foci (apexes at 0.76 s, SP 55, and 0.73 s, SP 90).*

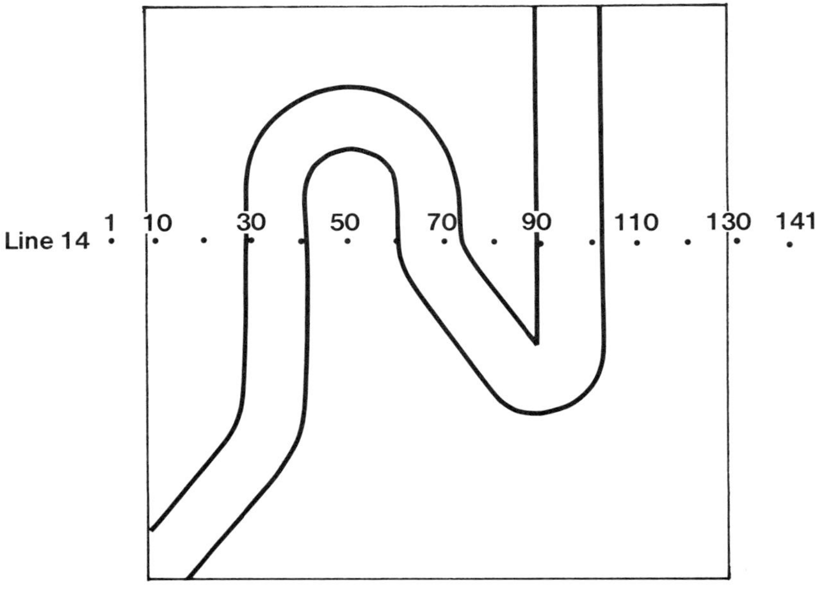

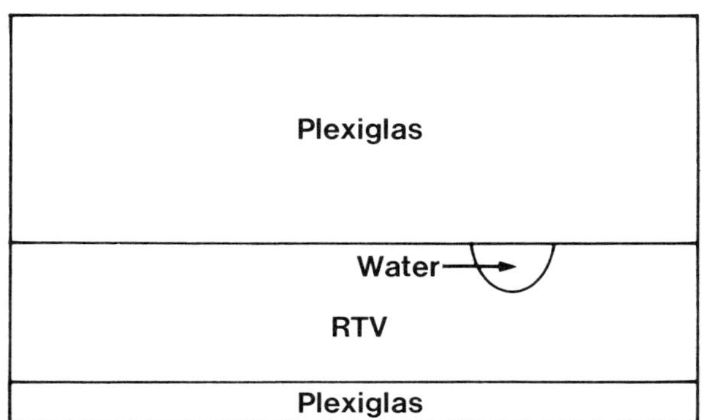

Figure 3.18. *Shotpoint map for SALCUT-2, line 14, with 141 common-offset (540 ft) shotpoints at 30 ft spacing.*

This amplitude phenomenon is a 3-D Fresnel zone effect. The actual portion of the 3-D Fresnel zone that the reflective body occupies determines the amplitude response. A one-wavelength Fresnel zone for this reflector depth and overburden velocity is a disc with a 1000 ft diameter. The width of the second undisturbed region is 450 ft, and that of the first region is 600 ft. Thus, the amplitude anomaly for the second undisturbed region is not so prominent.

The application of this amplitude method for differentiating undisturbed coals from channel-fill coals requires a locally homogeneous geologic environment. The amplitude anomaly for this lateral lithologic variation would be obscured by localized vertical changes in rock properties. For example, an increase in the

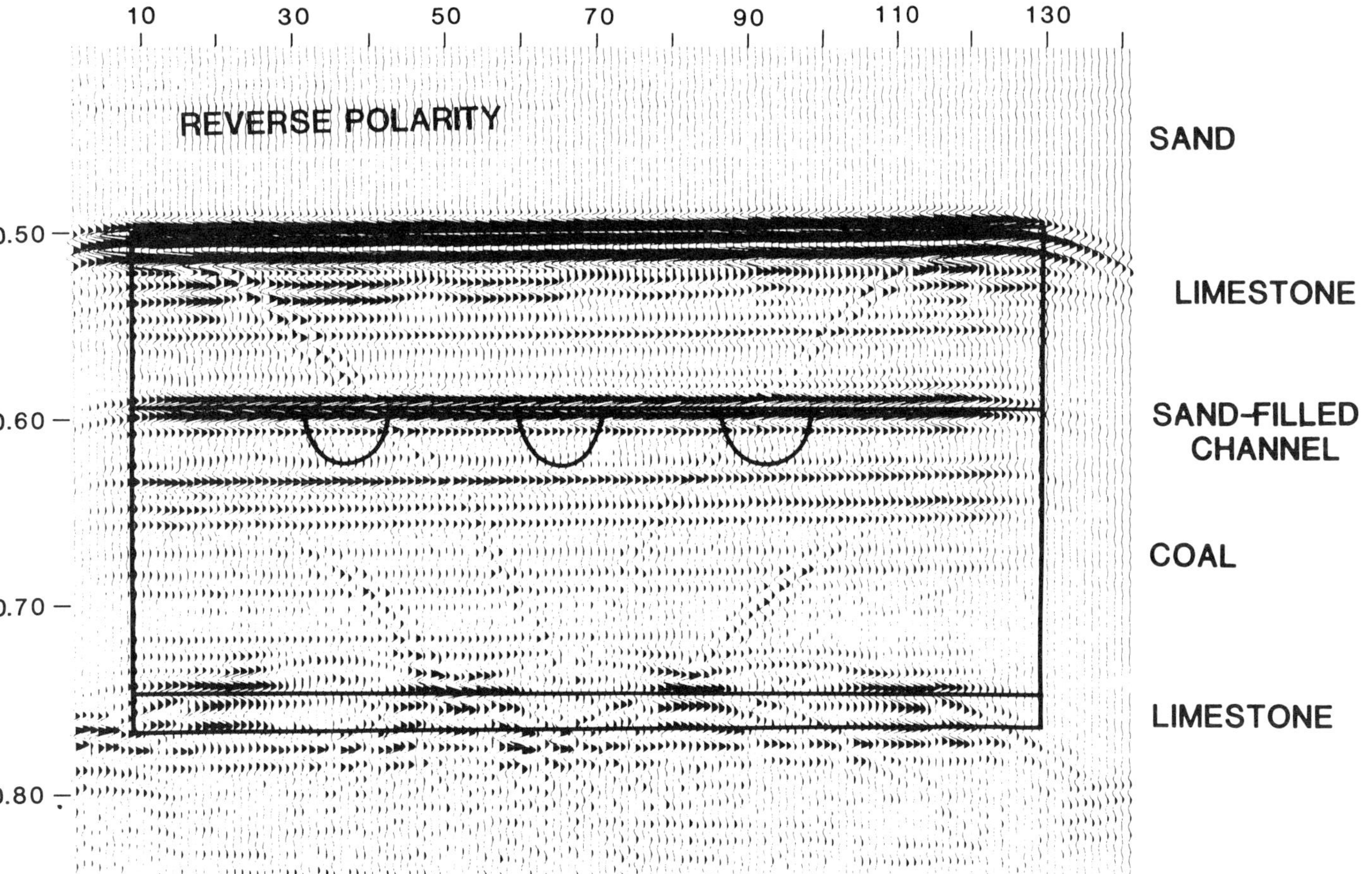

Figure 3.19. *The seismic response of SALCUT-2, line 14. Note the absence of the 2-D and 3-D foci seen on SALCUT-1, line 18, and the presence of the diffractionlike velocity pull-up underlying each channel.*

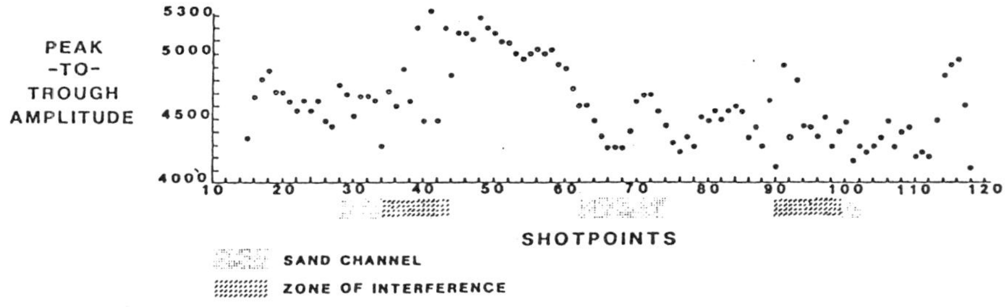

Figure 3.20. *Peak-to-trough amplitude of the basal limestone reflector (0.595 s) on SALCUT-2, line 14.*

number of mud-splits from neighboring crevasse splays could produce a similar amplitude anomaly.

PROGRADATIONAL DELTA COALS. The SALSTR-1 and SALSTR-2 physical models simulate a prograding delta. Thin and usually uneconomic coals accumulate along distributary levees and lagoons in the active lower delta and back-barrier depositional environments. The Clarion and Beckley Coals, discussed previously, exemplify the expected shape of the back-barrier coals.

The emphasis during these physical model runs was to illustrate the effects of a high acoustic impedance contrast on the coherence of the reflectors directly beneath. To maximize this contrast, Plexiglas represents the coal, and water represents the prodelta and deltaic sediments. A multilayered silicone model would fail to provide a sufficient velocity contrast. For thin coals, a 37 ft Plexiglas sheet has an effective one-way thickness of 0.29 wavelengths, while the objective coal, a 600 ft Plexiglas sheet, has an effective one-way thickness of 4.63 wavelengths. The reflections from the top and base of the thin coals constructively interfere, thus enhancing the amplitude of the seismic response. These lower delta and back-barrier coals are also laterally narrow. The widths of the lower delta and back-barrier coals are, respectively, less than the diameter of a 1/4 to 1/2 wavelength Fresnel disc.

In SALSTR-1 and SALSTR-2 (Figs. 3.21 and 3.22), there are six parallel back-barrier coals located 130 to 150 ft above the thick objective coal seam. These coals were assumed to be deposited in three progradational stages, approximately 500 ft apart. An inlet active during each of these stages separated the coals into six pod-shaped bodies, with 150 ft between progradational pairs. The shape of the resultant coals vary, with lengths of 1000 to 2000 ft and widths of 230 to 450 ft.

In SALSTR-1 (Fig. 3.21), there are four lower delta coals positioned orthogonally to the back-barrier coals 270 to 300 ft above the objective coal seam. These coals, grouped into pairs, were deposited along levees flanking two 300 ft wide distributary channels, which converge from 2000 to 1000 ft apart. One pair of lower delta coals maintains a width of 300 ft, but the second pair thins from 375 to 0 ft.

Twenty-five seismic lines were gathered over each physical model at 150 ft line spacing, with a 0.5 ms sampling rate and filter limits of 40 to 160 Hz. The lines contained 141 shotpoints, with 30 ft spacing. A common-offset source-receiver configuration was employed, with 540 ft between the source and the receiver.

Four characteristic lines, two from SALSTR-1 and two from SALSTR-2, are shown with the polarity reversed, so that a trough marks the top of the coal.

SALSTR-1, Line 5 (Figs. 3.23 and 3.24). This line traverses the four delta coals and the objective coal. The disturbed sequence of troughs at 0.590 s and the peaks at 0.680 s correspond to the top and base of the objective coal, respectively. The

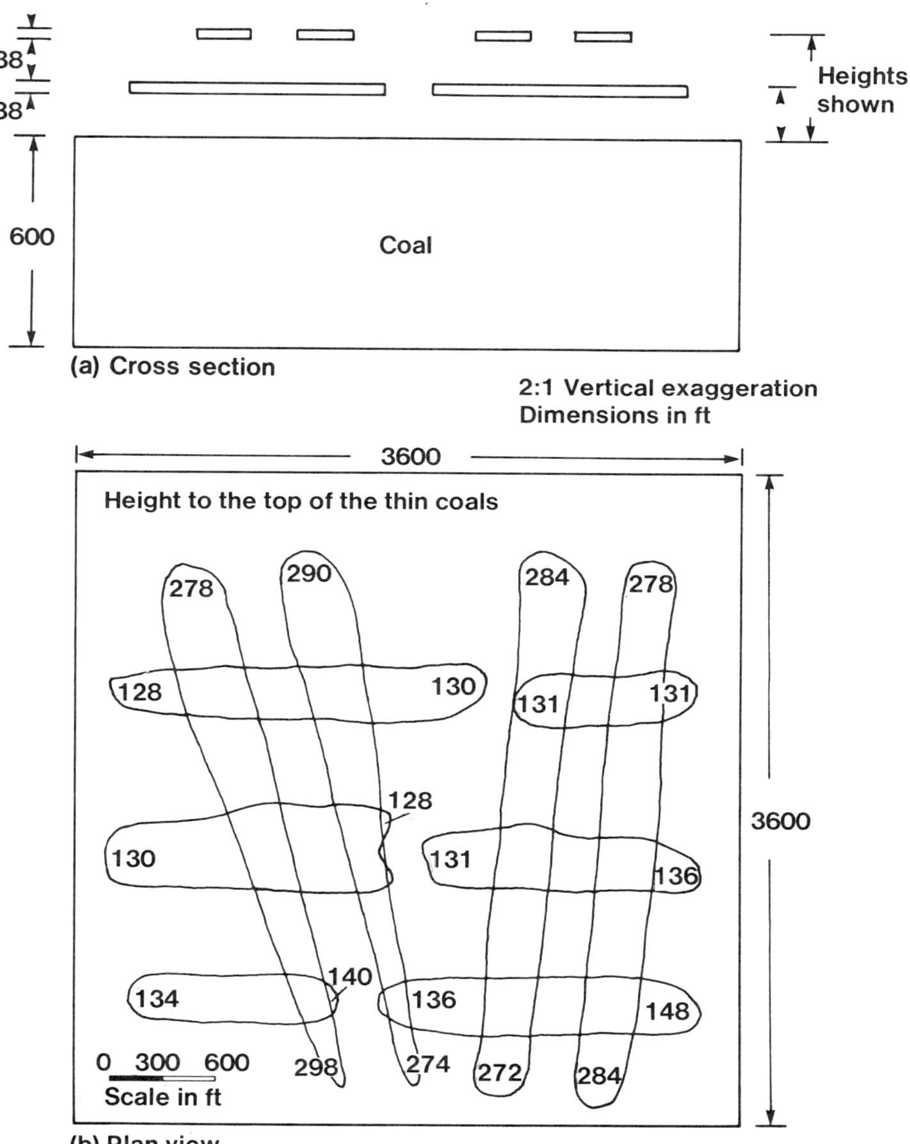

Figure 3.21. *Plan view and cross section of SALSTR-1.*

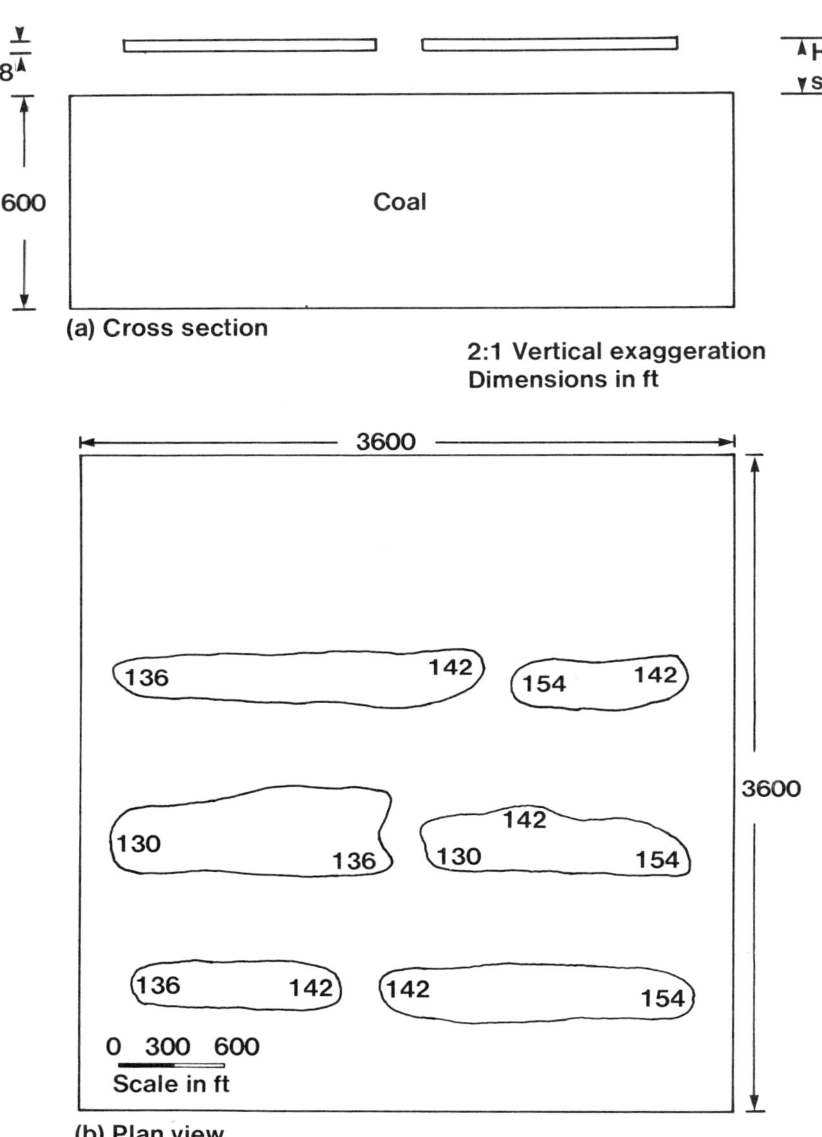

Figure 3.22. Plan view and cross section of SALSTR-2.

seismic response of the top and base of each of four delta coals is the pair of troughs and peaks at 0.515 s and 0.519 s (SP 26 to SP 38), 0.513 s and 0.517 s (SP 48 to SP 61), 0.514 s and 0.518 s (SP 92 to SP 104), and 0.510 s and 0.514 s (SP 110 to SP 118), respectively.

The discontinuity of the reflector demarcating the top and base of the objective coal is affected by shallower layering. The lower delta coals attenuate seismic energy, thereby reducing the amplitude of the objective coal reflections to create a dimspot immediately below. These overlying coals also produce a velocity effect on the objective coal. Raypaths that travel through the higher-velocity coal arrive

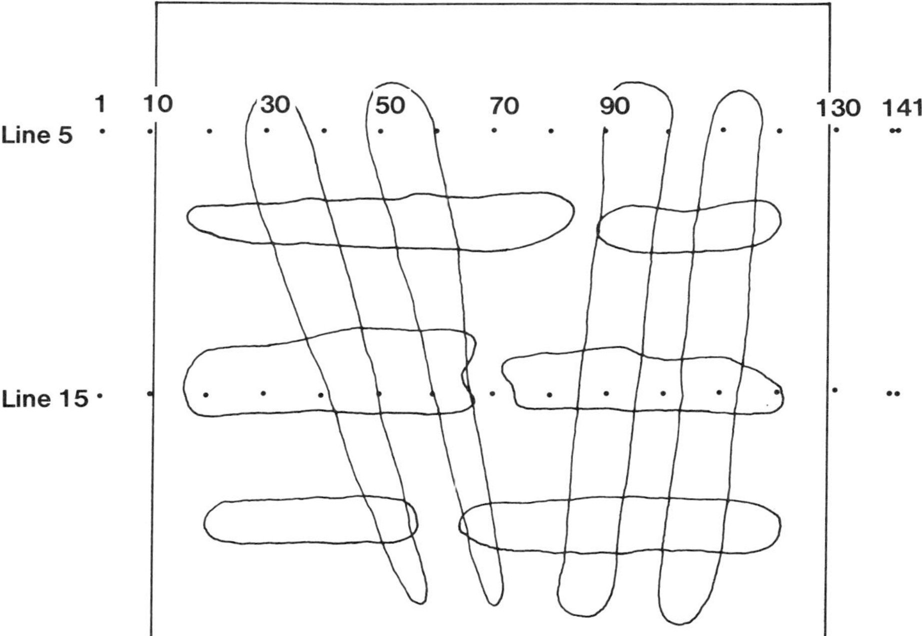

Figure 3.23. *Shotpoint map for SALSTR-1, lines 5 and 15, each with 141 common-offset (540 ft) shotpoints at 30 ft spacing.*

earlier than those that travel strictly within the deltaic sediments. The objective coal reflections below these delta coals experience a velocity pull-up of 5 ms.

Two types of diffraction features occur on this and other lines. The first is a classic edge diffraction, whereby energy returns from a location along the near edge of the body with a raypath perpendicular to the trend of the delta coal. The edge diffraction is steeper for the second pair of delta coals (SP 26 to SP 61) than for the first pair (SP 92 to SP 118), because the first pair is almost perpendicular to the seismic line, while the second pair is oblique. This relation is displayed by the two earliest diffractions between SP 61 and SP 78 from the first pair and between SP 78 and SP 92 from the more oblique pair.

The second type of diffraction results from 3-D focusing. For instance, line 5 crosses the second delta coal near its semicircular end, where energy is focused to make a higher-amplitude, leggy waveform. The small lateral extent of the delta coals creates an apparent point diffractor at the end of the body as energy is focused from around each end point. This is the steeper diffraction, prominent between SP 58 and SP 76, that is associated with the termination of each delta coal body and arrives later than the diffractions associated with the linear edge.

SALSTR-1, Line 15 (Figs. 3.23 and 3.25). This line crosses four lower delta coals and two back-barrier coals. The seismic response of the top and base of the back-barrier coals is the series of troughs and peaks at 0.558 s and 0.562 s (SP 16 to SP 67) and 0.555 s and 0.559 s (SP 74 to SP 122), respectively. The superposition of these six coals destroys the signal continuity from the thick objective coal. Even if there is some continuity, the objective coal appears highly faulted.

The only feature on this line of possible economic interest is the distinct high between SP 95 and SP 105 at the back-barrier coal's level. This high anomaly, which displays excellent closure, is in reality a secondary diffraction feature.

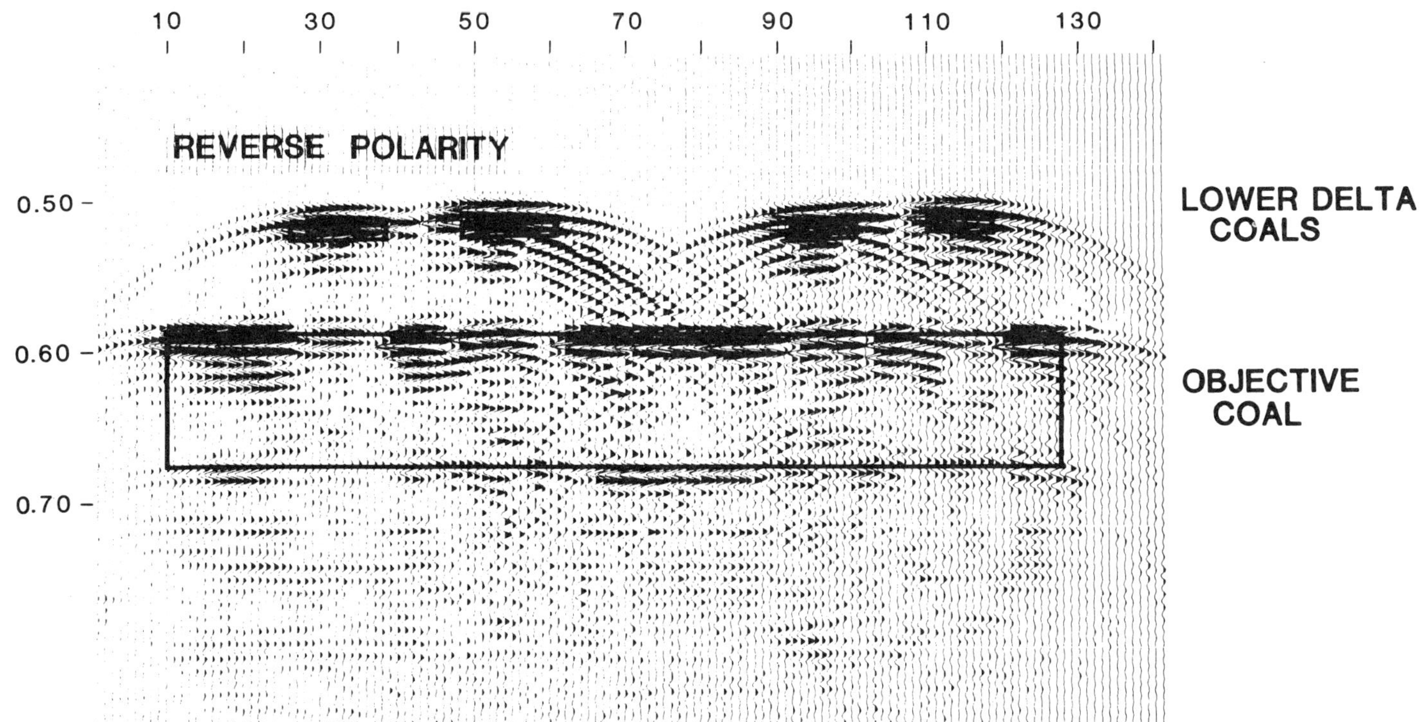

Figure 3.24. *The seismic response of SALSTR-1, line 5. Note both the velocity pull-up of the objective coal and the dimspots below each lower delta coal as well as the diffractions off the near edge of each lower delta coal (for example, at 0.513 s, SP 62, to 0.530 s, SP 75) and the steeper ones off the far end for example, at 0.530 s, SP 62, to 0.575 s, SP75).*

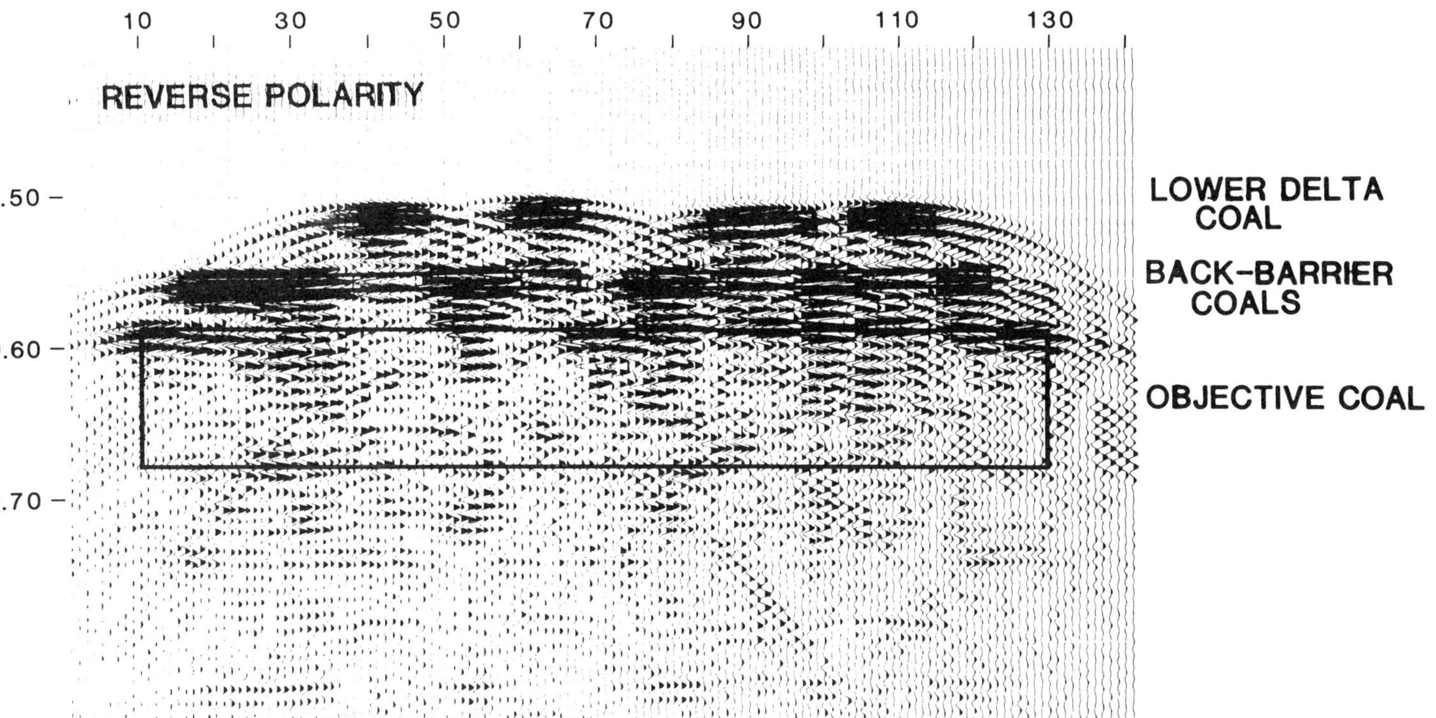

Figure 3.25. *The seismic response of SALSTR-1, line 15. Note the incoherent objective coal reflector and the apparent anticlinal feature (apex at 0.555 s, SP 98).*

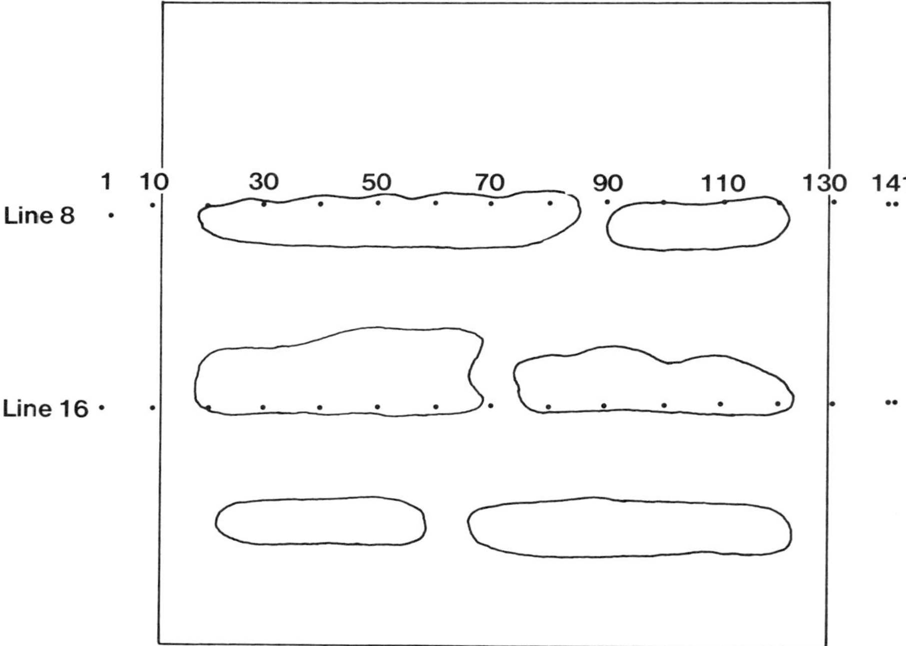

Figure 3.26. *Shotpoint map for SALSTR-2, lines 8 and 16, each with 141 common-offset (540 ft) shotpoints at 30 ft spacing.*

The pair of lower delta coals above the feature dip away from their adjacent sides. Besides the classical diffraction event generated from the lower delta coals, a secondary result is a downgoing ray that is refracted at the base of the lower delta coal surface before being reflected by the back-barrier coal. This creates an inward-facing set of diffractions, with an apparent onset at the deeper interface. These two diffraction limbs from the inner edges of each coal body imply false continuity between contrasting flat reflections.

SALSTR-2, Line 8 (Figs. 3.26 and 3.27). The sequences of troughs and peaks at 0.554 s and 0.559 s (SP 18 to SP 84) and 0.556 s and 0.560 s (SP 96 to SP 122) are the top and base of the two back-barrier coals, respectively. The sequences of discontinuous troughs and peaks at 0.590 s and 0.673 s are the top and base of the objective coal, respectively.

Line 8 skirts the irregular edge of the first back-barrier coal parallel to its trend. Approaching the body at this angle, the propagating signal is scattered by the edge of the body, and mapping of the objective coal without relying on faults would be difficult. The second back-barrier coal, also skirted by the line, has a crescent-shaped face, which selectively scatters the seismic signal. This yields the impression of roll-over in the objective coal below.

SALSTR-2, Line 16 (Figs. 3.26 and 3.28). The second back-barrier coal dips into the line, while the left coal is moderately flat. The dip on this right body differs from that of the objective coal below and creates the impression of a pinchout of the unit between the coals. The high amplitude at the top of the objective coal, where the back-barrier coals are absent, could easily be interpreted as a tuning phenomenon associated with the pinchout.

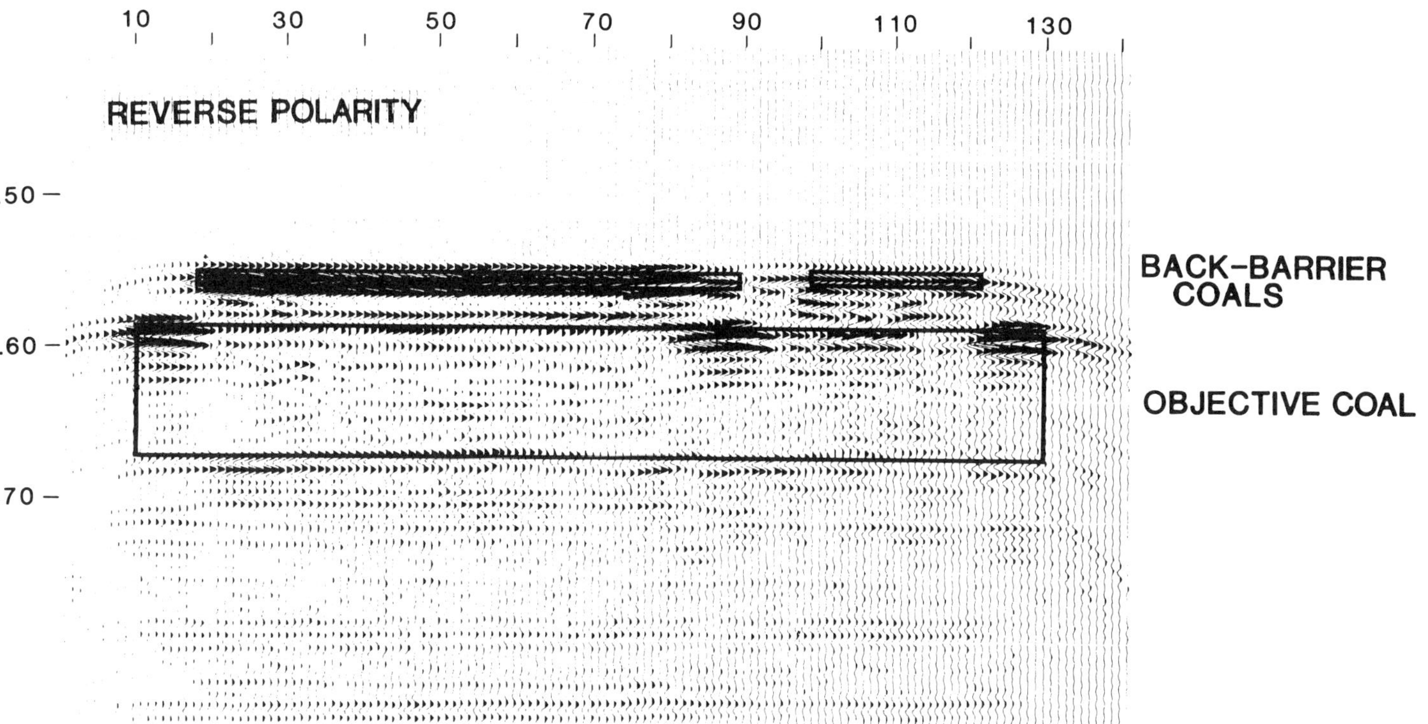

Figure 3.27. *The seismic response of SALSTR-2, line 8. The seismic signal is scattered by the edges of the back-barrier coals, yielding a faulted appearance for the objective coal's reflector.*

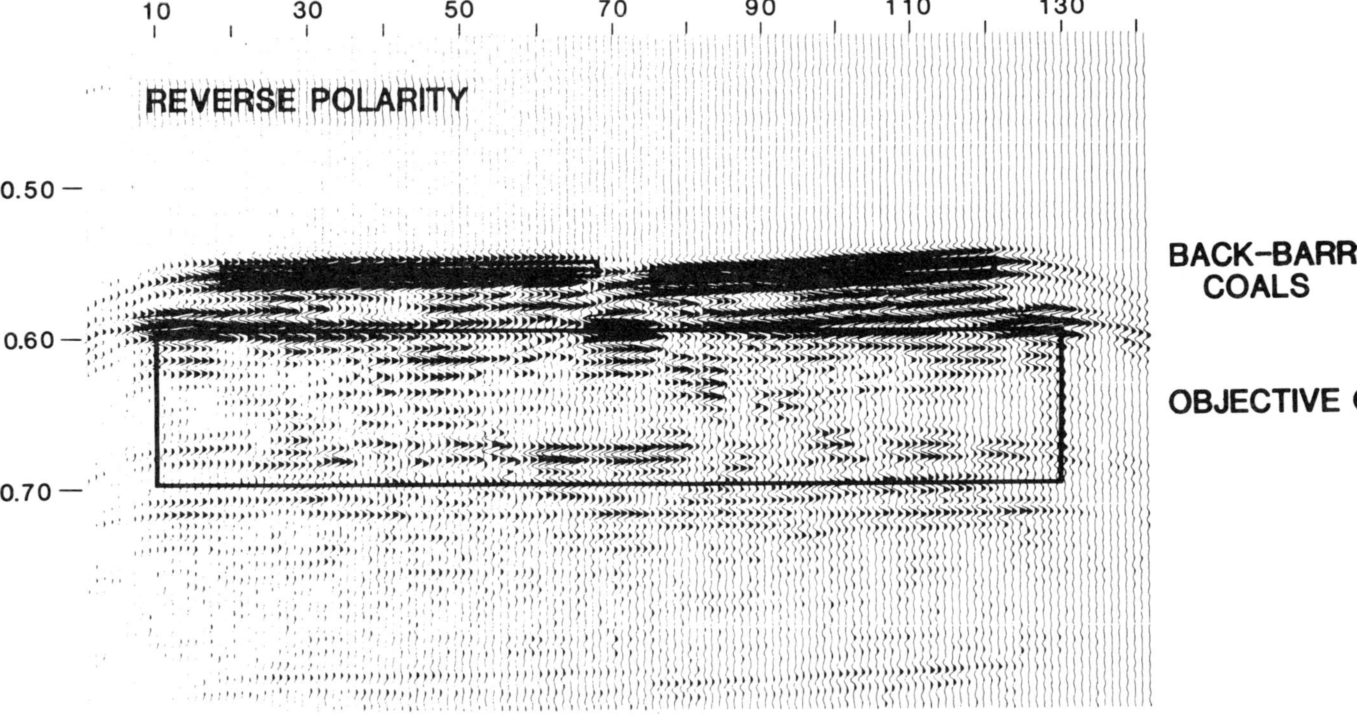

Figure 3.28. *The seismic response of SALSTR-2, line 16. Note the false pinchout on the right-hand side of the section, with an apparent amplitude tuning (0.592 s, SP 70).*

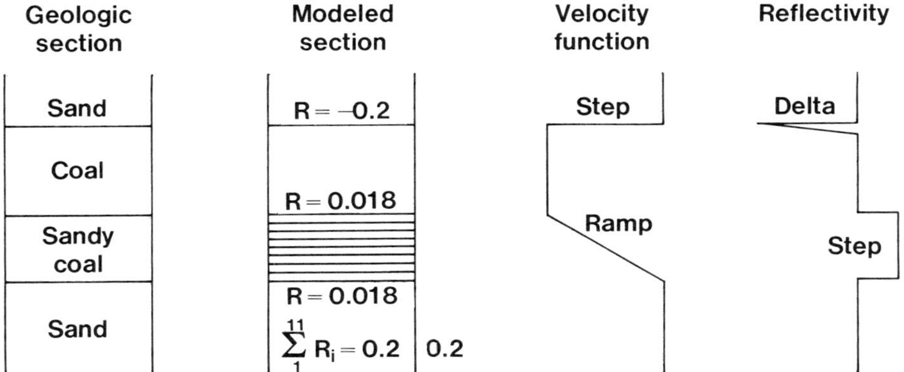

Figure 3.29. *Delineation of a coal seam encased in sand, with a sharp upper contact and a transitional layer for a lower boundary.*

DELINEATION OF THE STYLE OF A COAL'S CONTACT. With the advent of true-amplitude recording of seismic data and wavelet processing, many of the subtle, previously unnoticed features on conventional seismic sections are now being revealed. These and other exotic processing steps permit the seismic interpreter to stretch his limited knowledge of the subsurface to the edge of credibility. This is especially important if these high-resolution seismic techniques are to be applied to coal exploration.

Successful estimation of coal reserves requires accurate measurement of the thickness of a coal seam, which, as a stratigraphic unit, is often acoustically thin. Other problems include vertically thinning beds or pinchouts and transitional coal contacts. To be successful in estimating the amount of recoverable coal, both the amplitude and frequency response that characterize these coal seams must be understood.

In this section, the problem of the seismic resolution of these subtle features is approached, first, by expanding on the studies of Sengbush, Lawrence, and McDonal (1961), who modeled flat transitional boundaries. Then, the amplitude phenomenon associated with a thinning bed is reexamined and supplemented by observations of a similar frequency-thickness relationship. Finally, these two concepts are brought together to demonstrate that the interplay of amplitude and frequency can help define the style of the contacts of a vertically thinning coal seam.

Flat Transitional Boundaries. Most basal coal contacts are not sharp erosional boundaries, but, rather, are transitional. Deltaic inorganic sediments grade upwards into a soil profile containing root-filled underclays, which yield to a coal seam composed primarily of organic material (Figs. 3.2 and 3.29). If this transitional zone is very thin compared to the wavelength of the propagating seismic signal, the boundary can be approximated as a sharp contact. The seismic response of a sharp contact is simply a delta function convolved with the propagating wavelet, and for typical oil-exploration frequencies this has been an acceptable approximation. With the proliferation of high-resolution studies for coal or stratigraphic hydrocarbon accumulations, the validity of this approximation will be reexamined and the distinctive features of the seismic response at a transitional boundary classified.

Sengbush, Lawrence, and McDonal (1961) modeled transitional layers as consisting of a linearly increasing velocity function. This ramped velocity function yields a step function for the reflectivity response, and, consequently, the reflected waveform is basically the integral of the initial wavelet. This is true for the onset of the ramp function, assuming that it is large in extent. The onset of the reflection is at the change in slope of the velocity function, and its amplitude depends on the slope value.

A transitional layer with a linearly increasing velocity function generates two seismic events, one at the base of the layer and one at the top. If this layer is thin, the two integrated waveforms interfere, and the composite waveform is that of the input wavelet with an apparent onset at the center of the transitional zone (Fig. 3.30). The amplitude and frequency of this wavelet depend on the velocity contrast and the thickness of the transitional layer compared to the wavelength (λ) of the propagating wavelet.

Figure 3.30 shows the seismic response of a coal layer separated from the overlying sand by a sharp erosional contact and from the underlying sand by a transitional zone with a linearly increasing sand content. In this model and subsequent models, the seismic traveltime from the top of the transitional zone is kept constant at twice the wavelet period (T), while the thickness of the transitional zone varies.

In the lower portion of Figure 3.30, a transitional layer with a thickness of 1λ shows three separate seismic events. The first is a high-amplitude wavelet from the top of the coal, with a strong negative reflection coefficient; the other two events are the integrated waveforms associated with the top and bottom of the transitional layer. The amplitude of these integrated waveforms is small if the slope is gentle.

If the thickness of the transitional layer is $\lambda/4$, there are only two apparent seismic events (upper portion of Figure 3.30). The first is again the high-amplitude wavelet from the top of the coal. The second is a wavelet similar in appearance to the first, with opposite polarity and time onset at the center of the transitional zone. The amplitude of this second wavelet is slightly less than the first, and its frequency is lower. This second wavelet is the superposition of the two integrated waveforms, but the thinness of the transitional zone produces a derivative effect. The amplitude of these composite integrated waveforms is larger for the $\lambda/4$ case than for the 1λ case, because the slope of the transition is greater. Also, there is a relative amplitude tuning for the $\lambda/4$ thickness, as the two peaks are superimposed for maximum enhancement of the signal. This is similar to the classical thin-bed tuning, as described by Widess (1973).

Using a sequence of models similar to Figure 3.29 the seismic responses of ten transitional layers, varying in thickness from $\lambda/2$ to $\lambda/20$, were generated and analyzed. From the seismic responses shown in Figure 3.31, it is evident that the frequencies and amplitudes of the waves are inversely related to the thickness of the transitional layer. For very thin transitional cases, wavelets appear identical to those observed at sharp contact boundaries. Also, there is no dramatic amplitude tuning within this series. The tuning phenomenon is subtle and is evident only when the relative amplitudes are plotted against thickness of the transitional layer.

Figure 3.32 shows plots of transitional layer reflection amplitude and frequency against thickness. The amplitude anomaly is subtle, appearing as a bulge in the graph centered at approximately $\lambda/4$ thickness. This superposition, which is subdued compared to Widess's thin-bed tuning, results from a decrease in amplitude of the two integrated waveforms that comprise the measured wavelet with thickening of the transitional zone. There is also a nearly linear frequency decrease with thickening of the transitional zone. For transitional layers with a thickness of $\lambda/10$ or less, the frequency is that of the 25 Hz input frequency; at the tuning thickness, it is 21 Hz; and, finally, at a $\lambda/2$ thickness, the frequency is 15 Hz.

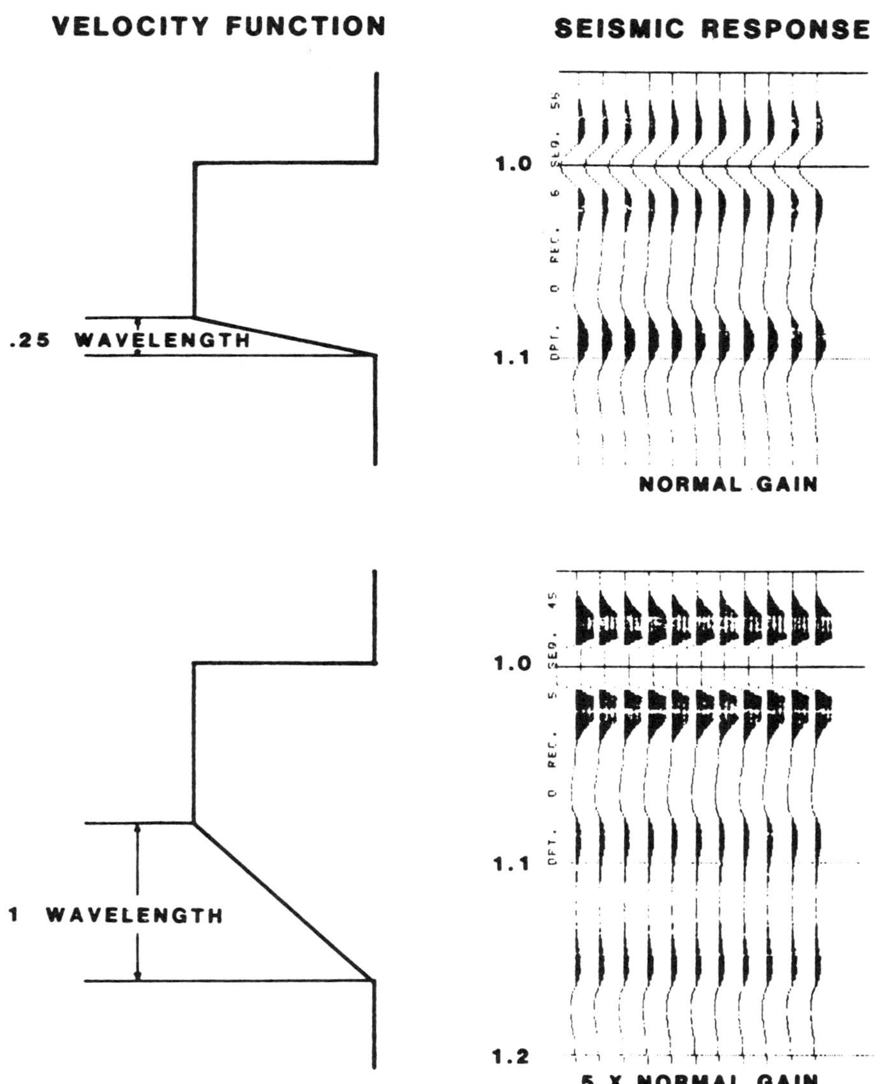

Figure 3.30. The seismic response and velocity function of two coals with transitional bases of 1λ and λ/4 thicknesses.

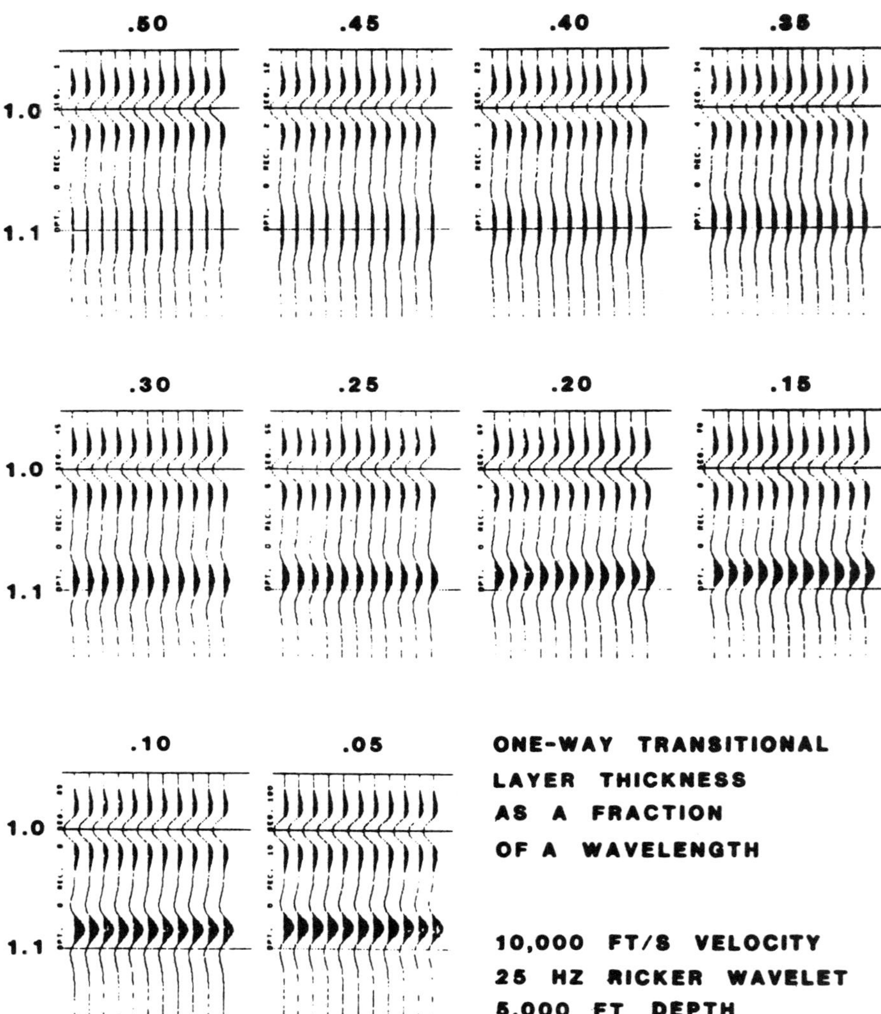

Figure 3.31. The seismic response of coals with transitional bases of various thicknesses.

Thus, only thin transitional boundaries, defined as those with a thickness less than λ/10, can be approximated by a sharp contact with no significant error. For transitional boundaries less than λ/2 and greater than λ/10, assumption of a sharp contact leads to an improper value for the amplitude and frequency of the seismic wavelet. The superposition wavelet would have the correct waveform and the amplitude could be scaled as a function of transitional zone thickness, but the frequency loss could not be compensated. Thick transitional boundaries greater than λ/2 should never be approximated with a sharp boundary, as the seismic response is really two separate integrated waveforms.

A Thinning Coal Seam with Sharp Contacts. Coal seams typically pinch out against, or interfinger with, adjacent sedimentary units. As the coal seam thins, it

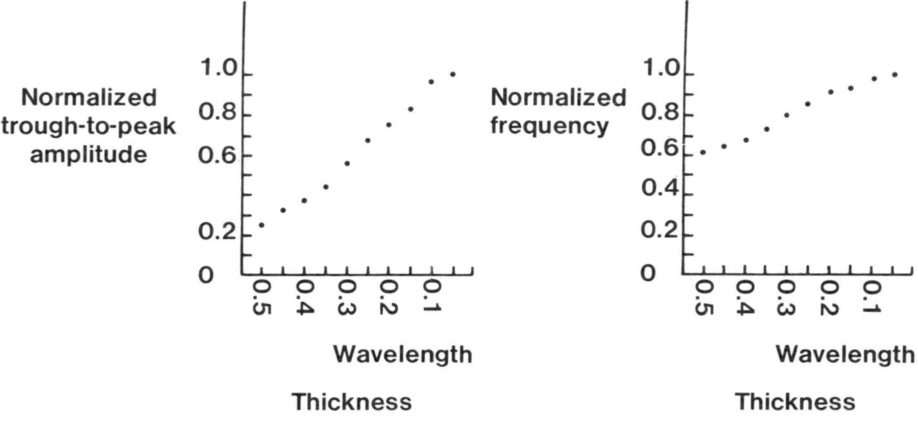

Figure 3.32. *Amplitude and frequency measurements of the seismic responses depicted in Figure 3.31.*

becomes impossible to separate the reflection derived from the top and base of the coal seam by direct observation. With today's improved resolution capabilities, a thinning bed can now be followed almost to its termination by indirect techniques.

Widess (1973) modeled a pinchout by deriving the observed waveform of the seismic response by graphical methods. The graphical convolution revealed that the seismic response was simply the superposition of the wavelets generated at the top and base of the thinning stratigraphic unit. Empirically, he determined that a thickness of at least $\lambda/8$ was needed to resolve the top and the base of a thinning bed.

Sengbush, Lawrence, and McDonal (1961) measured the amplitudes of the seismic response from flat layers at thicknesses that are various fractions of a wavelength. They demonstrated that an amplitude anomaly was associated with the thickness of $\lambda/4$ of their minimum-phase wavelet. The layer thickness of $\lambda/4$, at which the tuning phenomenon occurred, was consistent for all input frequencies. This $\lambda/4$ layer thickness was approximately where Widess noted that the observed waveform was the derivative of the input wavelet.

With improved processing capabilities, Neidell and Poggiagliolmi (1977) and Meckel and Nath (1977) presented a method for directly obtaining the thickness of a thinning bed from an amplitude graph, where amplitudes measured off a wavelet-processed section are plotted against thickness of the thinning bed. They also noted that the apparent bed thickness, determined by peak-to-trough time separation, became invariant at approximately the same thickness at which the amplitude peaked. However, this tuning thickness was different from that determined by Sengbush, Lawrence, and McDonal (1961) in their studies of flat beds. Figure 3.33 is a graph of amplitude and thickness, similar to that of Neidell and others, for a coal bed encased in a high-velocity sand. The top and base of the coal have sharp boundaries, with the base inclined at 10°. Also, the coal has a constant velocity of 10,000 ft/s, with a depth to the upper boundary of 5,000 ft. The horizontal axis of this graph displays both apparent thickness in milliseconds and relative amplitude, consistent with the format of other authors. However, the vertical axis differs, in that perpendicular thickness is plotted rather than actual thickness. Perpendicular thickness, as defined in Figure 3.34, is the difference

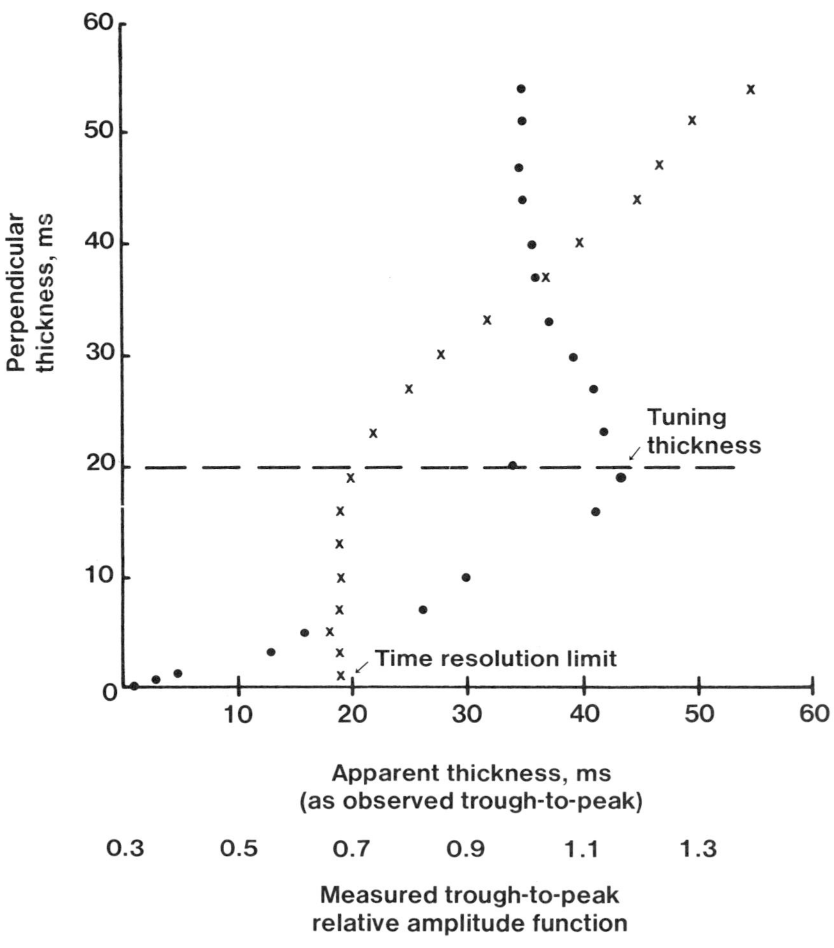

Figure 3.33. *Amplitude and thickness measurements of the 10° pinchout with sharp upper and lower contacts.*

between the traveltimes to the top and base of the coal from the surface along different raypaths, which are perpendicular to their respective boundaries. The seismic process measures this thickness, not the actual vertical thickness. Perpendicular thickness depends on both the angularity of the thinning bed and its depth of burial. For nearly flat beds, or surficial beds, vertical thickness may be an acceptable approximation; but it is not acceptable for the pinchout modeled in this section. (The discrepancies between vertical thickness and perpendicular thickness for this case are demonstrated later in Figure 3.36.)

In Figure 3.33, the amplitude tunes at a perpendicular thickness of 20 ms, or $T/2$. Figure 3.35, similar to Figure 9 of Sengbush, Lawrence, and McDonal (1961), demonstrates by graphical convolution why a thickness equal to the two-way time of $T/2$ or the one-way thickness of $\lambda/4$ should be the tuning thickness. The relationship of amplitude and thickness developed for a flat-bed study are

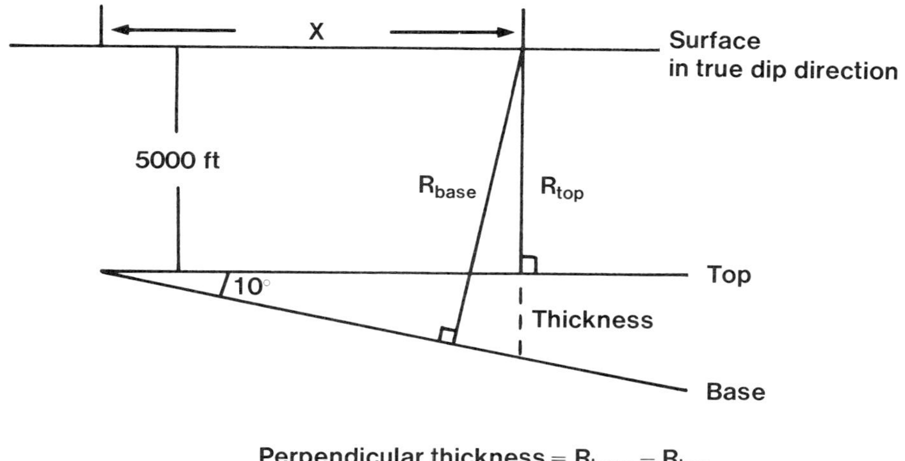

Source: Modified from Slotnick, 1959.

Figure 3.34. *Definition of perpendicular thickness.*

consistent with those of a thinning bed, provided that dip and depth of burial are compensated. The concept of perpendicular thickness, which compensates for dip and depth of burial, may be the specific mechanism that Neidell and Poggiagliolmi (1977) refer to as needing to be explained before this tool can have the profound economic impact on stratigraphic exploration that it should have.

A second feature that is diagnostic of thickness is the waveform itself. Figure 3.36 is the seismic response of the 10° pinchout. For thicknesses greater than twice the period, 80 ms (horizontal distance greater than 2700 ft), the two Ricker wavelets that are the reflections from the top and base are sufficiently separated in time to resolve both the top and the bottom. These two wavelets have the same amplitude and frequency but are 180° out of phase with each other. The interference of these waveforms is recognized between thicknesses of $1T$ and $2T$ (horizontal distance greater than 1600 ft but less than 2700 ft), with the adjacent side lobes disappearing at $1T$ separation. The trough-to-peak amplitude plotted in Figure 3.35 remains constant for thicknesses greater than 40 ms, or $1T$, and is unaffected by this overlap of the side lobes, as only the central lobe amplitude of each wavelet is measured.

The waveform at a thickness of $T/2$ (horizontal distance equal to 1000 ft), the amplitude tuning thickness, is the derivative waveform, or 90° out of phase with the wavelet from the top of the pinchout. Characteristic of this waveform are two high-amplitude lobes, a peak and a trough, rather than one central trough. At thicknesses less than $1T$, the superimposed waveform varies only in frequency and amplitude. The frequency of the resultant waveform is the inverse of twice the peak-to-trough time separation. Thus, apparent thickness, as plotted in Figure 3.33, is a measure of frequency for perpendicular thicknesses less than $1T$.

Consequently, the frequency is half the input frequency at a 40 ms, or $1T$, bed thickness and increases with thinning of the bed to equal the input frequency at $T/2$ thickness. For thicknesses less than $T/2$, the frequency of the waveform becomes nearly constant, remaining slightly higher than the input frequency. This frequency phenomenon provides a second criteria for determining perpendicular thickness of a thinning bed with a sharp upper and lower contact. Only when the bed is thick or when the perpendicular thickness equals exactly $T/2$, the

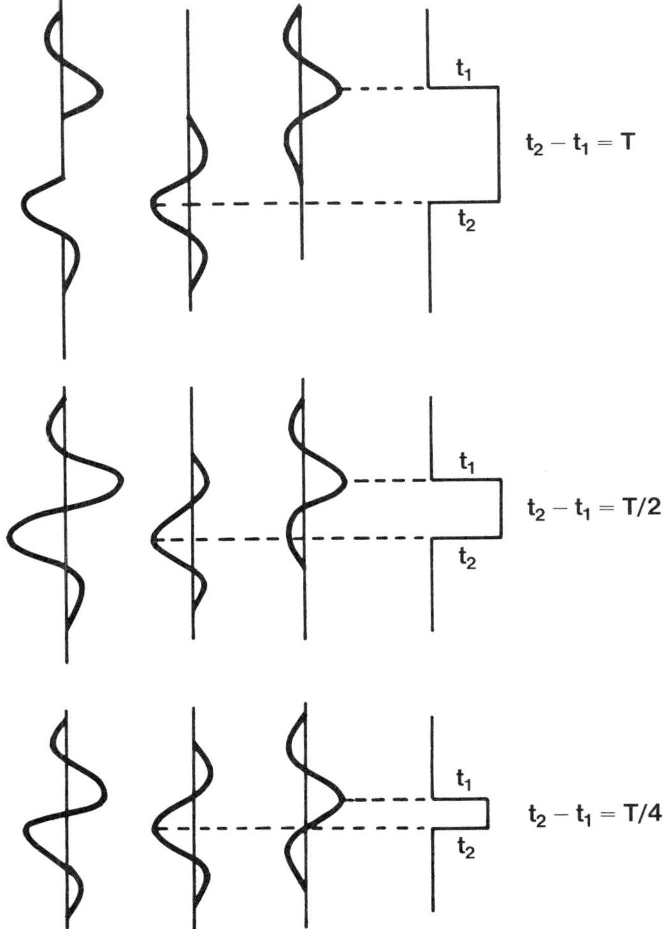

Source: Modified from Sengbush et al., 1961.

Figure 3.35. *Graphical convolution of Ricker wavelets at various time separations.*

amplitude tuning thickness, is the measured frequency equal to the input frequency.

A Thinning Coal Seam with a Transitional Lower Contact. A thinning coal seam was modeled with a sharp sand contact as an upper boundary and a transitional coal-sand lower contact, similar to the upper portion of Figure 3.30, except that the lower contact now dips 10°. Figure 3.37 is the seismic response for this model, with the width of the transitional zone T/2 and the input wavelet a 25 Hz Ricker wavelet. The wavelet associated with the basal contact has two-thirds the amplitude of the wavelet at the sharp upper boundary and a 21 Hz frequency. If the coal bed is thin, the superimposed waveform of the two Ricker wavelets of different amplitude and frequency lacks the clear definition that was present in the seismic response in Figure 3.36. Even though the waveform is dirtier, the tuning waveform is still 90° out of phase with the wavelet from the upper

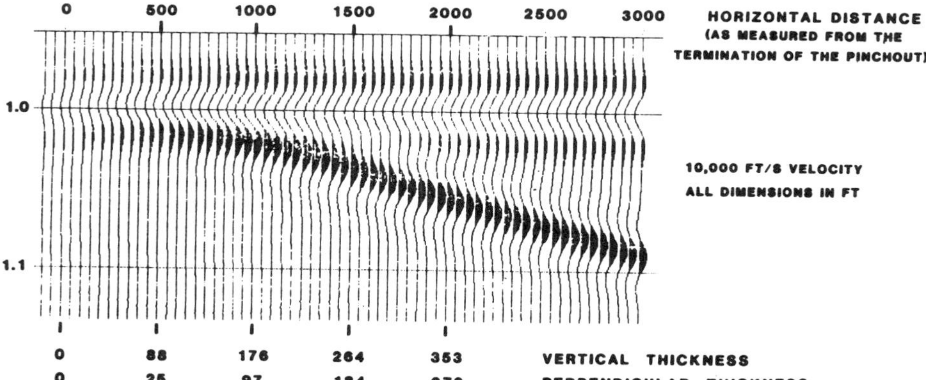

Figure 3.36. *The seismic response of a 10° pinchout with sharp upper and lower contacts.*

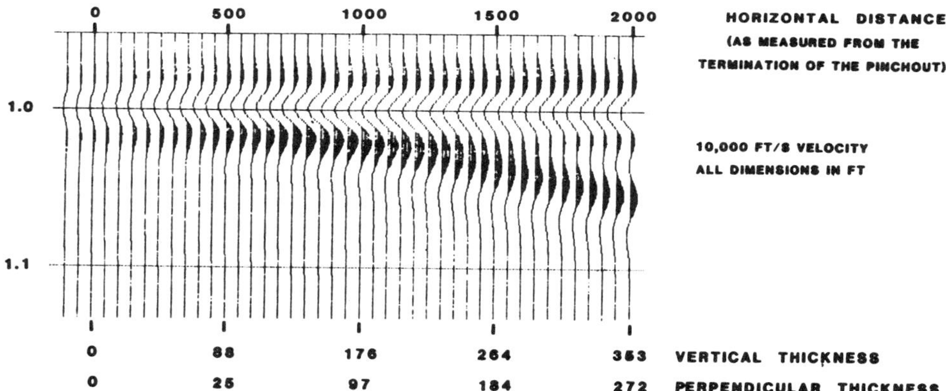

Figure 3.37. *The seismic response of a 10° pinchout with a transitional (T/2) lower contact.*

boundary, and the wavelets still start to separate at thicknesses greater than $1T$.

Figure 3.38 is a graph of amplitude and thickness similar to Figure 3.33, with perpendicular thickness measured to the center of the transitional zone. The amplitude again tunes at 20 ms perpendicular thickness, which is exactly $T/2$ of the higher-amplitude 25 Hz wavelet from the top surface. However, the frequency of this composite waveform is 23 Hz, as the trough-to-peak time separation is 21.5 ms at the tuning thickness, rather than the input frequency, as was the case for the pinchout with sharp contacts. The interference of the two different frequency wavelets results in a waveform whose frequency is the average of the two.

Conclusion. We have attempted to delineate transitional contacts and vertically thinning beds, and we have demonstrated that frequency is as powerful a tool in stratigraphic interpretation as amplitude is. The use of frequency as an indicator of thickness of a thinning bed may not be as reliable as amplitude, but the interrelationship of the two may provide more information than each parameter could alone.

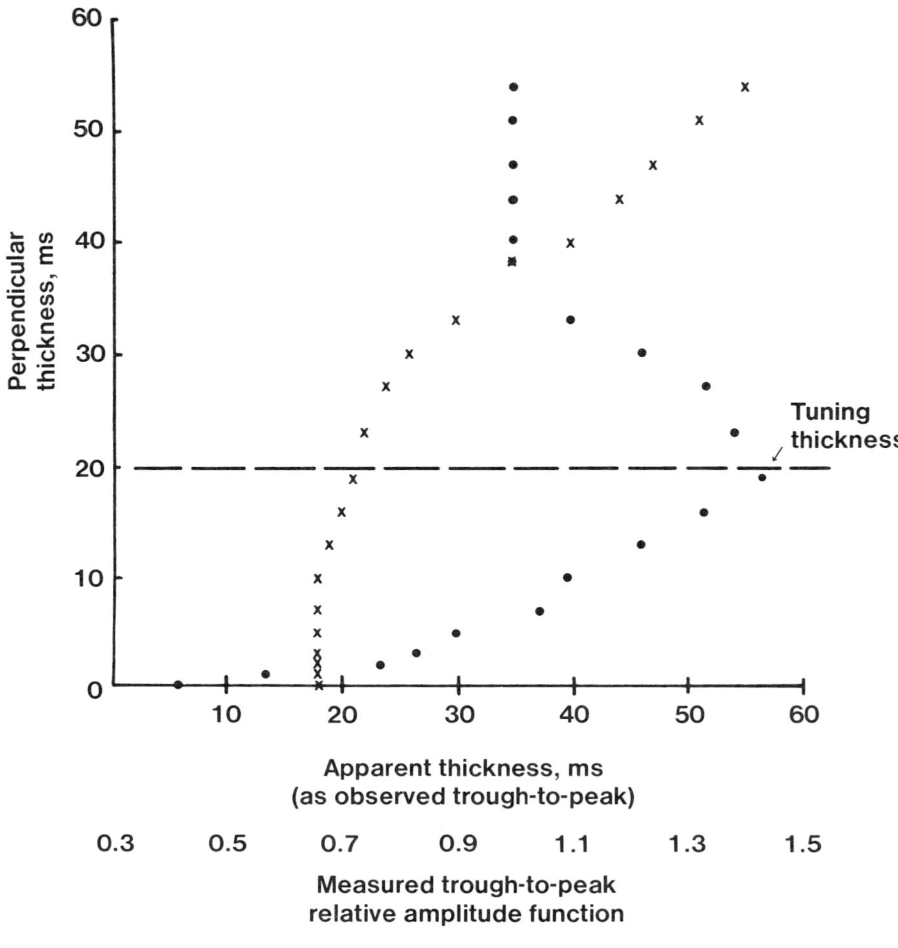

Figure 3.38. *Amplitude and thickness measurements of the 10° pinchout with a transitional (T/2) lower contact, depicted in Figure 3.37.*

SUMMARY. The seismic responses of several coal depositional environments are examined by several physical models. The meandering stream models provide tools for channel location by seismic methods; however, they also document several interpretational pitfalls. Undisturbed coal is discerned from channel-eroded coal by the 2-D and 3-D foci. The presence of these apparent anticlinal features, with their 90° and 180° out-of-phase waveforms, allows the channels to be charted accurately in most instances. The breadth of the 2-D focus measures apparent channel width, because the breadth is directly related to the azimuth of the seismic line and the trend of the channel. Also, the arrival time of the 2-D focus yields depth of the channel, but the arrival time of the 3-D focus yields the radial distance to the meander bend of the channel.

The presence of these 2-D and 3-D foci for a meandering channel creates a very complex seismic section. Their absence, however, does not guarantee undisturbed coal, because the apparent anticlinal features fail to develop when an interface is placed between the channel walls and their respective centers of curvature.

The most powerful indirect tool for mapping the sand channel is the velocity contrast between it and the coal. Velocity-induced sag in the underlying limestone is pronounced for the lower-velocity channel fill, and velocity pull-up is very evident for the higher-velocity channel fill. However, the velocity pull-up has a very unusual appearance because of the refractive properties of the channel. The pull-up displayed a diffractionlike appearance that was identical to the expected seismic response of an actual anticline with the same shape as the channel, but inverted. Thus, diffractions should not be used as the sole criterion for differentiating anticlines from velocity pull-up

A more dangerous pitfall associated with the refractive properties of the high-velocity sand channel occurs if there is channel sideswipe. In such a case, in which the channel is actually crossed by the seismic line, a high-amplitude 2-D focus is present. These same diffractions, or refracted reflections, are present when a meander bend is approached by the seismic line without necessarily being traversed. The strong 2-D focus is not present on the seismic section, as it is replaced by a faint sideswipe feature. Without this early event overlying the apparent anticlinal feature, it is very difficult to separate these events into true anticlines and velocity pull-ups.

Velocity contrasts also wreak havoc with the progradational delta coal models. The objective of such a seismic survey is to map a thick coal seam that is overlaid by thin lower deltaic and back-barrier coals. The superimposed coals have a sufficient velocity contrast with the surrounding deltaic sediments to yield a highly faulted appearance for the objective coal seam. Along with the velocity effects, strong diffraction events and dim spots below each thin coal prevents mapping of this thick coal seam.

In the theoretical modeling sections of this chapter, both amplitude and frequency are shown to be powerful tools for coal exploration. The modern amplitude studies of thinning beds are shown to be corollaries of the amplitude-tuning phenomenon associated with thin, flat layers, as described by Sengbush, Lawrence, and McDonal (1961). These contrasting studies indicate that the seismic process cannot measure actual bed thickness, except in the limiting cases of surficial or nondipping beds. In reality, seismic response is a function of perpendicular thickness, which takes into account depth of burial and dip of the interfaces before determining true bed thickness.

Another attribute of the waveform that is indicative of thickness is frequency. For coal seams with a one-way thickness of less than $\lambda/2$, frequency is defined as the inverse of twice the trough-to-peak time separation. At the amplitude-tuning thickness, the waveform is 90° out of phase with the input wavelet and has the same frequency.

Provided that the boundaries of the thinning coal are sharp contacts, these two characteristic phenomena always occur at a one-way perpendicular thickness of $\lambda/4$. If these boundaries are transitional, and if the thickness is greater than $\lambda/10$, the frequency of the tuning waveform is a measure of the thickness of the transitional zone, and its amplitude yields the actual perpendicular thickness of the coal.

Amplitude and frequency can also indicate areal dimensions. If the size of a coal body is less than that of a 1λ Fresnel disc, the amplitude consistently tunes, regardless of body shape, at a width equal to that of a $\lambda/2$ Fresnel disc. The tuning waveform is again 90° out of phase with the input wavelet, and the frequency is the same. This small-body phenomenon is identical to that of a thinning bed with Fresnel disc size comparable to a wavelength.

The determination of coal bed thickness in a small, thin coal body is complicated because amplitude and frequency depend on areal size. The amplitude-tuning thickness of a small coal body is not a constant perpendicular thickness of $\lambda/4$ of the input frequency. Instead, the thinning coal body tunes at $\lambda/4$ of the modified frequency, altered by the areal size of a small coal body. The tuning waveform of this small, thin bed is 180° out of phase with the input wavelet.

REFERENCES

Coon, J. B., Reed, J. T., and Dunster, D. E., 1978, Surface seismic methods applied to coal mining problems: 48th Annual SEG Meeting, San Francisco.

Dix, C. H., 1952, Seismic prospecting for oil: New York, Harper and Brothers.

Ferm, J. C., 1970, Allegheny deltaic deposits, *in* Deltaic sedimentation modern and ancient: J. P. Morgan, Ed., Soc. Econ. Paleont. and Mineral. Spec. Pub. No. 15.

———, 1974, Carboniferous environmental models in eastern United States and their significance, *in* Carboniferous of the southeastern United States: G. Briggs, Ed., Geol. Soc. Am. Spec. Paper 148.

Hilterman, F. J., 1970, Three-dimensional seismic modeling: Geophysics, v. 35, p. 1020–1037.

———, 1979, Interpretative lessons from three-dimensional modeling: Semi-Annual Progress Review, Seismic Acoustics Laboratory, Univ. of Houston.

Horne, J. C., Ferm, J. C., Caruccio, F. T., Cohen, A. D. Baganz, B. P., Cantrell, C. L., Corvinus, D. A., Geidel, G., Howell, D. J., Mathew, D., Melton, R. A., Pedlow, G. W., III, Sewell, J. M. and Staub, J. R., 1976, Depositional models in coal exploration and mine planning: Department of Geology, Univ. of South Carolina.

Meckel, L. D., Jr., and Nath, A. K., 1977, Geologic considerations for stratigraphic modeling and interpretation, *in* Seismic stratigraphy—Application to hydrocarbon exploration: C. E. Payton, Ed., Am. Assoc. Petrol. Geologists Mem. 26.

Neidell, N. S., and Poggiagliolmi, E., 1977, Stratigraphic modeling and interpretation—Geophysical principles and techniques, *in* Seismic stratigraphy—Application to hydrocarbon exploration: C. E. Payton, Ed., Am. Assoc. Petrol. Geologists Mem. 26.

Sangree, J. B., and Widmier, J. M., 1977, Seismic stratigraphy and global changes of sea level. Part 9, seismic interpretation of clastic depositional facies, *in* Seismic stratigraphy—Application to hydrocarbon exploration: C. E. Payton, Ed., Am. Assoc. Petrol. Geologists Mem. 26.

Sengbush, R. L., Lawrence, P. L., and McDonal, F. J., 1961, Interpretation of synthetic seismograms: Geophysics, v. 26, p. 138–157.

Slotnick, M. M., 1959, Lessons in seismic computing: Tulsa, SEG.

Spackman, W., Reigel, W. L., and Dolsen, C. P., 1969, Geological and biological interactions in the swamp-marsh complex of southern Florida, *in* Environments of coal deposition: E. C. Dapples and M. E. Hopkins, Eds., Geol. Soc. Am. Spec. Paper 114.

Thiessen, R., 1947, What is coal?: U.S. Bur. Mines Info. Circ.

Trorey, A. W., 1970, A simple theory for seismic diffractions: Geophysics, v. 35, p. 762–784.

Wanless, H. R., Baroffio, J. R., Gamble, J. C., Horne, J. C., Orlopp, D. R., Rocha-Campos, A., Souter, J. E., Trescott, P. C., Vail, R. S., and Wright, C. R., 1970, Late Paleozoic deltas in the central and eastern United States, *in* Deltaic sedimentation modern and ancient: J. P. Morgan, Ed., Soc. Econ. Paleont. and Mineral. Spec. Pub. No. 15.

Wanless, H. R., Baroffio, J. R., and Trescott, P. C., 1969, Conditions of deposition of Pennsylvanian coal beds, *in* Environments of coal deposition: E. C. Dapples and M. E. Hopkins, Eds., Geol. Soc. Am. Spec. Paper 114.

Widess, M. B., 1973, How thin is a thin bed?: Geophysics, v. 38, p. 1176–1180.

4. *TWO- AND THREE-DIMENSIONAL VELOCITY ESTIMATION PROBLEMS*

J. K. Owusu and G. H. F. Gardner

INTRODUCTION. Seismic velocity is an all-important parameter in seismic data processing and interpretation. Velocity estimates obtained using multifold seismic data are employed in stacking data, in migrating data, and in computing intrinsic interval velocities. Proper selection of velocities affects the quality of the stacked migrated data. Seismic velocity estimates constitute the primary tool for predicting abnormal pressures and, hence, drillability profiles in designing a drilling program for a virgin basin.

Many commonly used seismic velocity estimation schemes are intimately related to the stacking of common-depth-point (CDP) data and are based on a layered-medium assumption (Schneider and Backus, 1968; Taner and Koehler, 1969.) Because of this assumption, their performance tends to degrade as the earth becomes nonlayered and more complex. The stacking process assumes that the reflection point is common to all traces in a gather, regardless of offset, and that the common source and receiver midpoint is its epicenter. This can be true if and only if the reflector is horizontal. For dipping reflectors, the reflection points are located at different positions, displaced more in the updip direction as the offset increases (see Fig. 4.1). Stacking velocity is affected not only by the dip of a reflector but also by the azimuth of the profile, with respect to the dip direction of the reflector. Further complications occur when the reflector is curved. In such a case, the stacking velocity is dependent on the curvatures of the reflector as well as on its depth of burial.

Diffraction events also provide an additional complication. The apparent dip of the limbs of the diffraction event on a stacked section might be used erroneously to correct the overburden velocity in order to obtain a normal moveout (NMO) velocity.

Many of the difficulties associated with the estimation of velocities from seismic data recorded over nonhorizontal and sometimes nonlayered media can be overcome if the earth is modeled as consisting of point scatterers. This makes it possible to model the recorded seismic wave field as diffracted or scattered energy rather than as specularly reflected energy. By so doing, the subsurface points can be imaged by focusing the energy at the apex of the diffraction hyperbolas. Using imaging as a preprocessor, the aforementioned problems will be overcome and a better estimation of velocities can be obtained. Also, the velocities obtained will be independent of the reflector geometry.

Seismic waves propagate in three dimensions (3-D), and the subsurface features they encounter scatter them in all directions. As a consequence, the interpretation of conventional seismic sections based on two-dimensional (2-D) profiles may provide an erroneous description of the subsurface. The 2-D conventional profiles contain information not only from within the vertical plane of the profile but also from out of the plane. In some cases, the information that pertains to a particular vertical plane is nonexistent on the seismic section taken along the profile. In those cases, the conventional seismic profile constitutes an insufficient sampling of the wavefields. It is well known that 2-D migration of a conventional stacked section very often contains sideswipe. However, 2-D migration of a 3-D stacked section through the same line, obtained by partially migrating areal data, will not contain sideswipe. The fact is that the stacked section resulting from the partial migration of the areal data is without sideswipe

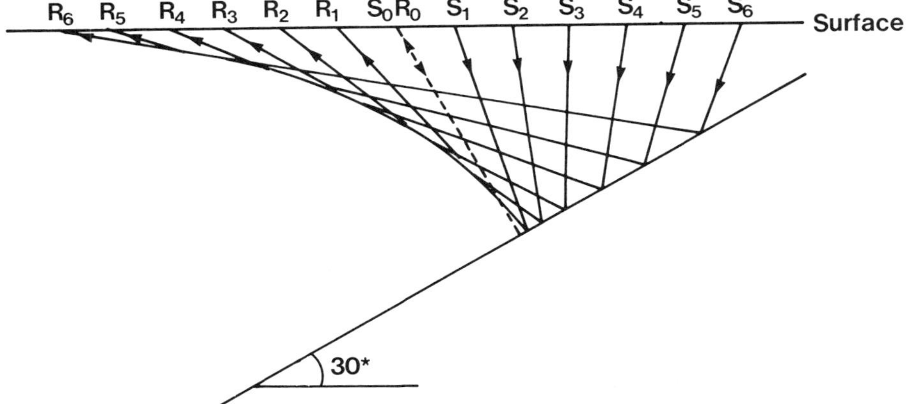

(a) CDP shooting geometry over a dipping bed

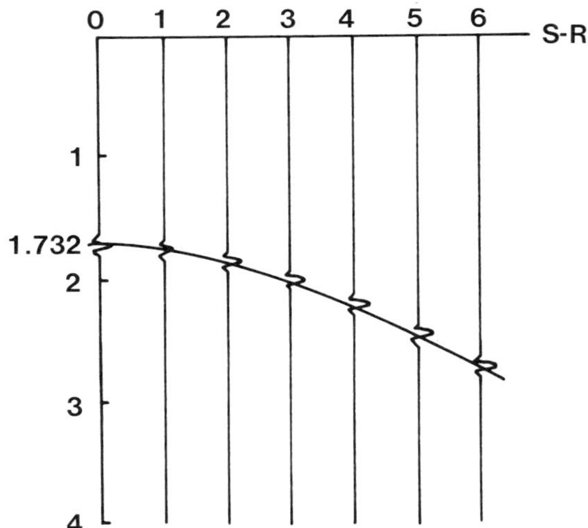

(b) CDP gather for a dipping layer

$$T_x^2 = T_o^2 + \left(\frac{x}{v/\cos\theta} \right)^2$$

Figure 4.1. *Distribution of points in a CDP caused by dip.*

but contains all the information from within the vertical plane of the line. Areal coverage therefore constitutes a sufficient sampling of the wavefield. We will demonstrate here its use in improving the resolution and the signal-to-noise ratio of the stacked and migrated section.

In a previous presentation (Owusu and Hilterman, 1980), the effect of profile azimuth, reflector depth, reflector dip, reflector curvatures, and velocity gradients were demonstrated using synthetic data. The model used was a single reflecting surface of second degree in x and y, and the stacking velocity for CDP gathers was computed for a range of these parameters.

In this chapter, a 3-D stacking velocity for arbitrary 3-D areal coverage is defined, based on the first step in a two-step approximation to 3-D migration. This 3-D stacking velocity reduces to the usual CDP stacking velocity if the coverage is simply a single, multifold line. This method of velocity analysis eliminates many of the difficulties associated with a 3-D subsurface, since it basically extracts a true 2-D multifold profile (without sideswipe) from 3-D data and computes the corresponding stacking velocity. The feasibility of the analysis is demonstrated with 3-D tank data.

The velocity obtained is still dependent on the dip and curvature of the reflector in the plane of the profile, although cross-dip effects have been eliminated. The next objective is to surmount this obstacle and obtain a velocity that is completely independent of dip, curvature, and all geometric properties of the reflector. It is hoped that this velocity can be related to intrinsic velocities of formations and can most accurately predict high-pressure zones and other drilling hazards.

VELOCITY ESTIMATION FROM AREAL DATA, USING MIGRATION AS A PREPROCESSOR. The discussion in the aforementioned presentation (Owusu and Hilterman, 1980) pointed out some of the difficulties and inaccuracies of conventional velocity estimation in the presence of structure. Many of the difficulties can be traced to the fact that the seismic data sometimes do not resemble, in detail, the earth structure over which they were recorded. The reflection point is no longer the same for all shot and receiver pairs in a gather (Fig. 4.1). Diffraction and scattering also contribute to the difficulties and inaccuracies. Migration is an operation that suppresses diffraction and scattering effects by reorganizing the recorded seismic energy so that the data will resemble the reflectivity map of the subsurface over which it was recorded. The process is successful because it focuses the energy at the apex of the diffraction pattern, corresponding to the scattering point. The migration of properly sampled areal seismic data enhances the resolution needed to distinguish subsurface scattering points. This property can be exploited to improve velocity estimates. Thus, there is no longer any diffusion of reflection points caused by structure to contaminate the velocity estimates. These concepts are applied here to the data from a model tank experiment.

Method of Imaging. The method of migration used here is the Kirchhoff integral approach, which can be implemented in three dimensions by summing the time derivative of the seismic data over diffraction surfaces. The diffraction surface is determined primarily by the root-mean-square (rms) velocity. With some approximations, rather than doing a complete 3-D migration, the process can be done in two steps (Gardner et al., 1978; Larner and Gibson, 1979). The first step is to image the prestacked seismic data into equally spaced output points on a reconstruction line in a direction perpendicular to the line. The second step is to image the output stacked data of the first step along the reconstruction line. The scheme is illustrated in Figure 4.2, where S and R indicate the source and receiver locations, M is their midpoint location, P is the projection of M onto the reconstruction line, O is any output on the reconstruction line, T_0 is any sample point on the output

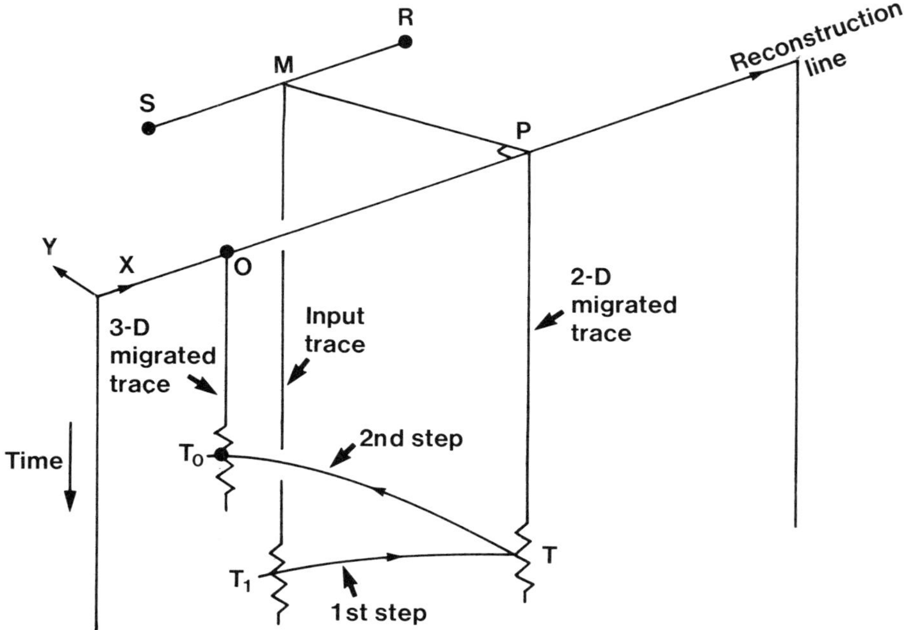

Figure 4.2. Perspective diagram to illustrate migration of 3-D data

trace at O, T is the time on the zero-offset trace at P corresponding to T_0, and T_1 is the time on the input trace corresponding to T.

The moveout equation relating T_1 to T is assumed to be given by

$$T_1 = [(T/2)^2 + (SP/V)^2]^{1/2} + [(T/2)^2 + (RP/V)^2]^{1/2}, \tag{4.1}$$

where V is the rms velocity at time T at location P.

Velocity Estimation. Data were collected over a physical model in the experimental water tank. The physical model was a 3-D oblong basin made of RTV silicone rubber. (A schematic diagram of the model is shown in SAL Catalog No. 1, November 1979, under the name SALBAS.) The model was placed in the water tank, and the source and receiver transducers were positioned at a scaled height of 5000 ft above it. The major and minor diameters of the basin were 6400 ft and 4700 ft, and the major and minor radii of curvature were 7000 ft and 4000 ft. Figure 4.3 illustrates how the data were collected. Along a single line off the center of the model and oblique to the axes of the basin, eightfold CDP gathers were collected perpendicular to the line. This was done to simulate a first-step subset of a 3-D multifold data set. The near offset of each gather was 1500 ft and the far offset 5700 ft. The midpoint spacing was 50 ft. There were a total of 351 midpoints along the line.

Using the diffraction pattern moveout equation (4.1), all 2808 input traces from the experiment were imaged into an output point at midpoint location 176, indicated by the arrow in Figure 4.3. The 351 input traces for each offset were imaged using a suite of five constant velocities: 11,000, 11,500, 12,000, 12,500, and 13,000 ft/s. These traces are displayed in Figure 4.4. Each panel corresponds to a constant velocity; each trace within a panel corresponds to a constant offset.

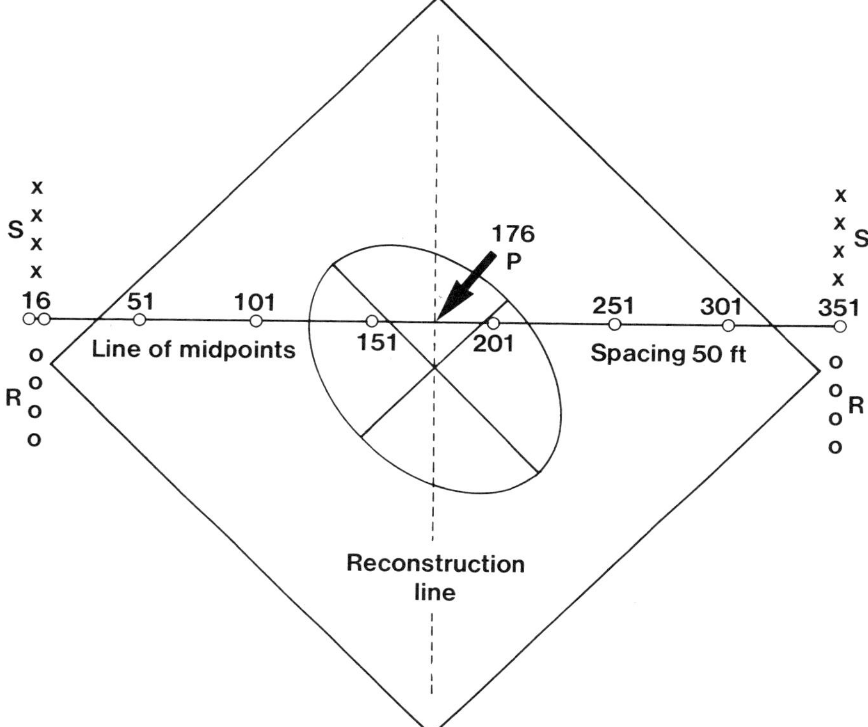

Figure 4.3. *Velocity analysis over a basin model*

The vertical stack of the eight traces in each panel will give the output for the first step of the migration for the corresponding velocity. Hence, the velocity for the events within a panel that are horizontally aligned gives the appropriate velocity for performing the first step. The imaged events curve upward when a lower-than-correct velocity is used, indicating overmigration, and curve downward when a higher-than-correct velocity is used, indicating undermigration.

Figure 4.5 shows the result of stacking the panels. The resulting five traces form a crude velocity analysis spectrum. Local maxima in amplitude can be used to pick velocity and time for discrete events.

Comparison of 3-D and CDP Stacking Velocities. The eight traces from the original data set at midpoint 176 form a CDP gather. A standard velocity analysis for this gather can be compared with the 3-D analysis just described. Figure 4.6 shows the eight-trace gather, and Figure 4.7 shows these traces moveout-corrected with the five constant velocities used previously.

A comparison of Figures 4.7 and 4.4 indicates several differences. Perhaps the most striking difference is the prominent event on the 3-D analysis (Fig. 4.4) between $T = 1.0$ s and $T = 1.1$ s, which is hardly observable on the 2-D analysis (Fig. 4.7). This event is the reflection from the first interface at the curved surface inside the basin. The 3-D analysis indicates a velocity in excess of 13,000 ft/s. This is higher than the water velocity of 12,000 ft/s, as expected, because the vertical cross section of the basin is inclined to the horizontal at the reflection points. The

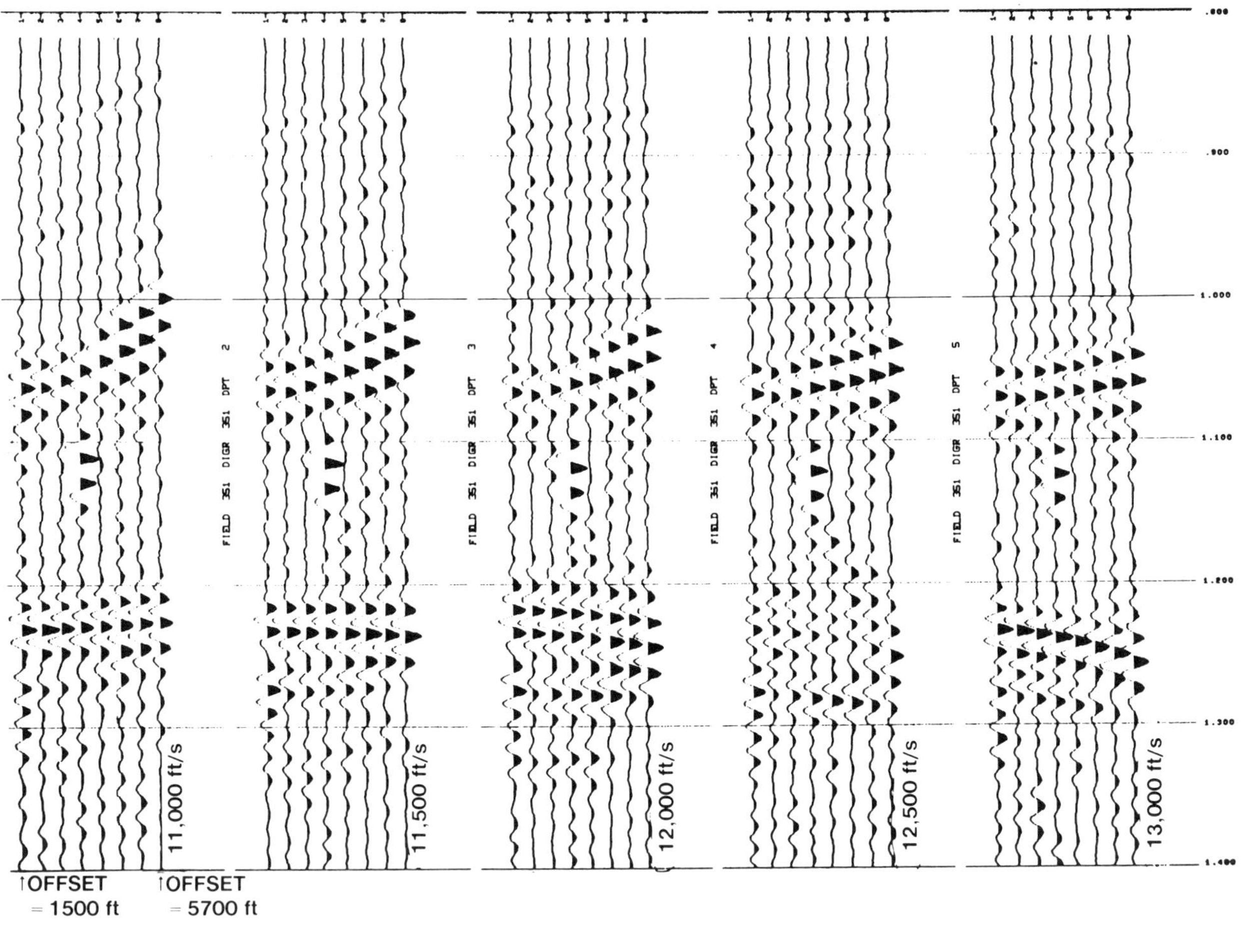

Figure 4.4. *3-D velocity analysis panels.*

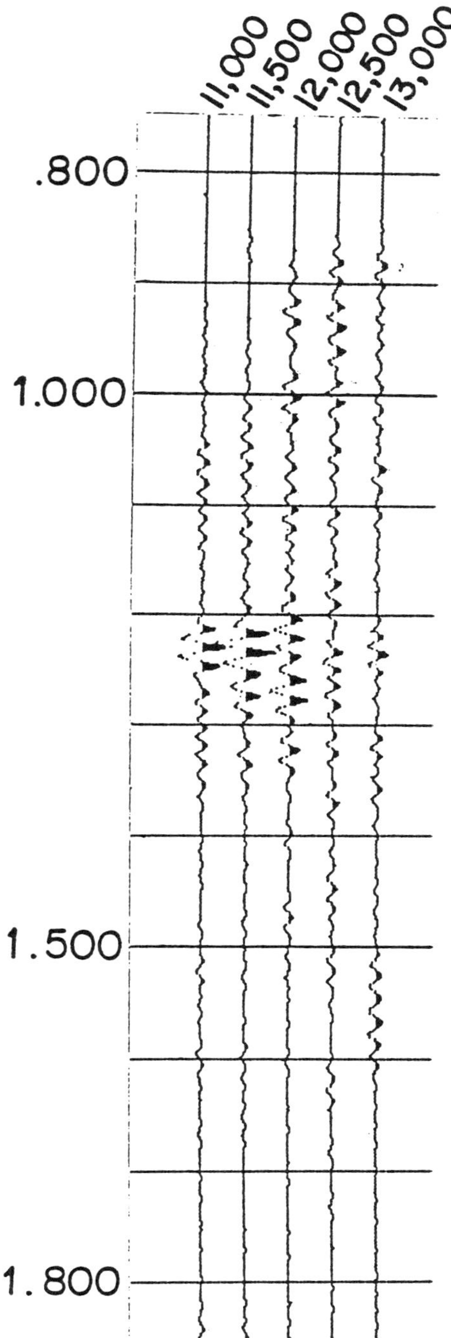

Figure 4.5. *Summed trace velocity spectrum.*

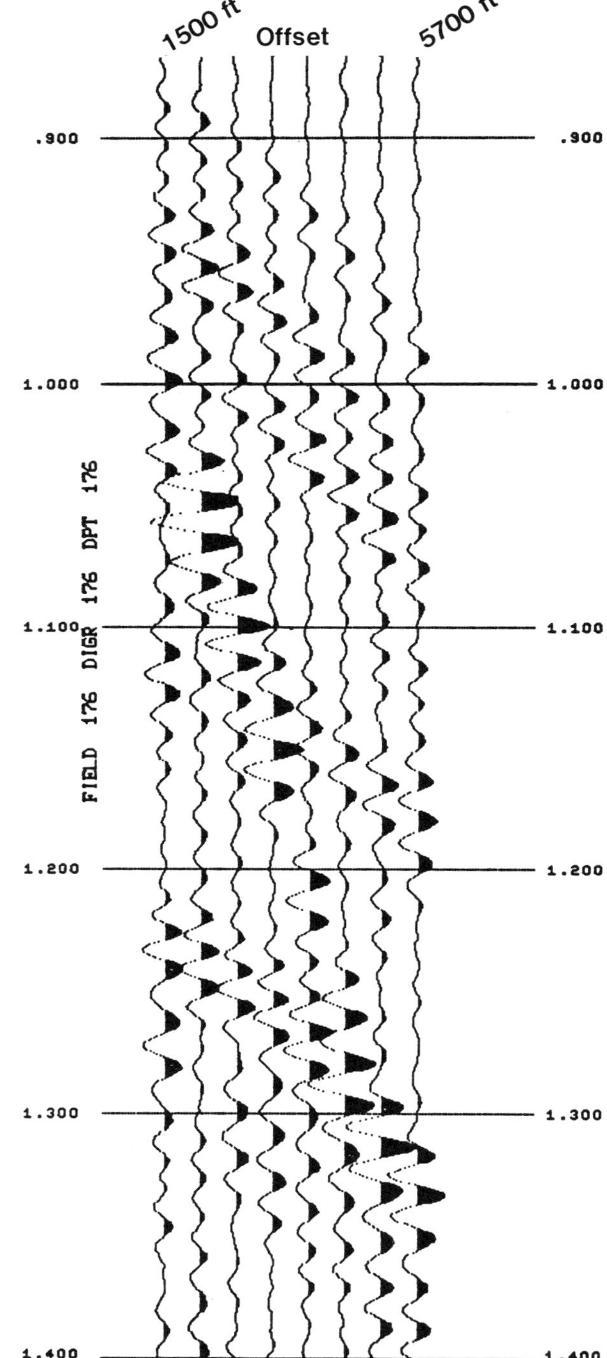

Figure 4.6. CDP gather at midpoint 176.

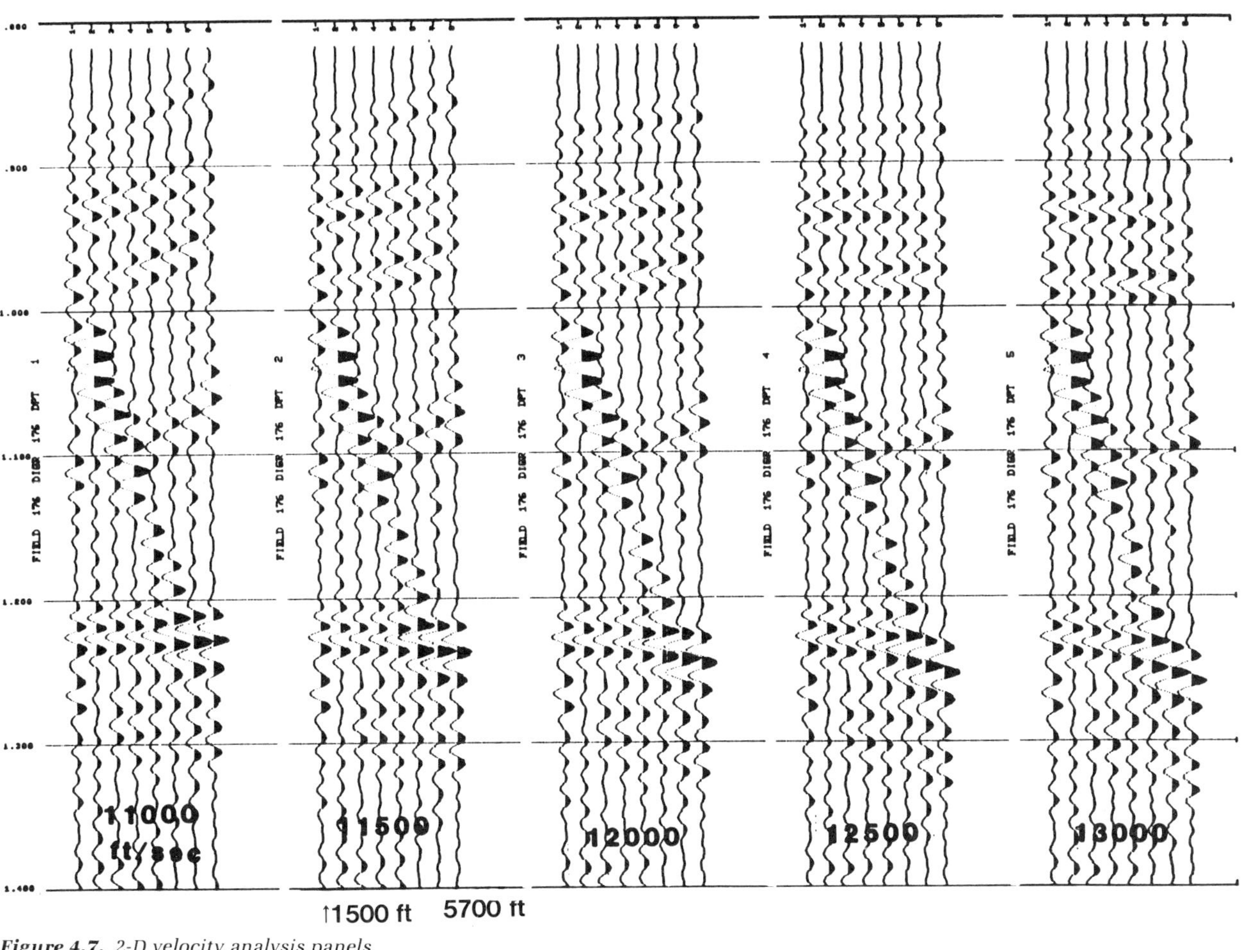

Figure 4.7. 2-D velocity analysis panels.

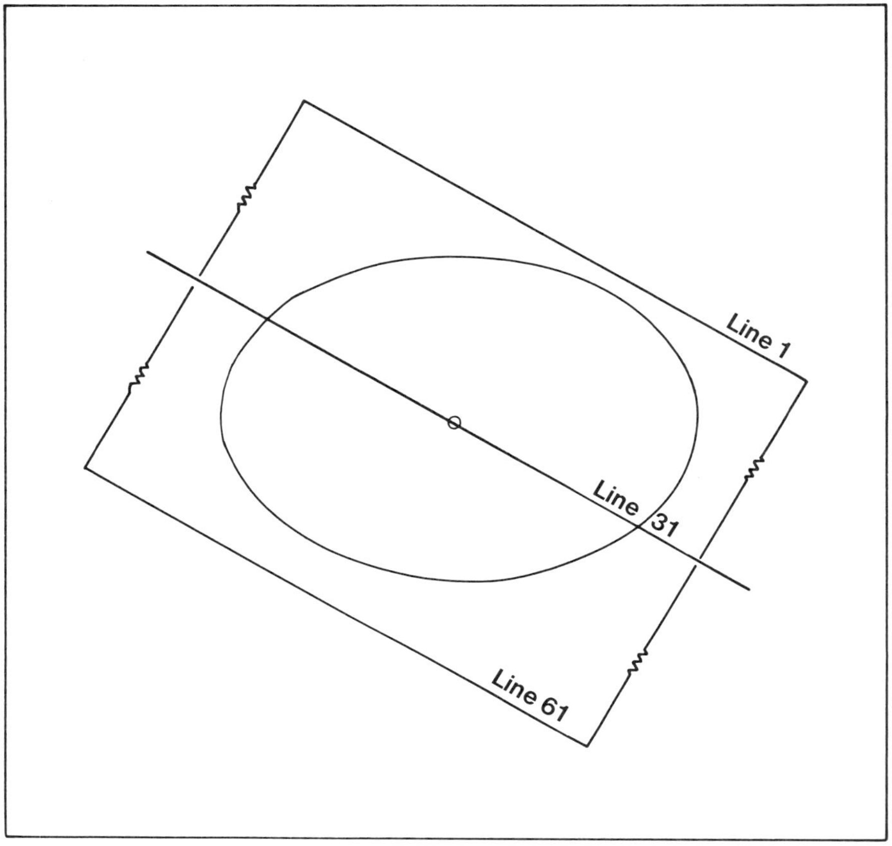

Figure 4.8. *Common-offset data shot over the basin model.*

spread in the reflection points with offset presumably accounts for the curvature of the alignment. The small offsets indicate a lower velocity than is indicated by the large offsets.

The 2-D velocity analysis (Fig. 4.7) does not show a clear alignment between 1.0 and 1.1 s but indicates a confused zone. This may be attributed to the effects of the cross-dip and curvature of the synclinal surface. One benefit from 3-D data is that this confused zone is clarified.

The reflection from the flat bottom of the model occurs between 1.2 and 1.3 s. The alignment of this event is clear on both the 2-D and 3-D analyses at around 11,500 ft/s. The 2-D result is satisfactory because the apparent anticlinal surface (caused by velocity pull-up) does not cause much diffusion of the reflecting points. Events between 0.8 and 1.0 s are diffractions from the rim of the basin and do not show much difference between the 2-D and 3-D analyses because the reflecting points are fairly close in both cases.

This example illustrates that, when 3-D effects in a 2-D CDP gather degrade the velocity analysis, a better result can be obtained by using a 3-D stacking velocity analysis program. The example also indicates some remaining difficulties. Note that, in the time interval 1.1 to 1.2 s in Figure 4.4, a large-amplitude event is concentrated on the fourth offset trace for all velocities. This appears to be a

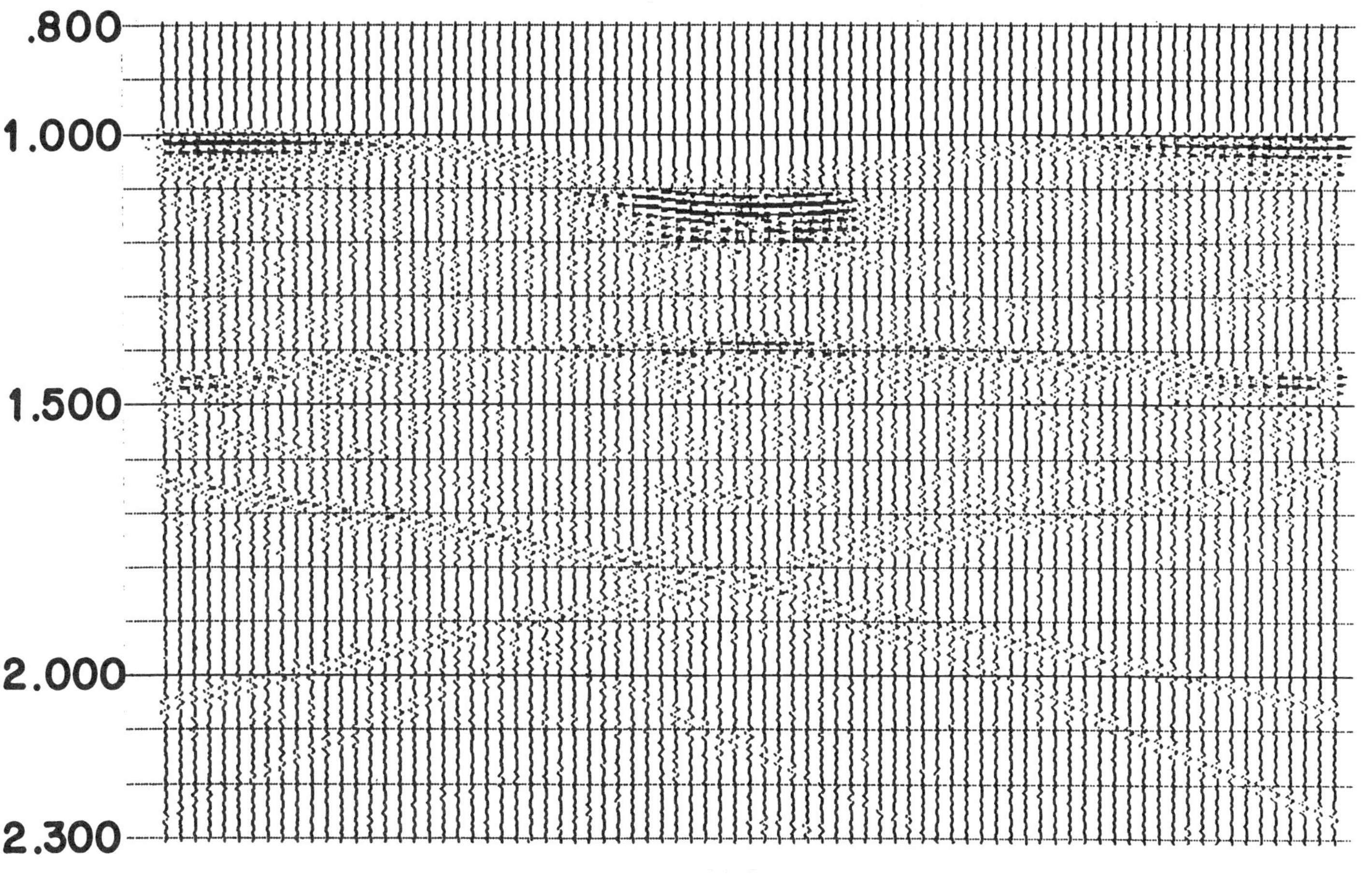

Figure 4.9. *Common-offset data along line 31.*

focusing effect, and it can be seen on the 2-D data in Figure 4.7 as well. This problem would be resolved by 3-D migration.

One benefit that can be anticipated from the use of areal data for velocity analysis is not demonstrated by this example. The combination of hundreds of traces improves signal-to-noise ratios. Hence, CDP data that would be of marginal value because of noise can be replaced by data of greatly improved quality.

Effectiveness of the Migration Scheme. Figure 4.8 illustrates 61 common-offset lines, with 81 midpoints per line, shot over the basin model at 100-ft midpoint spacing and a line spacing of 100 ft. (The data are from SAL Catalog No. 1, SALBAS 3.) Figure 4.9 shows the common-offset data along the central line 31 with an offset distance of 1500 ft. Line 31 goes over the center of symmetry of the basin but at an angle of about 30° with the major axis. The raw data of this line and its 2-D migration output are compared to the outputs of steps 1 and 2, respectively, of the 3-D migration scheme discussed earlier.

Figure 4.10 gives the output of step 1, using all 61 lines. A comparison of Figures 4.9 and 4.10 shows a number of significant features. First, the well-known bow-tie feature of 2-D synclines can hardly be seen on Figure 4.9, whereas, by focusing the areal data in the correct spatial location, the bow-tie feature shows very prominently in Figure 4.10. This should make a significant difference in the 2-D migration of the two sections.

It is difficult to explain the 2-D section in detail because the reflection points do not necessarily lie vertically below line 31. However, the algorithm for the 3-D stacked section constrains the reflection points to lie exactly below line 31, and the section therefore resembles the familiar pattern for a 2-D syncline.

Another significant feature can also be seen. All the diffraction effects from the edges of the model, which show on Figure 4.9, tend to cause interpretation problems. Most of them are from out of the vertical plane of the line. The imaging process removes all the out-of-plane information and, as Figure 4.10 shows, only the in-plane edge diffractions remain.

Figure 4.11 shows the 2-D migration of the common-offset line 31 shown in Figure 4.9. Although the basin has been defined by the process, the flanks have a phase discontinuity at the rim as shown on the section. This is probably caused by the wandering of the reflection points away from line 31. Figure 4.12 shows the 2-D migration of the output in Figure 4.10. This section shows a significant improvement in the continuity of phase at the rim of the basin. Also, all the edge diffractions have disappeared in the areal data output (Fig. 4.12) as compared to the migrated output of the common-offset data (Fig. 4.11.)

CONCLUSIONS. The assumptions of the CDP method and, hence, conventional velocity analysis, fail as the subsurface geologic features become more complex, deviating from a horizontally layered medium. The reflection points are no longer coincident for all the source-receiver pairs with a common midpoint and are not situated at a point vertically below the common midpoint. This tends to degrade the quality of the velocity estimates obtained, thus affecting the quality of the stacked or migrated section. Velocity becomes dependent on the structure of the reflector—the dip of a plane reflector, the radii of curvature of a curved boundary, and its depth. Velocity also becomes dependent on the azimuth of the profile.

It has been demonstrated that, by using areal data, velocity estimates may be obtained with a two-step, 3-D migration as a preprocessor. These velocities are less dependent on the geometry of the reflectors than are standard CDP velocities because the effects of cross-dip and cross-curvature have been removed. However, the effects of dip and curvature in the profile section remain and must be allowed for in estimating intrinsic velocities.

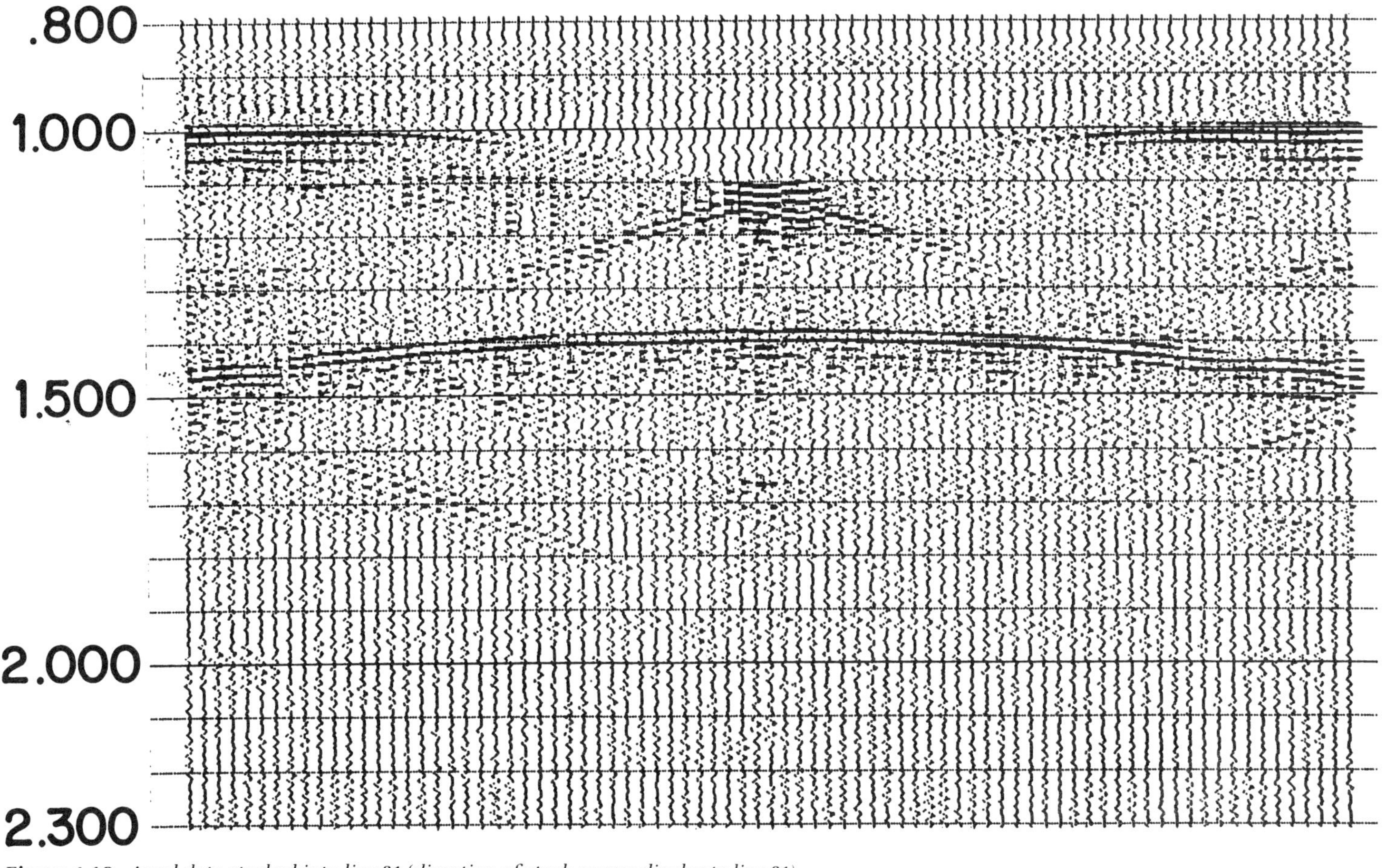

Figure 4.10. *Areal data stacked into line 31 (direction of stack perpendicular to line 31).*

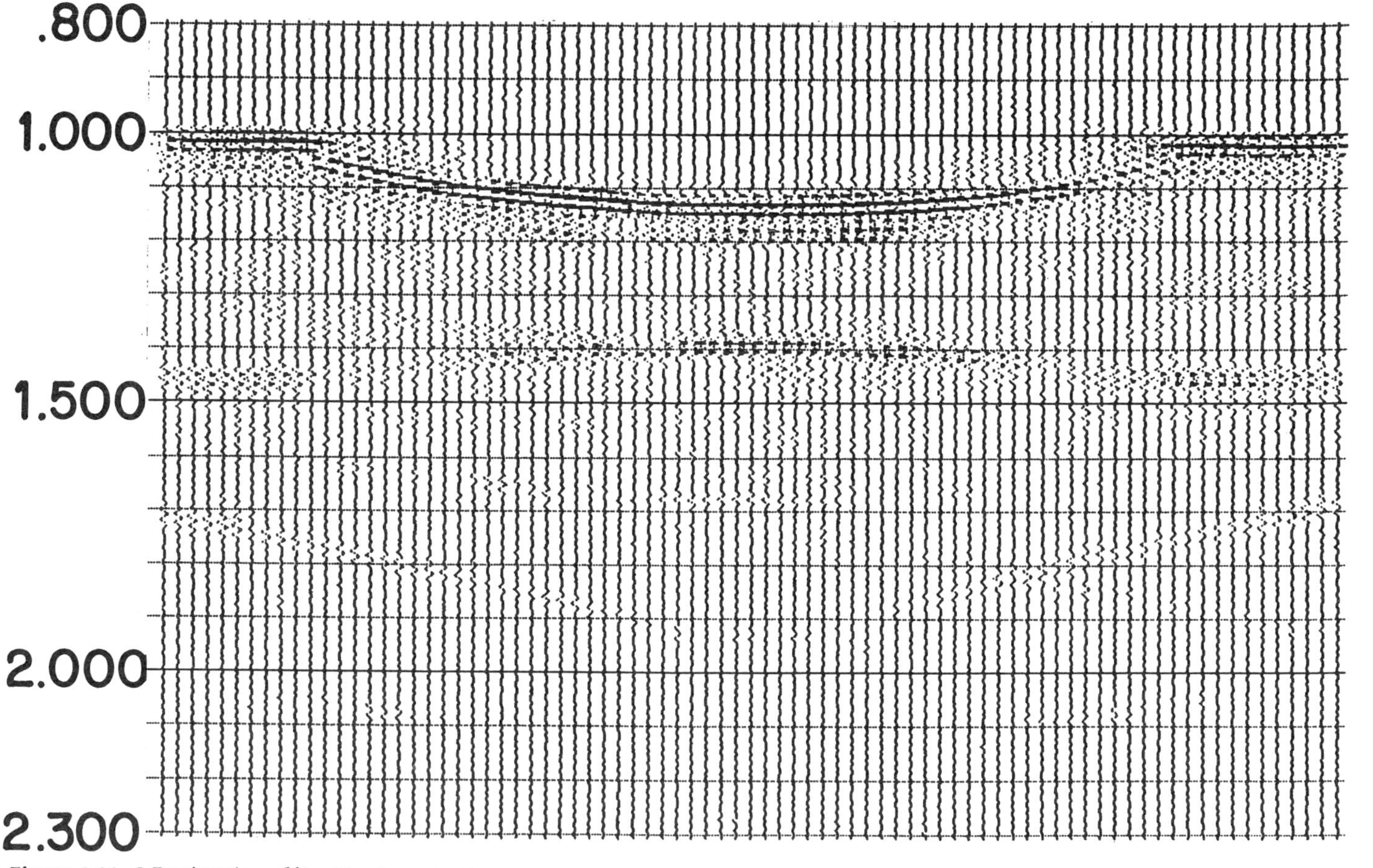

Figure 4.11. *2-D migration of line 31, shown in Figure 4.9.*

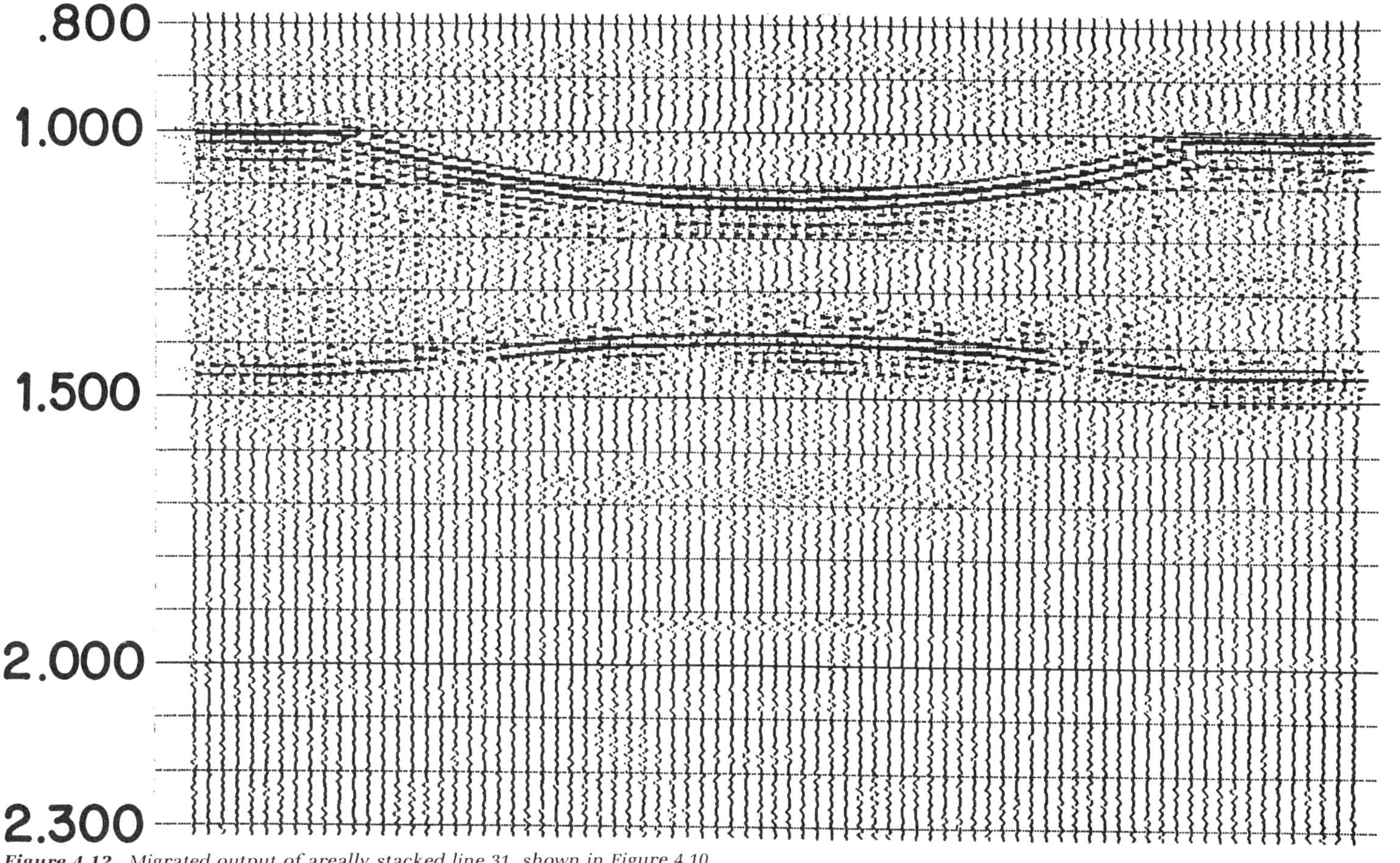

Figure 4.12. *Migrated output of areally stacked line 31, shown in Figure 4.10.*

REFERENCES

Gardner, G. H. F., McDonald, J. A., Watson, T. H. & Kotcher, J. S. 1978, An innovative 3D marine survey: Presented at European Association of Exploration Geophysicists Meeting, Dublin. (See also abstract, Geophys. Prosp., v. 26, p. 674, 1978.)

Larner, K., and Gibson, B., 1979, Course notes for the SEG Continuing Education Short Course "3-D Seismic Techniques": Houston, SEG, September.

Owusu, K., and Hilterman, F., 1980, Velocity analysis over a 3D curved boundary: 23rd Annual Midwestern Exploration Meeting, March. (See also Seismic Acoustics Laboratory Annual Progress Review, v. 4, p. 385–406, 1979.)

Schneider, W. A., and Backus, M., 1968, Dynamic correlation analysis: Geophysics, v. 33, p. 105–126.

Taner, M. T., and Koehler, F., 1969, Velocity spectra-digital computer derivation and applications of velocity functions: Geophysics, v. 34, p. 859–881.

5. *EXPERIMENTAL SEISMIC SURVEYS OVER A SIMULATED PETROLEUM FIELD*

Thomas A. Smith

As the incentives to find oil fields have increased in recent years, geophysicists have been challenged to explore much more carefully than was ever done in the past. New energy sources, super multifold data, and pseudoacoustic logs are but a few examples of ways in which new ideas are brought to bear on old problems.

New techniques often lead to new questions. Vertical seismic profiling (VSP) is a method that allows us to follow seismic events back into the earth. A set of one or more downhole receivers records time traces at a number of different depths. The method brings to mind a great number of interesting questions, although only two will be mentioned here.

First, VSP data are very nearly a true realization of downward continuation. There are now wavefield extrapolation methods to project data into the earth in a fashion consistent with the wave equation. The techniques that are applied numerically may be tested against measurements of the real wavefields. The question is how successful we are at continuing the waves. If the answer is that we are very successful, then we will, of course, move on to other questions. If we are only partially successful, however, then what limited success we enjoy undoubtedly will stimulate us to devise extrapolation methods for wave equations that are more consistent with our recorded data.

Another important potential for the VSP method lies in the intuitive notion that what lies close at hand is better resolved than what lies afar. Petroleum fields certainly are features that need to be studied in more detail. One important question posed by the method asks whether or not VSP data will allow us to better resolve the shape of a reservoir. It is this question that leads us to the present study.

This chapter will describe a Plexiglas model that represents a field of five high-velocity reservoirs. The choice of materials and the selection of scaling parameters were made to simulate a limestone layer in shale. Because of the anticipated wide angles of incidence, model materials were used that responded like real-earth materials.

Conventional sixfold lines were run over the model to mimic the types of data that might be used to first discover the field. A surface three-dimensional (3-D) survey was also collected, with two purposes in mind. In a hypothetical exploration scenario, the survey might have been collected to assist in the design of a well program for field development. Another purpose was to provide a benchmark data set against which the VSP data might be tested. The VSP data were gathered at a hypothetical discovery well. The location of this well was chosen in a manner that was consistent with data available at the time of the initial discovery. Two types of data were collected. For one type, there were four downhole receiver positions and a surface source grid. For the other, the source was fixed at the surface of the well site and the receiver scanned along a line.

MODEL MATERIALS. Experimental modeling for the VSP technique requires careful selection of model materials and a judicious choice of scaling factors. These are necessary because many of the waves measured in VSP surveys will have interacted with the model at wide angles of incidence. It is particularly important to avoid using model materials that could generate wide-angle reflections with unrealistic amplitudes.

The commonly used SAL scaling factors are listed in Table 1.6 in Chapter 1. There are two choices of mass scaling factors, which lead to density factors of 2.4 and 1.9, respectively. Figure 5.1 shows the trend relations between bulk density and compressional velocity for several common rock lithologies (Gardner, Gardner, and Gregory, 1974). The dashed line plots the quarter-power relationship that generally follows the trends for sandstone, shale, limestone, and dolomite. Using the first scaling factor, water falls directly on the quarter-power line and scales lithologically to a high-velocity sandstone. Using the same scaling factors, Plexiglas scales to a dense limestone or dolomite.

When the first scaling factor is used with the RTV silicone rubbers, the scaled densities are unrealistically high. Using the second scaling factor, the RTV 3110 and RTV 184 scale to shale and sandstone, respectively. Water, however, then scales to a material with a density less than that of salt. It was concluded that a water-over-Plexiglas interface resembles a shale-over-dolomite contact. It must be assumed that the acoustic wavefield, supported by the water, is suitable to model the subsurface.

Theoretical plane wave reflection and transmission coefficients have been computed for incident and reflected compressional waves over a water-Plexiglas interface (Cerveny and Ravindra, 1971). The displacement coefficients in the direction of propagation are plotted in Figure 5.2. Starting from normal incidence, the reflection coefficient remains near 0.36 out to the critical angle of 33.5°. Beyond this angle, the reflected wave is very weak and is altered in shape by the phase of the reflection coefficient. At the grazing angle, the magnitude approaches 1.0 and the wave is out of phase by 180°, so that the reflected wave approaches that of a polarity-reversed version of the incident wave. Values of the plane wave reflection coefficients are not valid in the vicinity of the critical angle.

RECONNAISSANCE SURVEY. Five irregularly shaped tubular Plexiglas bodies were mounted on pins over a substratum block of Plexiglas. These simulated reservoir bodies are 125 ft thick. The space between the bottom of the bodies and the substratum is 625 ft; the block is 1000 ft thick. The top of the bodies is approximately 10,000 ft below the source-receiver observation plane. A map view of the bodies is presented in Figure 5.3.

Three sixfold reconnaissance lines were run over the model, using a line spacing of 2500 ft. Each line of 121 common-depth-point (CPD) data was occupied, with the source and receiver aligned in the direction of recording; CDP spacing was 100 ft. Offset separations ranged from 2000 ft to 7000 ft. The unprocessed inside traces from the first line are shown in Figure 5.4. The basic system wavelet is long and oscillatory. Reflections from two bodies may be seen at 1.7 s, and reflections from the top and bottom of the substratum may be seen to arrive near 1.8 and 1.9 s.

The data were wavelet-filtered (Fig. 5.5), with a 20–80 Hz filter. In Figure 5.5, the pattern of interfering diffractions between the two bodies is easily seen.

The data were stacked using a velocity function designed to enhance the edge diffractions. The three stacked sections are shown in Figures 5.6, 5.7, and 5.8. The water velocity of 11,928 ft/s was used to stack the body reflections and the reflection from the top of the substratum. Between 1.789 s and 1.874 s, the velocity increased linearly to the root-mean-square (rms) velocity of the bottom of the substratum. Using the Plexiglas velocity of 21,600 ft/s, the stacking velocity for the bottom of the block is 12,580 ft/s.

The reflections from the bodies and from the top and bottom of the substratum have all been stacked coherently. Note, in all three sections, the profusion of edge diffraction below the reservoir bodies. The reflections from the top of the block under the bodies arrive earlier than those to the side, because the Plexiglas has a velocity greater than that of the water. This velocity pull-up may be seen under the right body in the gather of Figure 5.5.

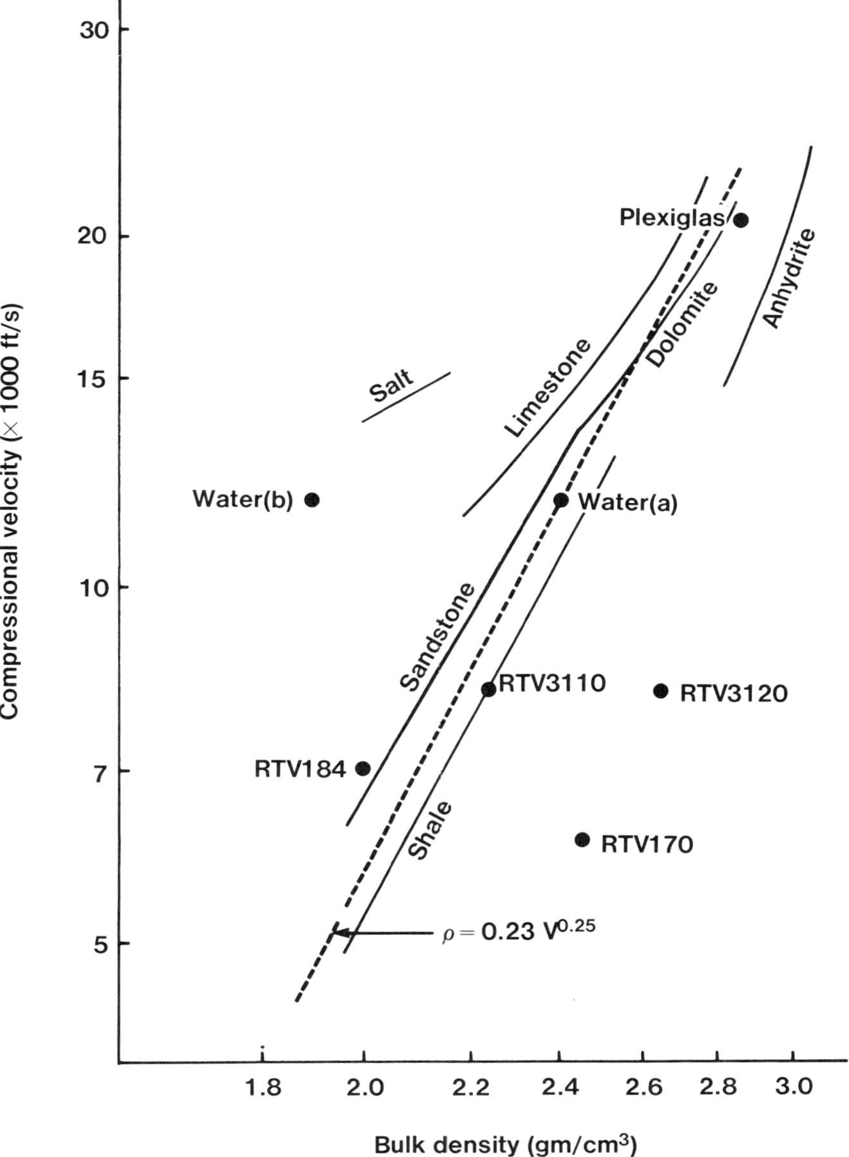

Figure 5.1. *Compressional wave velocity as a function of bulk density for common rocks compared to RTV silicone rubber.*

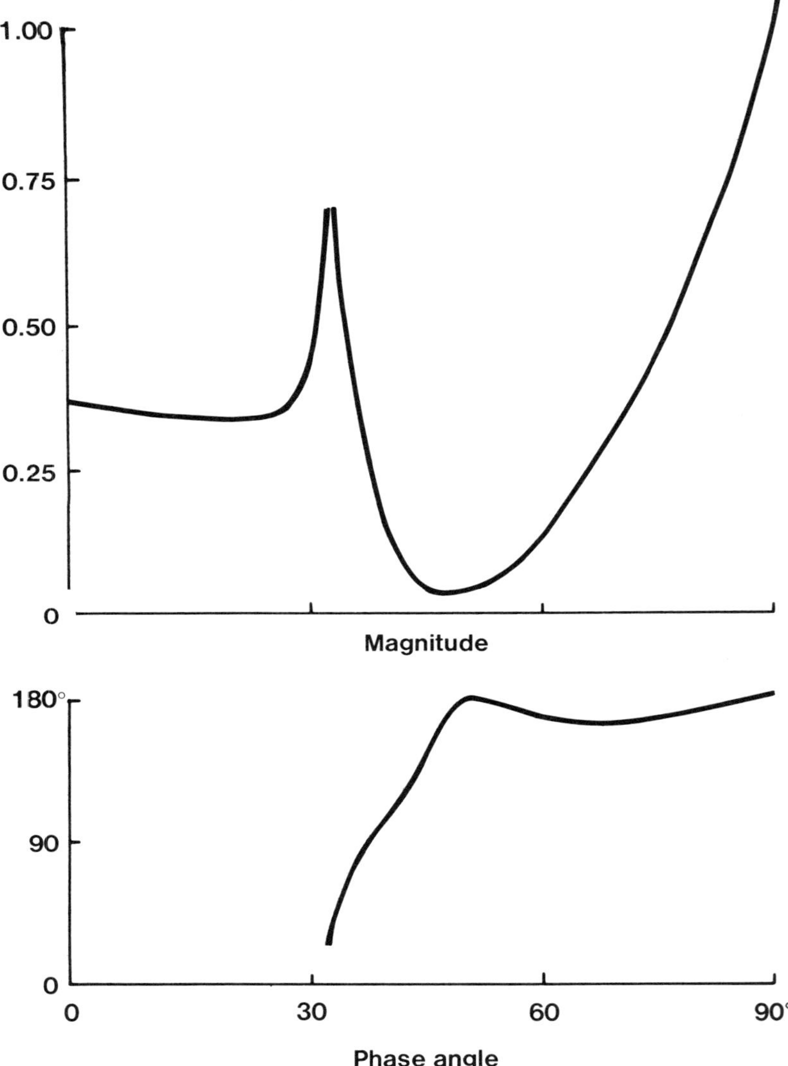

Figure 5.2. *Plane wave displacement reflection coefficients.*

Because the velocity function was based on the section away from the field, the stack in the vicinity of the field is not correct. The presence of the Plexiglas plate raises the normal incidence time and flattens the moveout curve for the event. In these sections, the block reflections have been overcorrected and do not stack in-phase properly. On the stack section of line 1 (Fig. 5.6), it is tempting to interpret the top-of-block reflection as a velocity push-down rather than a pull-up, since the more coherent energy is later in the section. This erroneous interpretation would lead to a conclusion that the overlying body was low-velocity rather than high-velocity. A look at the inside trace gather (Fig. 5.5) would clear up this misidentification immediately.

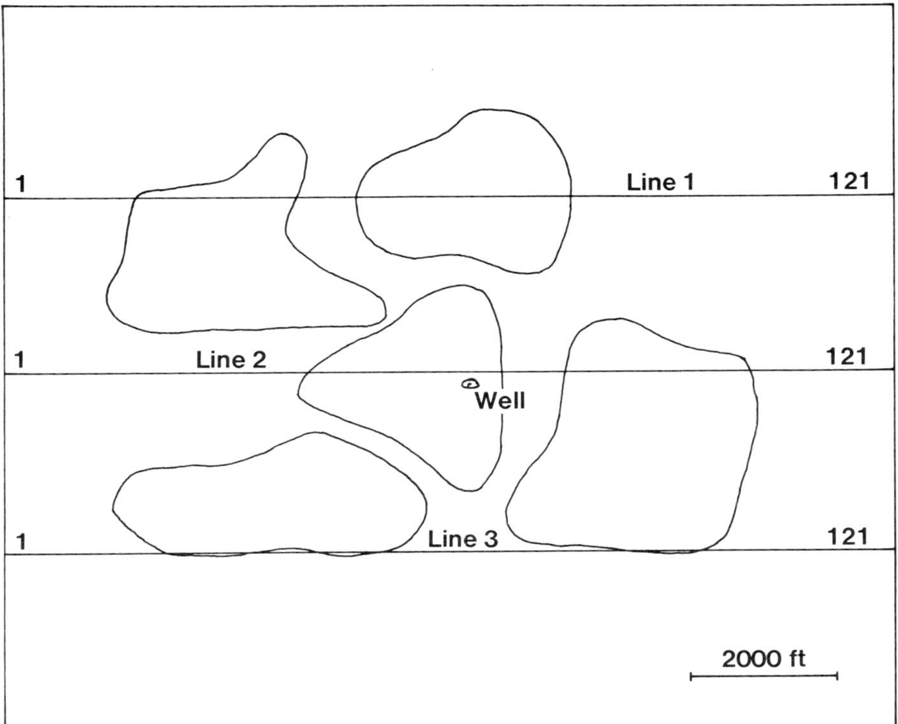

Figure 5.3. *Reconnaissance lines.*

CONCEPTS. When analyzing diffractions from 3-D structures, it is helpful to bear in mind two simple concepts. The first of these helps identify the location of a diffraction. First, consider a simple reflection on an infinite plane. The only raypath that carries energy is the one in which the incident and reflected rays of the path make equal angles with the normal. For more complicated but smooth surfaces, Fermat's principle is used to show that energy is carried along raypaths for which traveltime is stationary. This is a geometric optics approximation.

The same principle may be applied to diffraction raypaths. If the infinite plane is cut into a closed surface, there are an infinite number of diffraction raypaths. Each raypath points from the source to a point on the edge and then to the receiver. Again using Fermat's principle, the only diffraction raypaths that carry energy are those for which the traveltime is stationary. For the strongest energy-carrying diffraction raypath, the incident and diffracted rays make equal angles with the edge.

Energy-carrying reflection raypaths have incident and reflected rays that fall in a plane that passes through the normal to the reflector, but energy-carrying diffracted raypaths are not constrained in this manner. They need only be time stationary. Whereas the reflected ray leaves the stationary time point on the plane in only one direction, the diffracted rays leave the stationary point as a bundle. The rays fall on the surface of a cone, and the vertex of the cone is the stationary time point. When source and receiver are coincident, the stationary time points on the edge are easy to pick—they meet the edge at right angles.

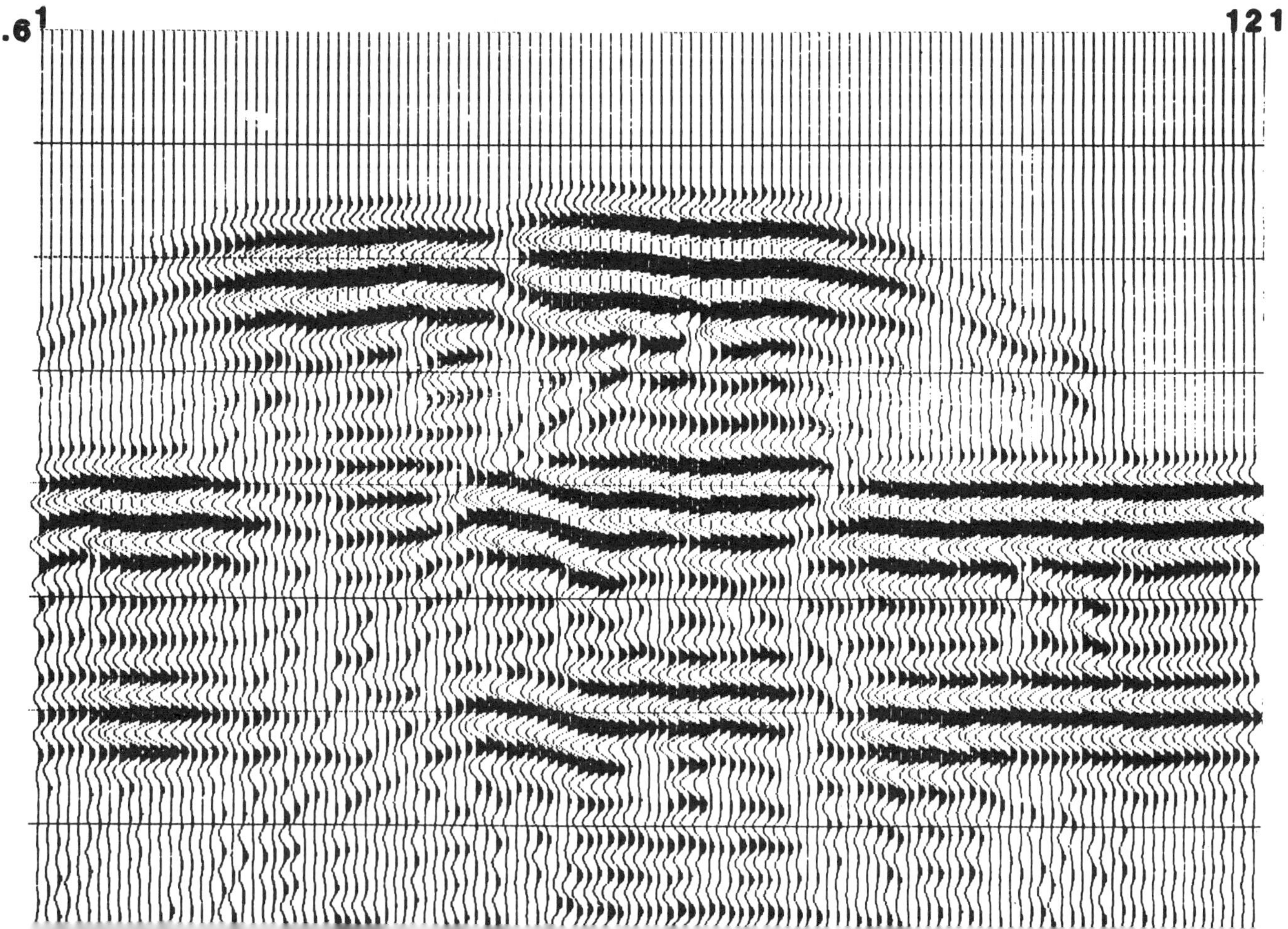
1.6
1
121

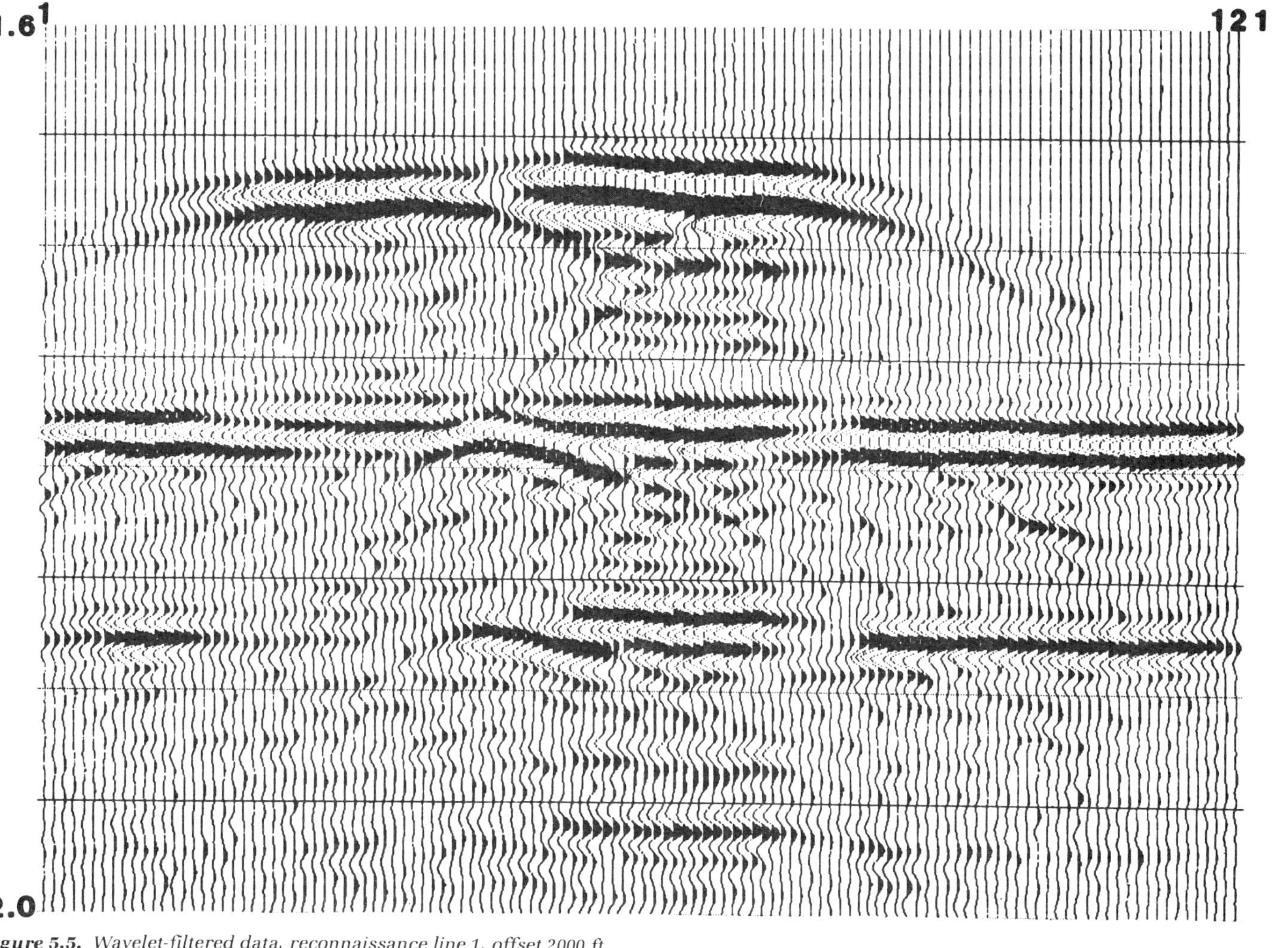

Figure 5.5. *Wavelet-filtered data, reconnaissance line 1, offset 2000 ft.*

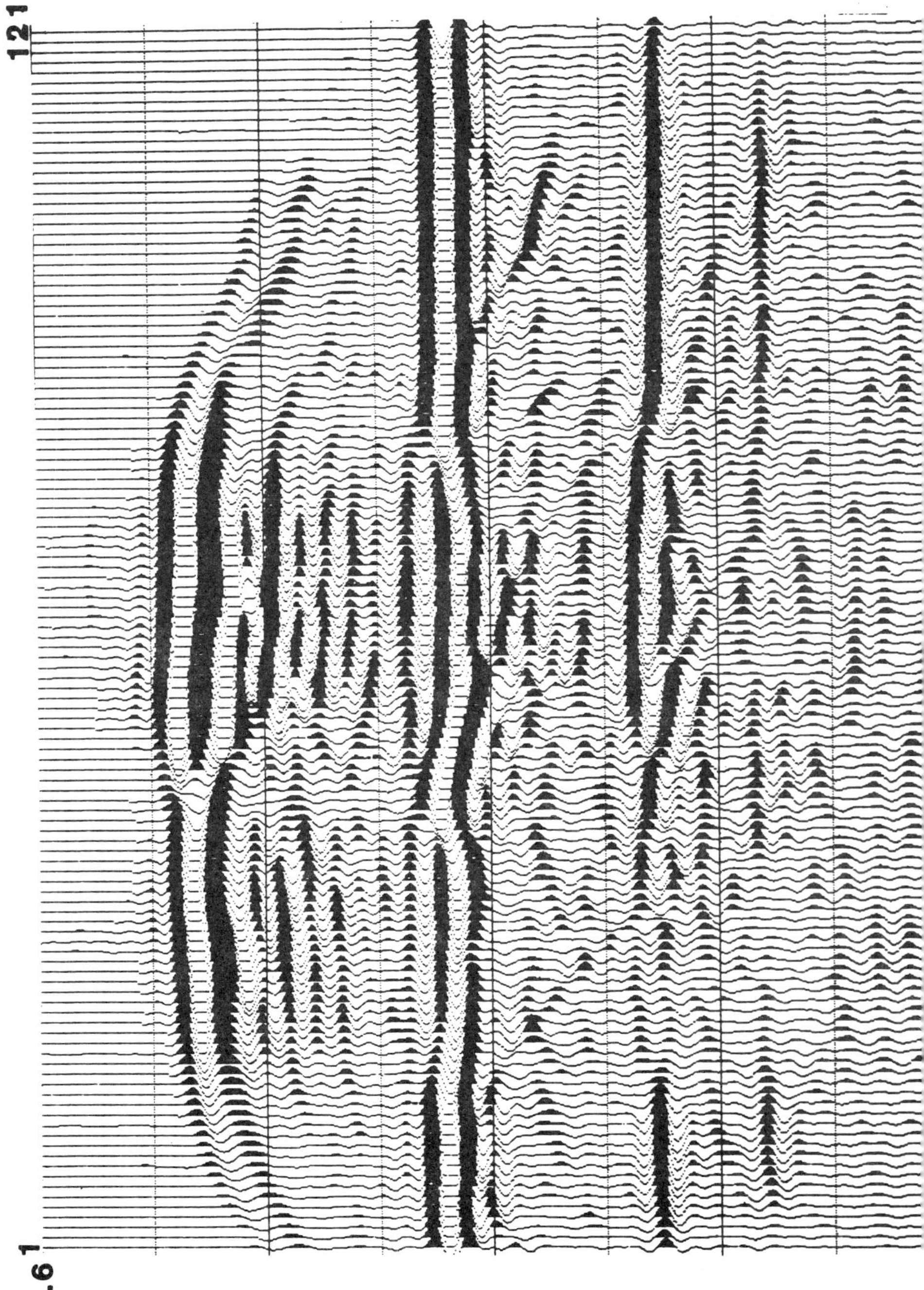

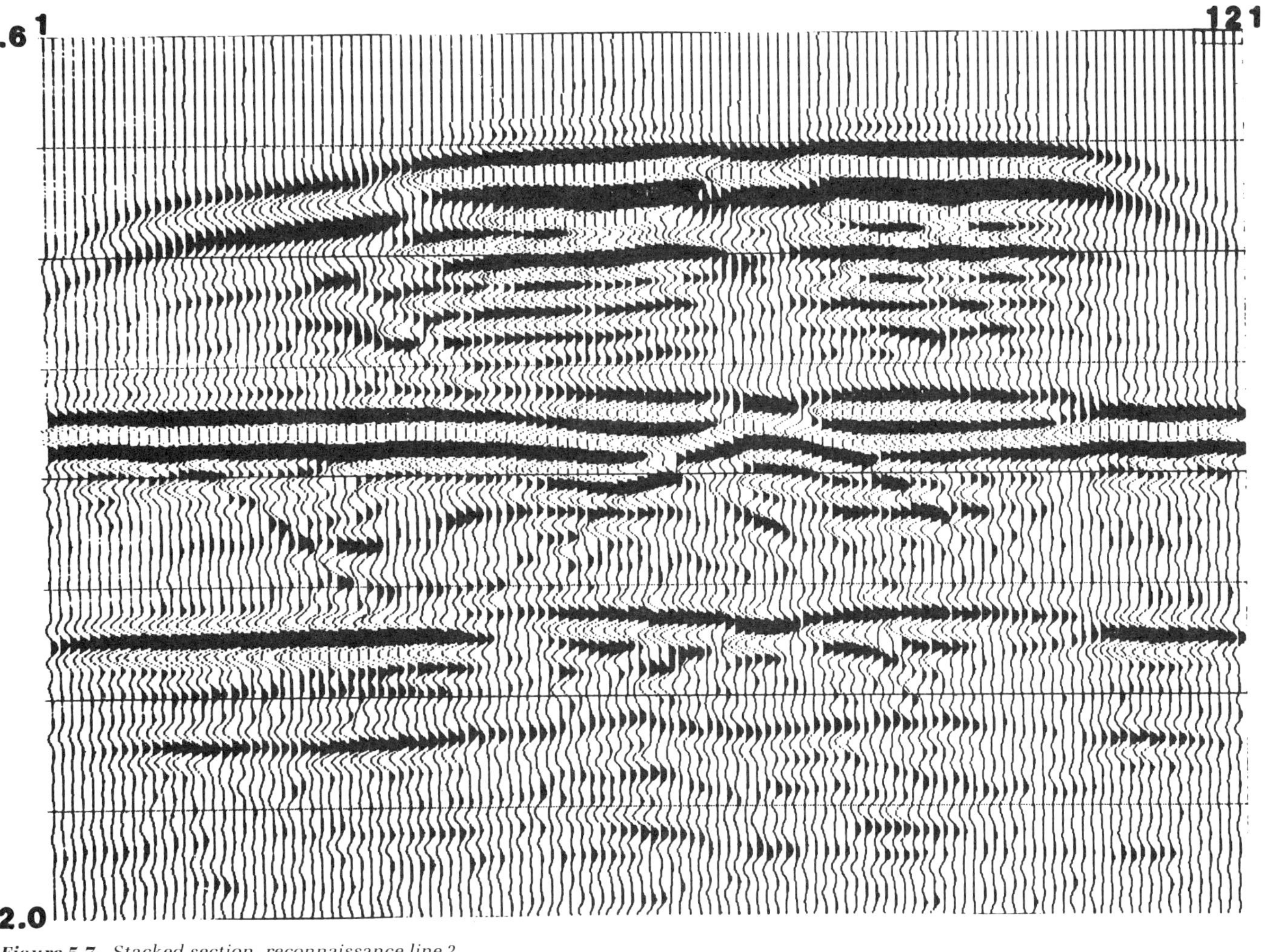

Figure 5.7. *Stacked section, reconnaissance line 2.*

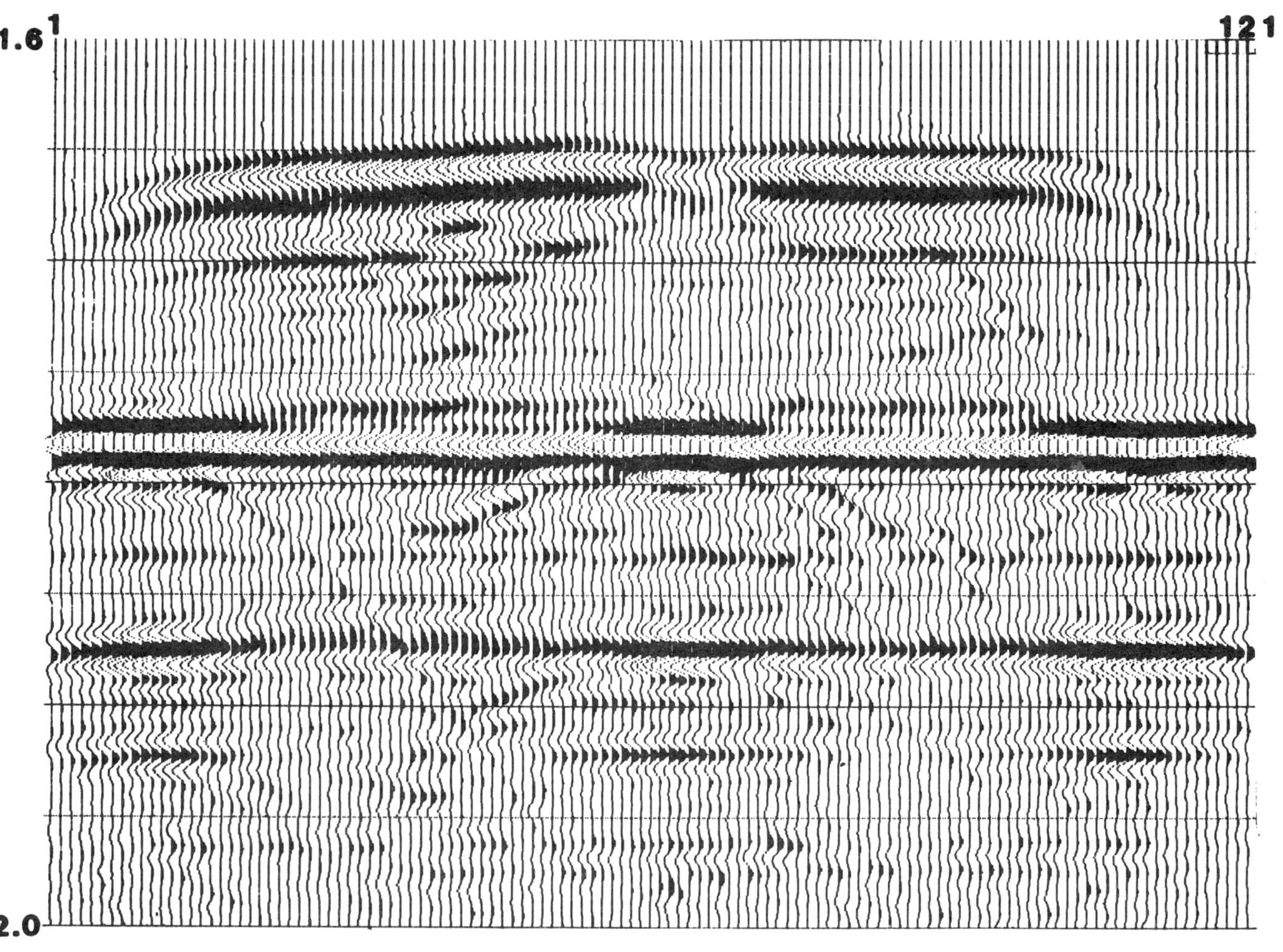
1.6
1
121
2.0

Treating diffractions in terms of raypaths is central to the geometric theory of diffraction (Keller, 1962). Like geometric optics theory, diffraction theory is a high-frequency approximation. In both theories the raypaths are completely independent of the shape of the basic wavelet.

The downfall of the first concept comes about when the raypath concept is applied to smooth, undulating surfaces that have no edges. When diffraction theory is applied to such surfaces, the scope of the solution becomes unwieldy. The stationary time point occurs when the incident ray is tangent to the surface. Thus, the diffracted rays do not form a conical sheet, as they do for edges, but spread out as a volume.

The second concept, physical optics, views diffractions as being generated by changes in the rate at which a wavefront sweeps past a diffracting body. When source and receiver are coincident, the strength of the diffraction is proportional to the rate of change of the solid angle that the wavefront makes with the diffracting body.

According to physical optics, sharp edges are seen as strong diffraction sources, because the solid angle function changes abruptly. Curved surfaces (or surfaces that appear curved because of lateral velocity changes) may also be sources of diffractions, but they are weaker in strength. The diffractions generated by this model are all caused by sharp edges or changes in lateral velocity. Their relative strength is controlled by the radius of curvature of the edge in the vicinity of the stationary time point. When the edge is very sharp and the surface area small, it is useful to view the reflection and diffraction together. The strength of the event is proportional to the ratio of the surface area of the reflector to the area of the first Fresnel zone.

AREAL SURVEY. A fixed-offset areal survey was made over the area bounded by the outline of the block substratum (Fig. 5.9). In each line, the 81 recording points were spaced 150 ft apart. The 68 lines were spaced 150 ft apart. The source-receiver offset was fixed at 1500 ft and aligned in the direction of the line. The observation plane was again 10,000 ft above the top of the bodies.

The odd-numbered lines 5 through 19 are presented in Figures 5.10 through 5.17. These data are presented to show how diffractions are generated near curved edges. Line 5 is well away from the bodies, and isolated edge diffractions may be seen from the upper left and right borders. Below the top reflection from the block, an artifact of the original wavelet can be seen. It is caused by a nonstationary part of the source waveform that was not removed by the wavelet filtering process. The fluctuating part of the source waveform is identified by the undulating event in the middle of the block. The event is also present below the bottom substratum, but because the polarity is negative, the event is not pronounced.

On lines 5 and 7, the diffraction from the convex edge of the right body gives the appearance of an anticline. The concave edge of the left body is the source of the isolated event at 1.7 s on these two lines. Line 9 is not quite over the right body. The event closure is caused by the convex edge of the body. The concave event from the left body is stronger than it was on line 7.

By line 11, the first true specular reflections may be seen over the right body. Three events are found to the right of the edge. One is the specular reflection from the top of the substratum; one is an edge diffraction from the body; and one is an edge diffraction that has been reflected from the top of the substratum. The two diffraction events have the same time moveout. The diffracted reflections also have the same polarity as the edge diffractions. These interfere with the specular reflections, so that the top of block reflections appears to have a synclinal bow-tie.

On the left side of line 11 is an interesting split-edge diffraction. One has a shallow time curve and the other has a steep curve, and both bend down to the left. The flatter diffraction is caused by the portion of the edge nearly parallel to

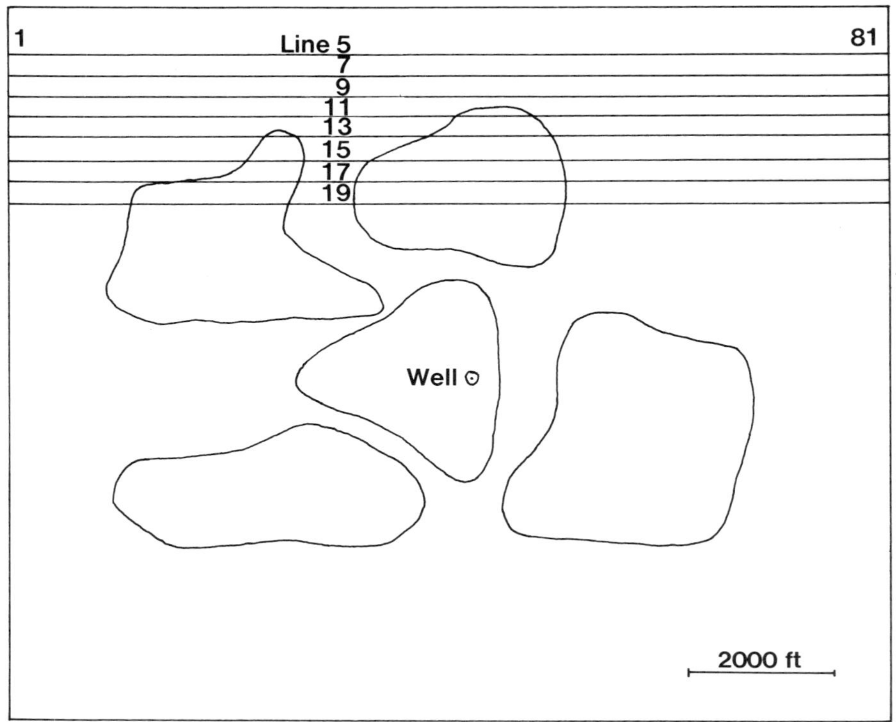

Figure 5.9. *Areal survey lines.*

the line, and the other is caused by the edge more oblique to the line. The shallower event retains its basic shape and grows in strength between lines 13 and 17. By line 19, it has become a specular reflection.

On lines 13 through 19, the right edge of the right body holds its position in the line, while the left edge changes from nearly parallel to perpendicular. The right edge is normal to the line, and its position may be picked by noting the location of the dim spot in the top-block reflection. Whereas the tops of the right-edge diffraction and dim spot align vertically, the left-edge diffraction and its apparent dim spot do not. On line 13, a line connecting these two features is swung down to the right at about a 150° angle; this feature could easily be mistaken for a fault. From lines 13 through 18, the angle of the alignment changes to vertical. Across this sequence of lines, the stationary points of both edges have shifted to the left.

There is also a curious effect due to the interaction of edge diffractions from both bodies. On line 19, the gap between the bodies is noted by an apparent dead zone, caused by distructive interference of the two diffraction patterns. This gap occurs because the right body is approximately a half-period shallower than the left body. Below the gap is an apparent buried focus whose origin is due to diffracted reflections from the two bodies. The gap and buried focus remain together vertically between lines 13 and 19.

COARSE MIGRATION GRID. A coarse-spaced grid of migration lines was used to delineate the edges of the model (Fig. 5.18). In all, six rows and nine columns of

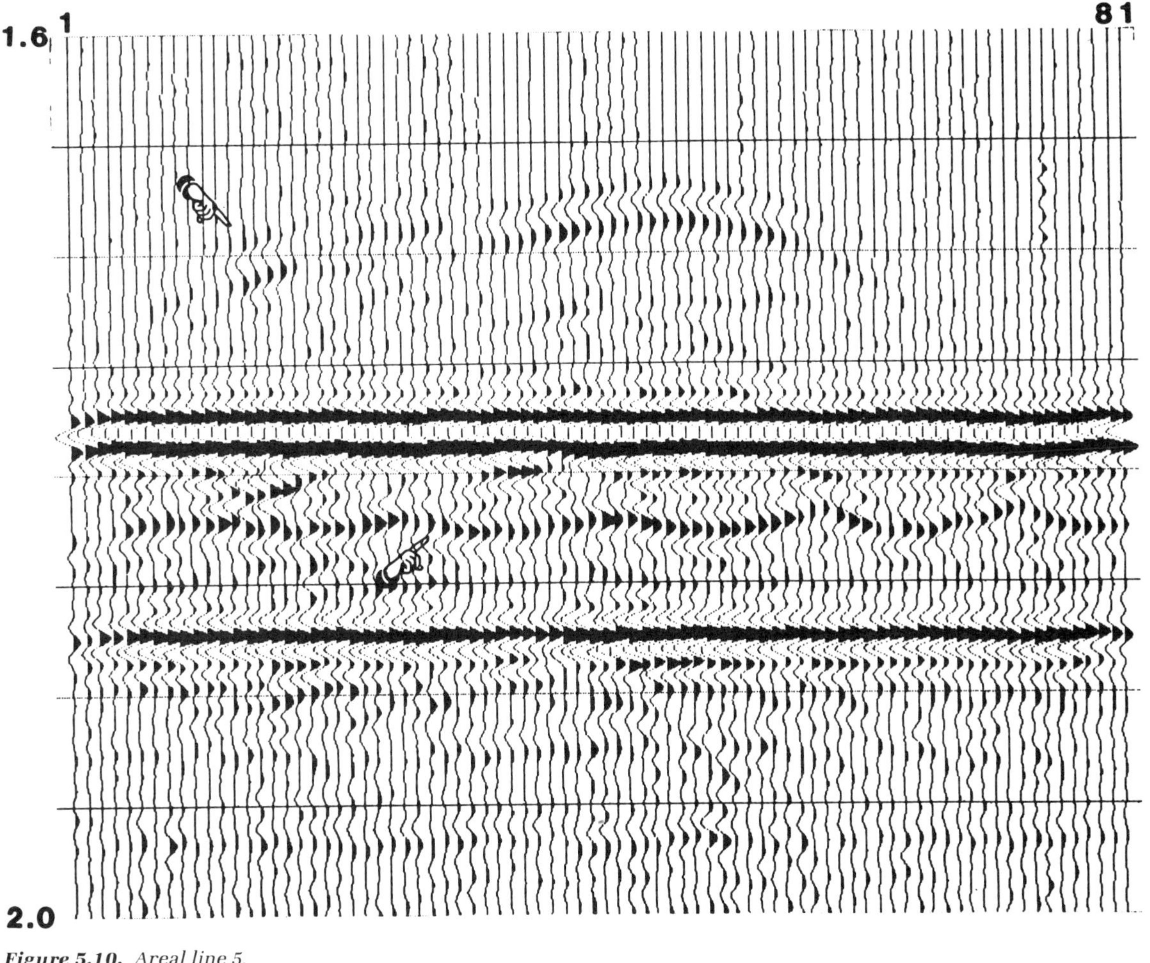

Figure 5.10. Areal line 5.

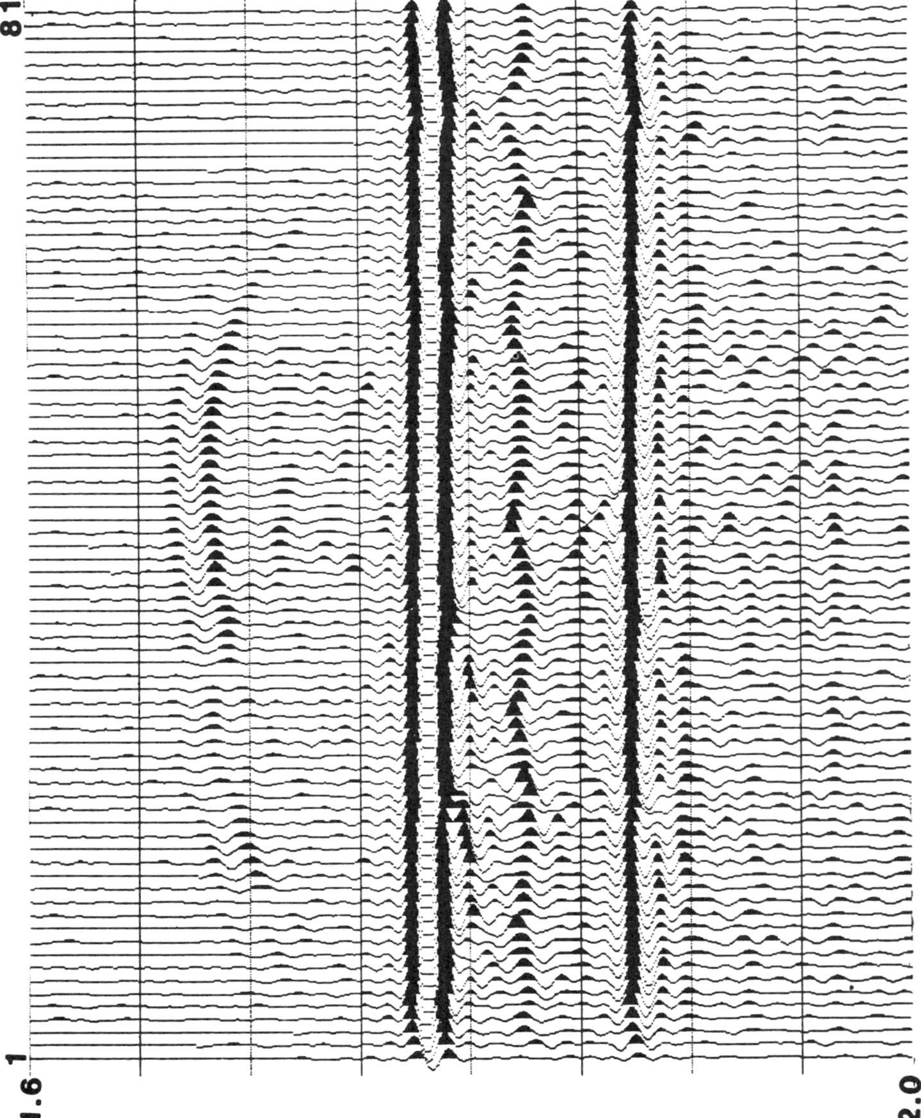
81
1
1.6
2.0

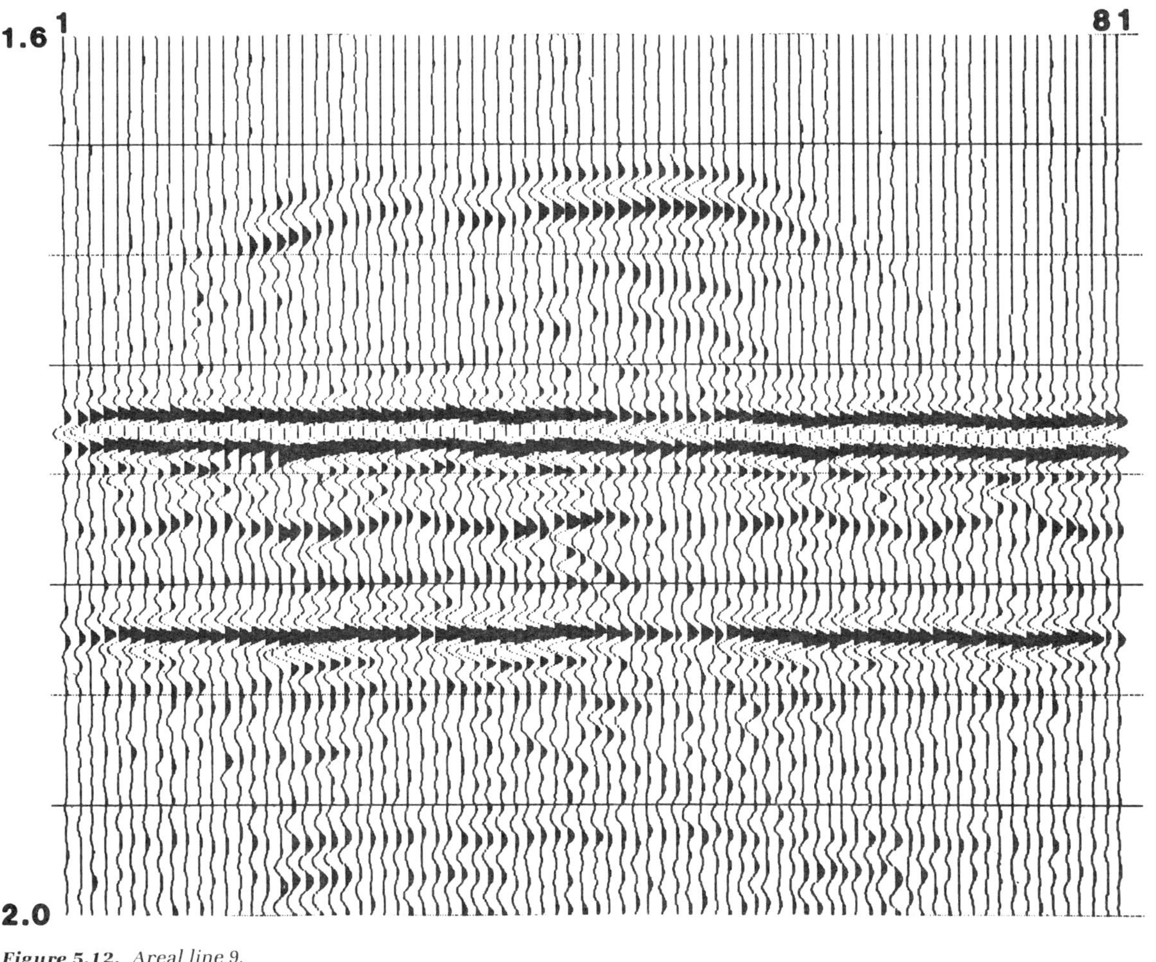

Figure 5.12. Areal line 9.

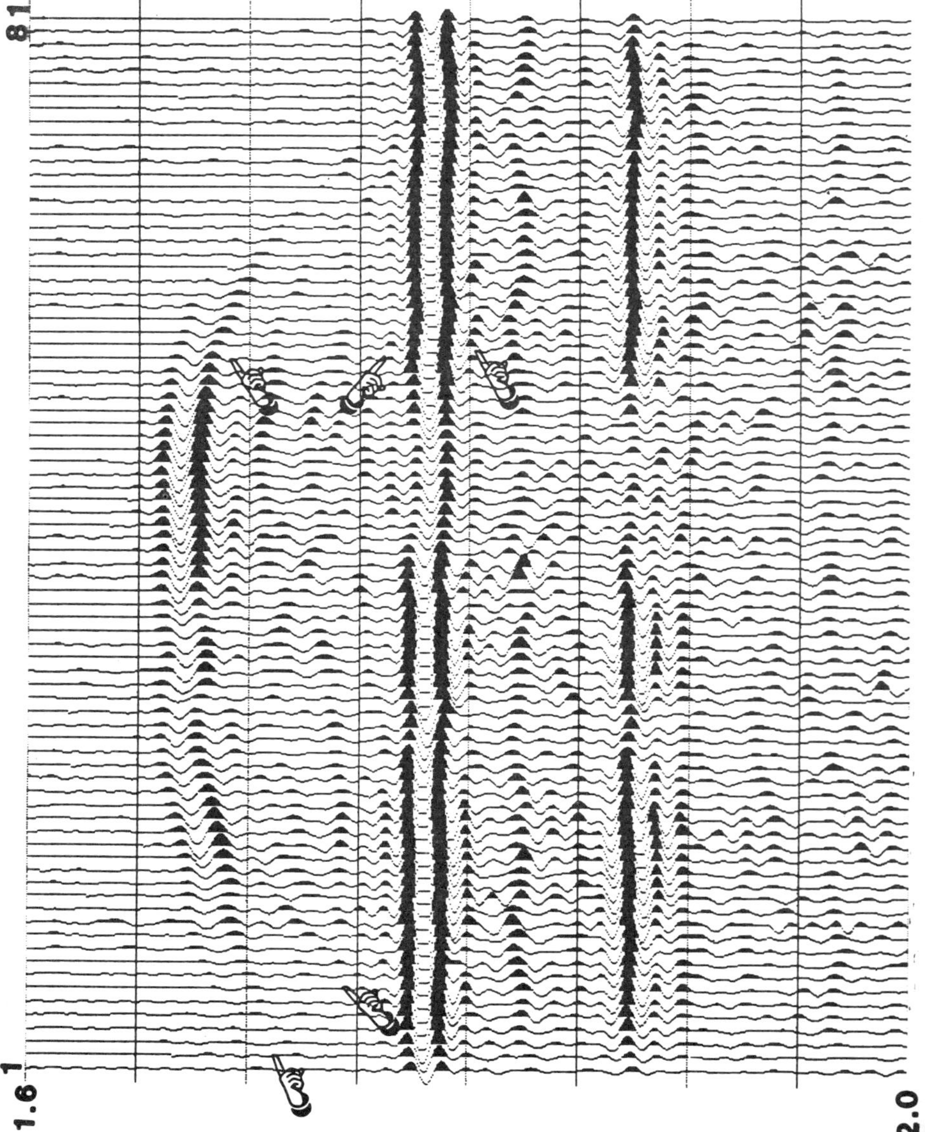
81
1
1.6
2.0

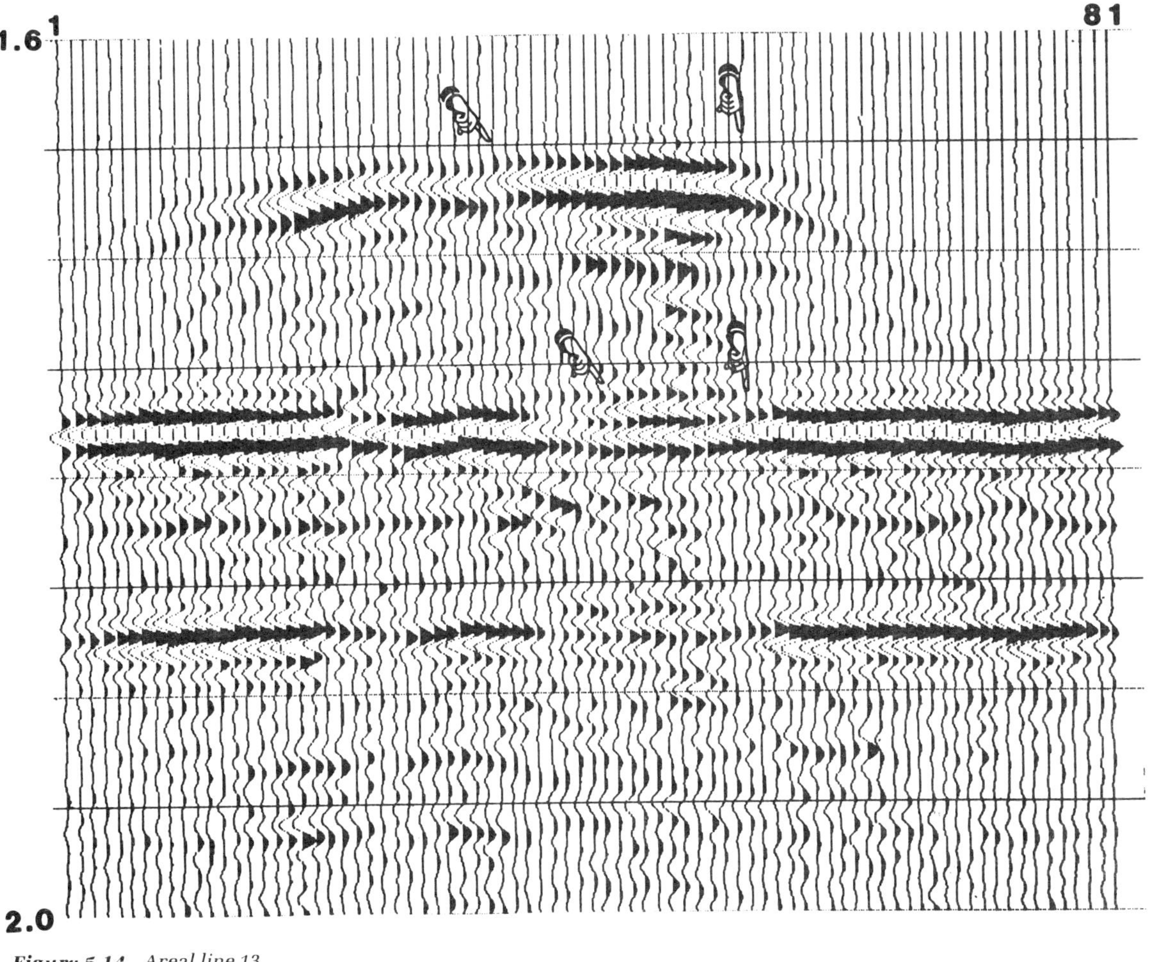

Figure 5.14. Areal line 13.

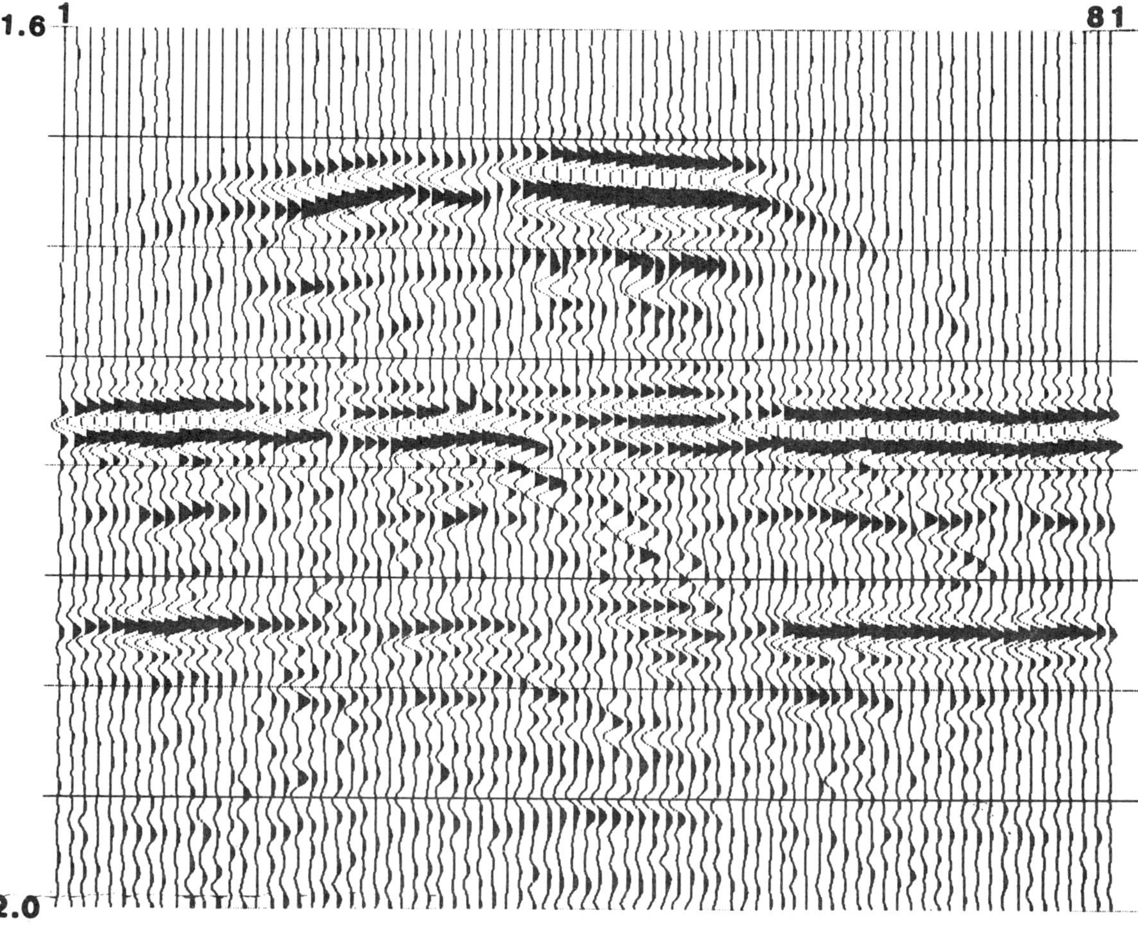
1
81
1.6
2.0

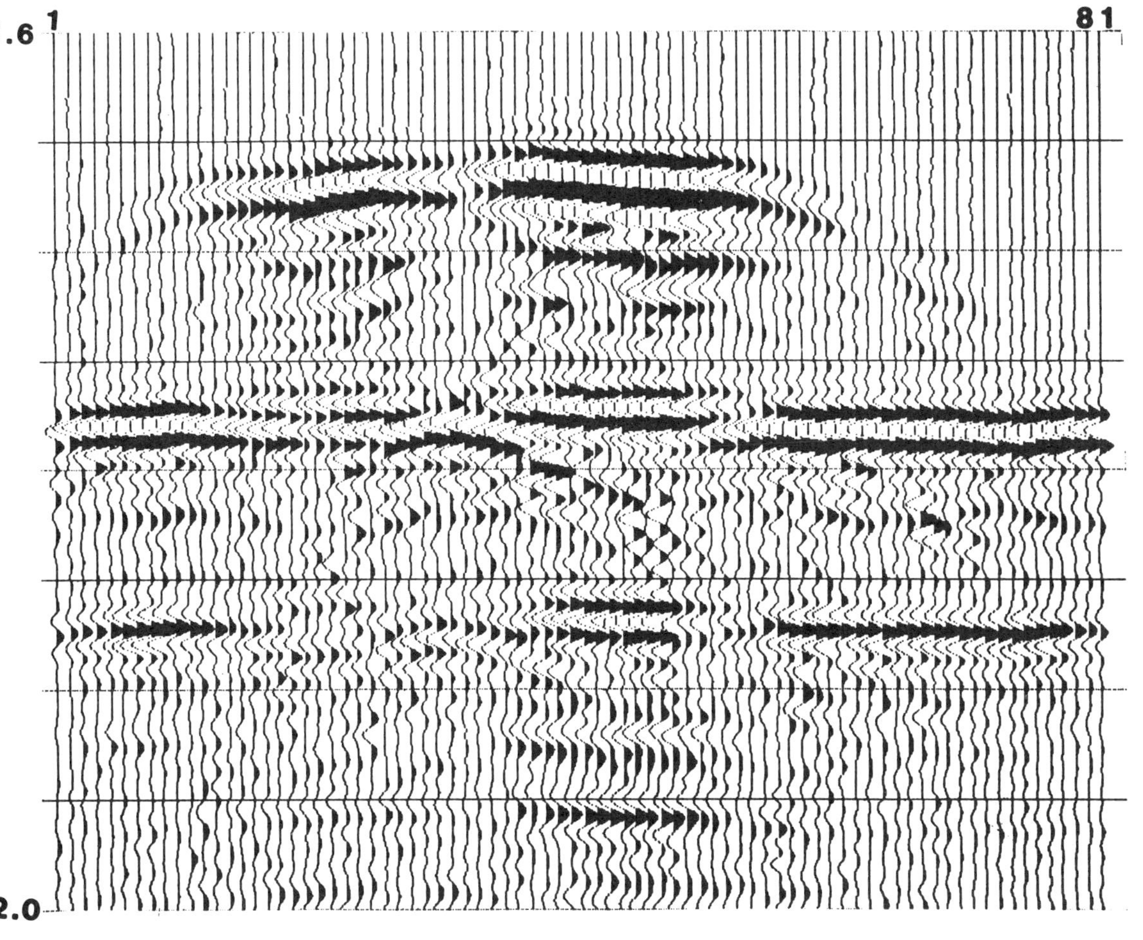

Figure 5.16. *Areal line 17.*

81

1

1.6

2.0

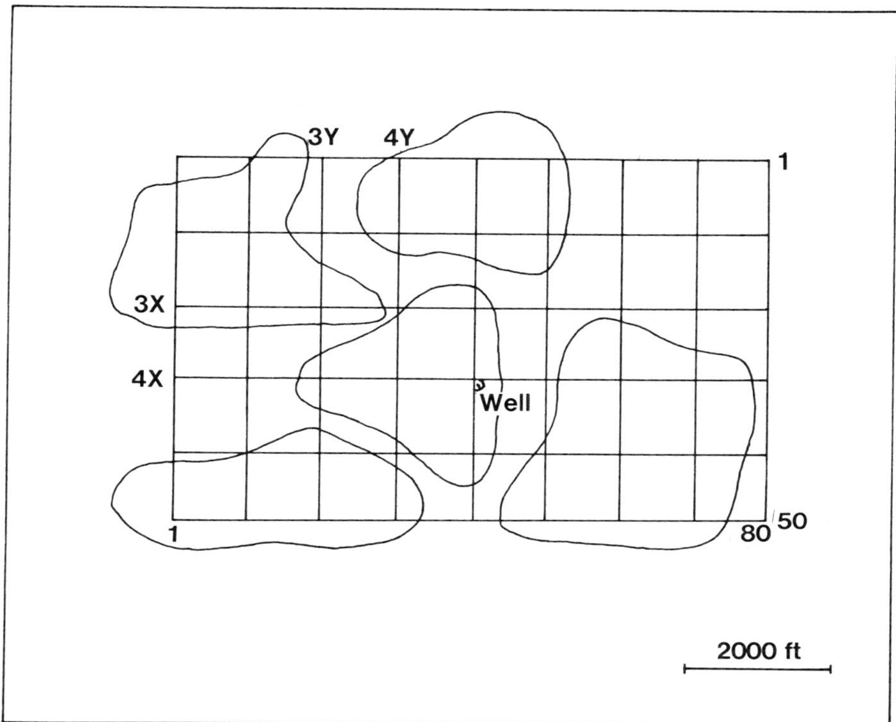

Figure 5.18. Coarse migration grid.

the grid were completed. The migrated trace spacing was 100 ft, and the depth-sampling interval was 20 ft. Time traces were migrated using a Kirchhoff integral method, with a circular aperture of 2000 ft radius. A triangular aperture function was used to taper the effects of the edge. Water velocity was used for the migration. Row line 3X and column line 4Y are presented in Figures 5.19 and 5.20.

The velocity pull-up of the top-of-block reflection under the bodies is very apparent in these figures. The source-generated artifact, previously noted, follows the main signal by approximately 0.040 s. The time signal event may be seen as a spurious reflector that is approximately 240 ft below the real image.

Line 3X crosses the upper left and center bodies. Line 4X crosses the center and lower right bodies. In this line direction, the upper left body dips to the left. The center body also dips to the left, but not at such a steep rate. The lower right body does not appear to dip in this direction. The substratum has no perceptible dip in either direction.

Measurements were made of the dip components, using line 3X (Fig. 5.19) and 2Y (not shown). They indicate that the upper left body is dipping to the left at an angle of 0.82° and toward the bottom at an angle of 0.38°. The resultant dip component of 0.91° has not been verified on the real model.

It is clear that, although migration has collapsed the diffraction events so that the edges are easier to pick, there is still a limitation to their location accuracy. The amplitudes in the migrated image are still controlled by the principles described in the previous section.

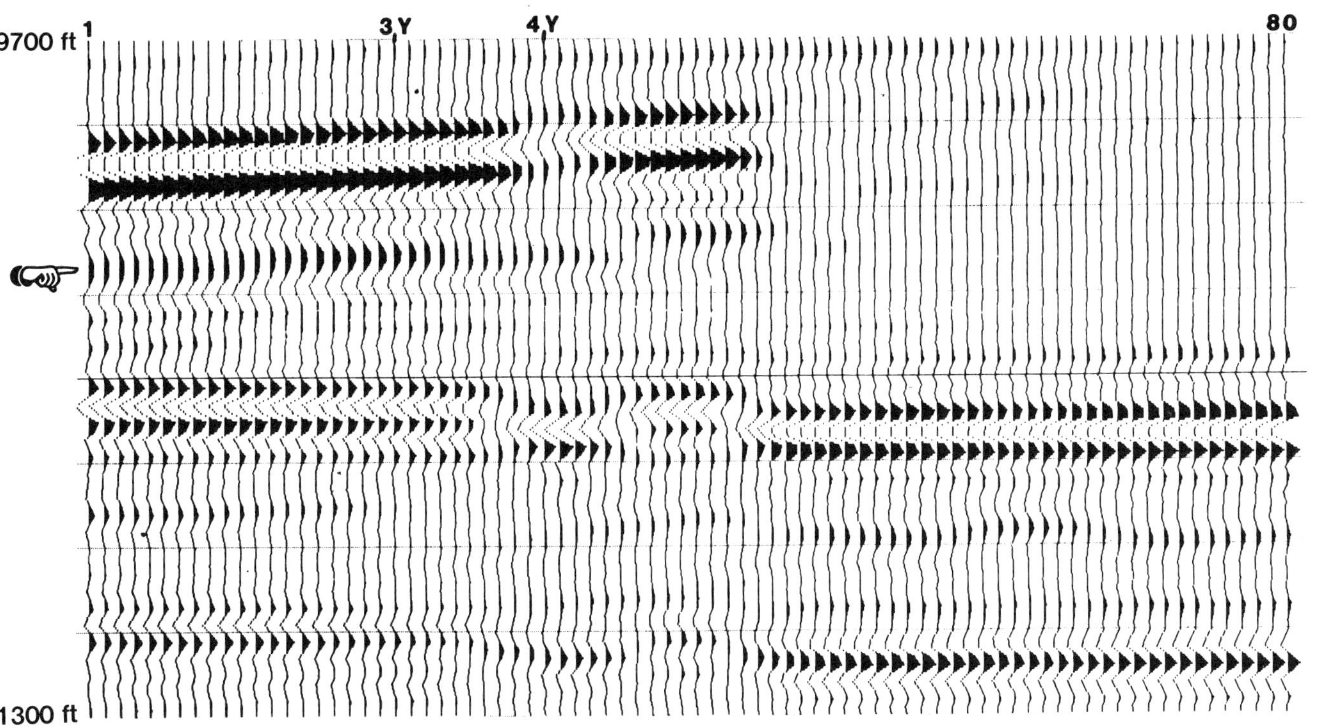

Figure 5.19. *3-D migration line 3X.*

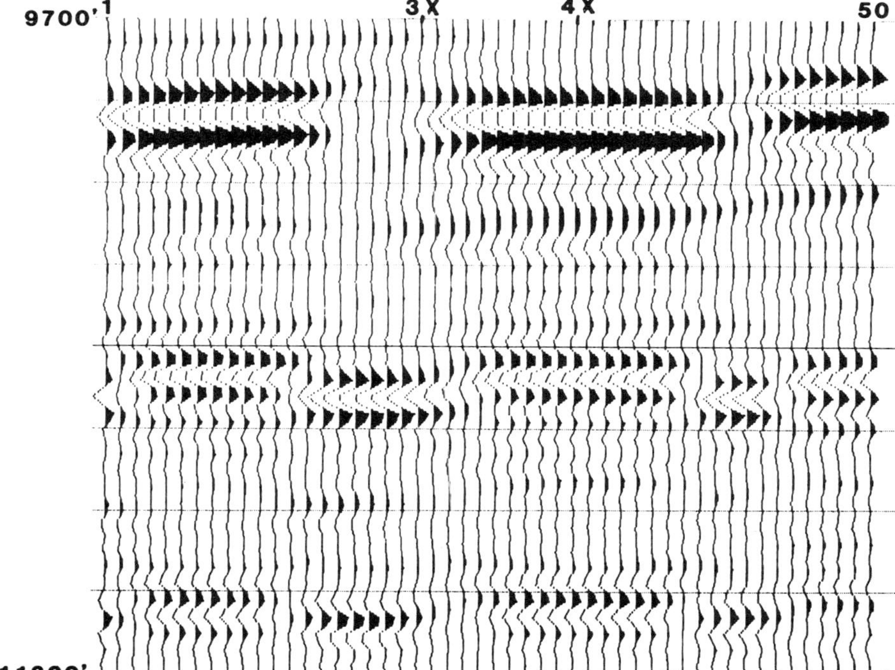

Figure 5.20. 3-D migration line 4Y.

Images of four of the five reservoir bodies may be seen along line 4Y (Fig. 5.20). The very slight dip component of the center body may be seen in the figure. In Figure 5.20, the left edge of the image of the center body is affected by the proximity of the nose of the upper left body.

The model is basically the source of two types of waves—specular reflections from the reservoir bodies and diffractions from the edge. In a sense analogous to resolution in optical instruments, the lateral resolution of sharp edges, such as faults and other sharp-edged geologic features, is controlled by the size of the aperture and by the wavelength (Berkhout, 1980, ch. 12). Lateral resolution is improved by either increasing the size of the aperture or decreasing the wavelength. However, there is no substantial improvement of lateral resolution of the specularly reflected waves beyond a certain aperture radius. If the end of the reflected ray is sufficiently within the aperture for the traces in the first Fresnel zone to be included, then the greater part of the reflected waves will be migrated. Increasing the aperture radius any larger than this will not strengthen the image significantly.

DETAILED MIGRATION GRID. A very closely spaced migration grid was used to identify the shape of the boundaries of two adjacent bodies. Figure 5.21 presents the location map of the survey. The survey area includes the tip of the upper left body and a portion of the center body. The depth-trace spacing is 20 ft, and the depth interval is 10 ft. The areal data, described earlier, were used to compute the images. Sections along columns 1 and 21 are presented in Figure 5.22. The data in this section are all plotted with reverse polarity.

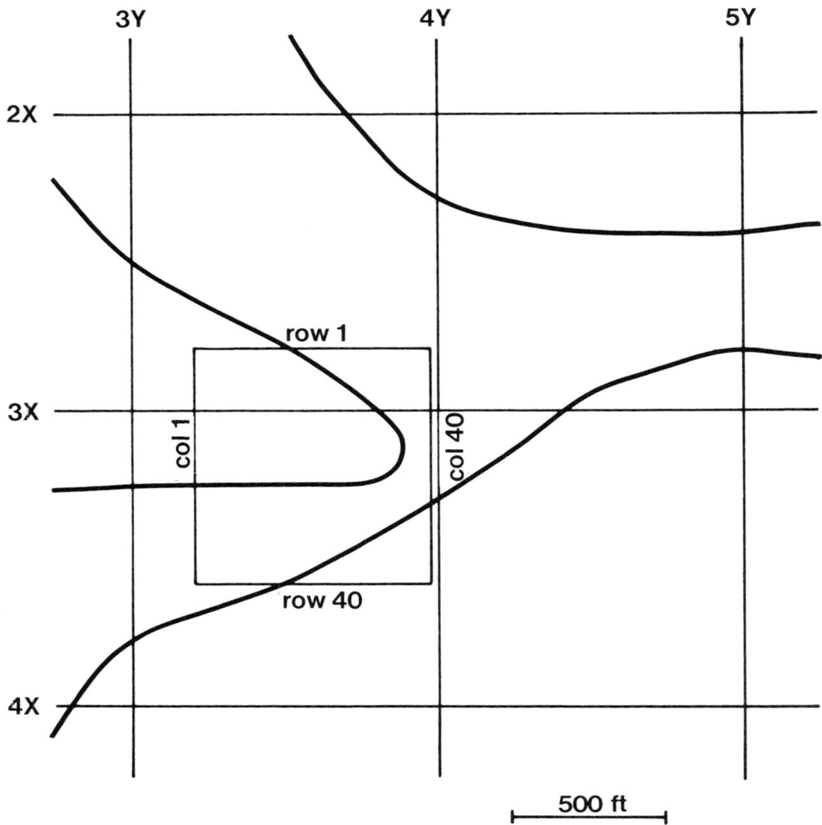

Figure 5.21. *Detailed migration grid survey map.*

The image of the upper left body may be seen along column 1. The top and bottom of the reservoir body are not separated, and the composite waveform gradually decays in amplitude off the edge. Near the end of the line, the amplitude starts to increase as the center body is approached. Along column 21, the gap between the two bodies is 340 ft, and the wavelength is 200 ft. Note that the center body is slightly shallower than the upper left body. In this section, the edges are difficult to pick.

The migrated data were organized into depth slices, several of which are presented in Figures 5.23 through 5.29. The plot-scale factor is constant for these figures. At a depth of 9920 ft (Fig. 5.23), the peak of the image of the center body is seen as a dark area in the lower right portion of the figure. At a depth of 9930 ft (Fig. 5.24), the dark area has grown, indicating the peak of the image wavelet has not been reached. Figure 5.25, at 9940 ft, the image of the upper left body becomes visible as a dark area, indicating that the polarity has changed. At a depth of 9950 ft (Fig. 5.26), the dark area from upper left body fills much of the slice. In the next deeper slice (Fig. 5.27), the dark area fills, because the center body is contracting while the area for the upper left body is still growing. This confirms the earlier observation that the center body is slightly above the upper left area. In the final two figures, at depths of 9970 and 9980 ft, respectively, the polarity of

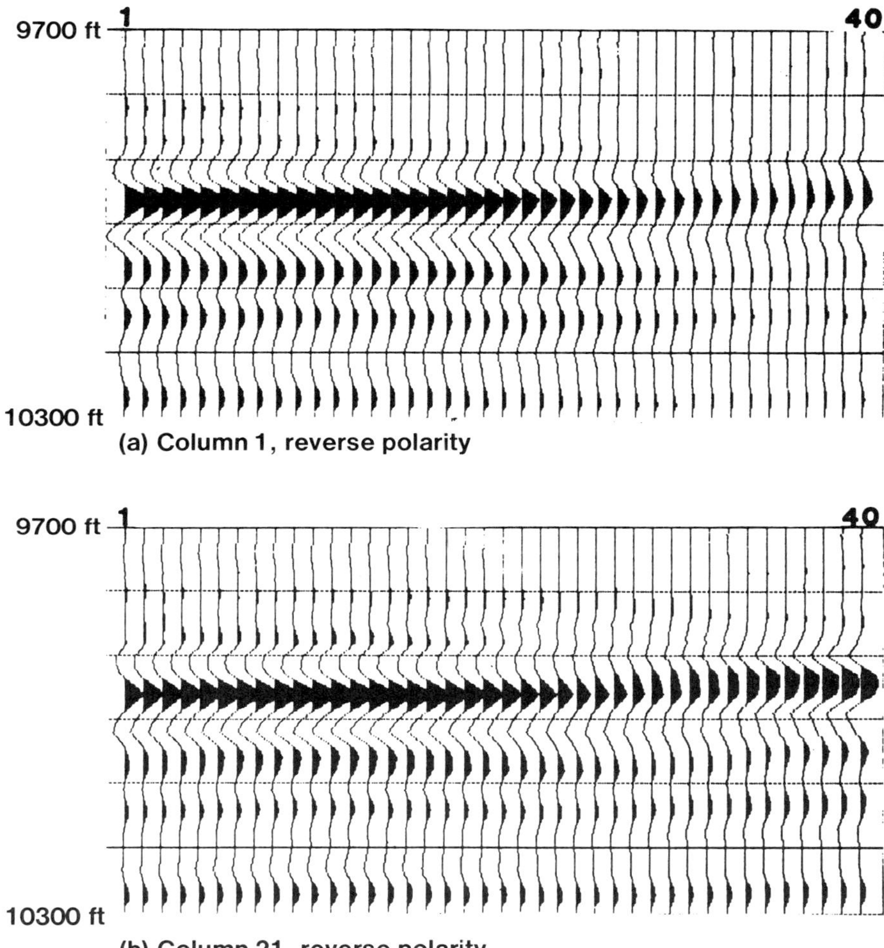

Figure 5.22. *Detailed migration grid sections.*

the image of the center body has changed, while the polarity of the other body remains the same.

AREAL VSP SURVEY. A hypothetical well site was selected near reconnaissance line 2. Because the center body is dipping slightly to the left, the well was placed near the updip edge. The receiver was lowered in this hypothetical well to depths of 0, 2500, 5000, and 7500 ft. The tops of the bodies are at a depth of 10,000 ft. A rectangular grid of surface source locations was chosen to illuminate the model to the left of the well location. The side lengths were 2,550 ft by 11,250 ft. The trace spacing within the 76-row by 18-column grid was 150 ft. The grid box was offset 2450 ft to the left of the well.

Figure 5.30 shows a map view of the rectangular boxes that bound the specular reflection points at the 10,000 ft level. These bounds were computed assuming that the reflecting plane was flat. See Appendix 5A, at the end of this chapter, for a

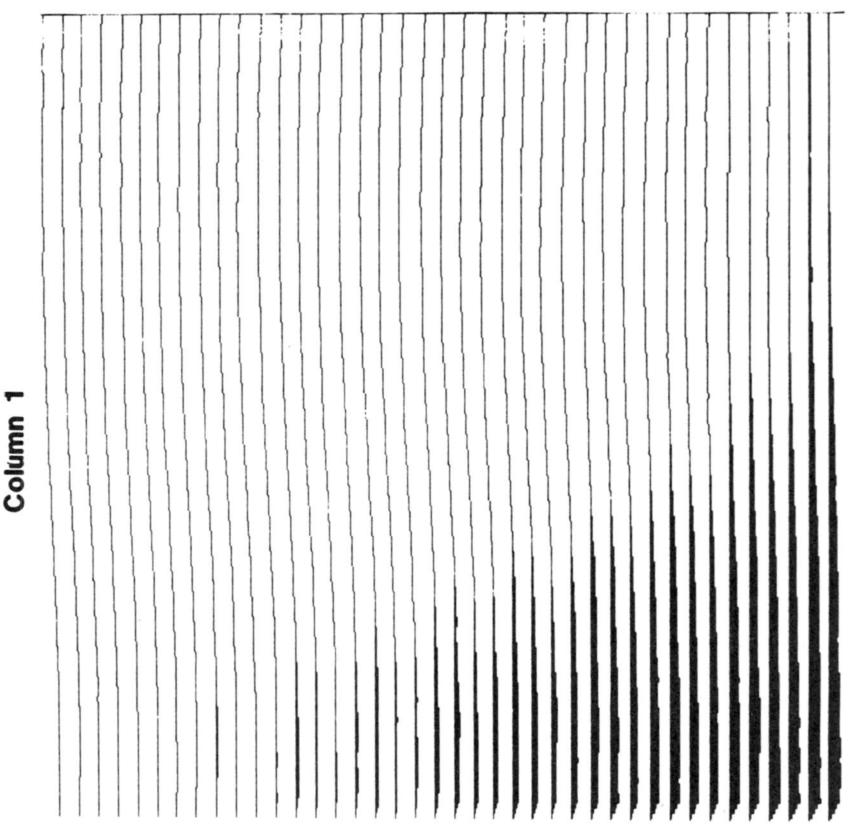

Figure 5.23. *Horizontal depth slice, 9920 ft, reverse polarity.*

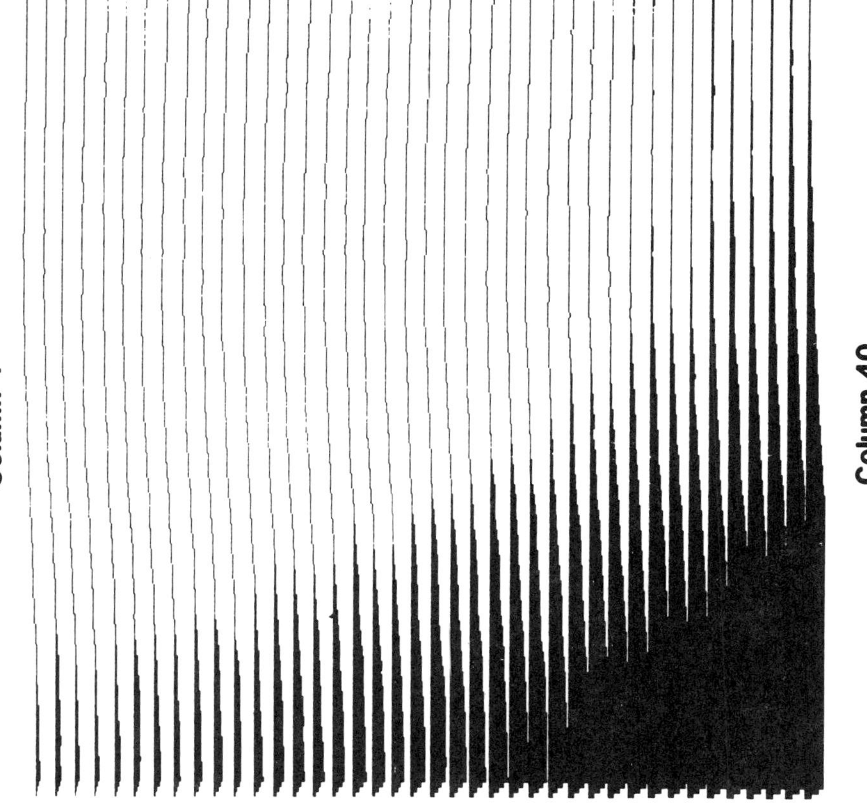

Figure 5.24. *Horizontal depth slice, 9930 ft, reverse polarity.*

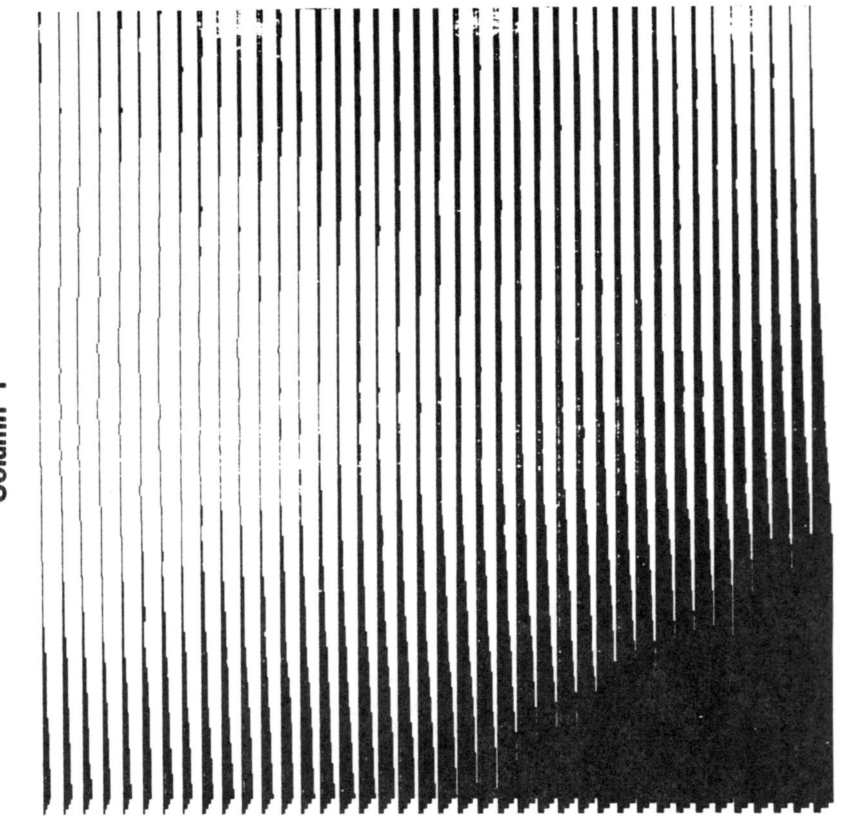

Figure 5.25. Horizontal depth slice, 9940 ft, reverse polarity.

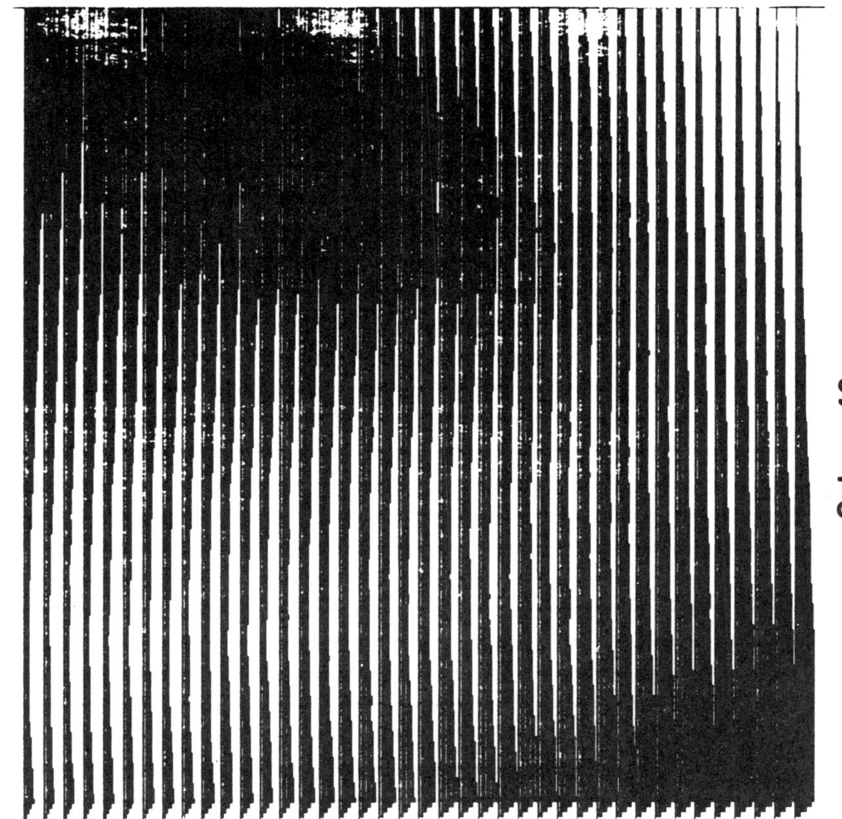

Figure 5.26. Horizontal depth slice, 9950 ft, reverse polarity.

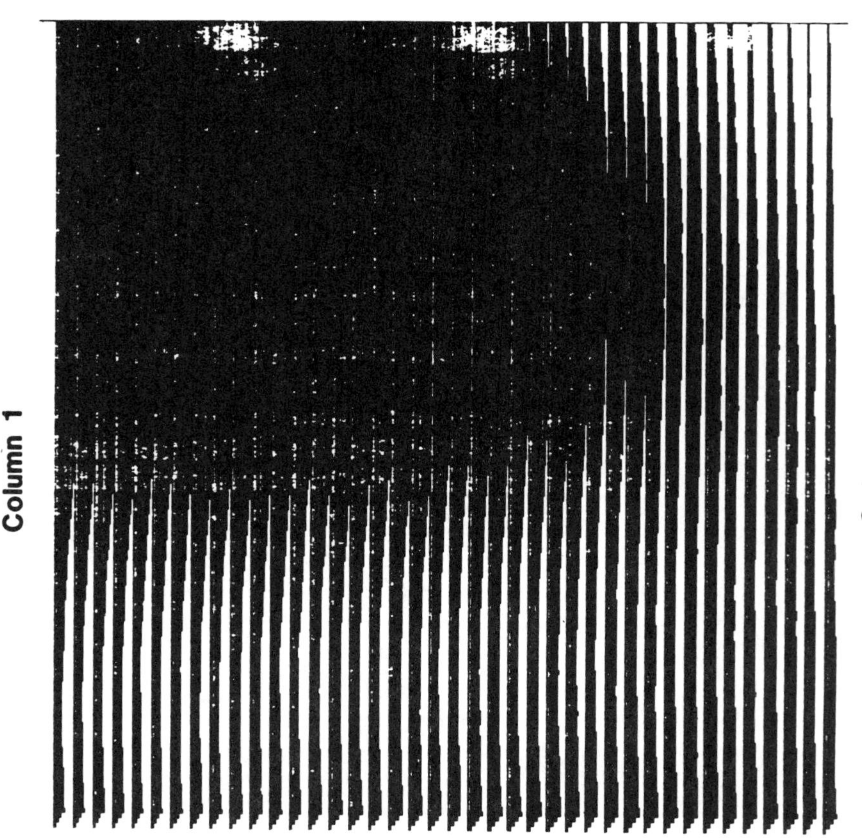

Figure 5.27. Horizontal depth slice, 9960 ft, reverse polarity.

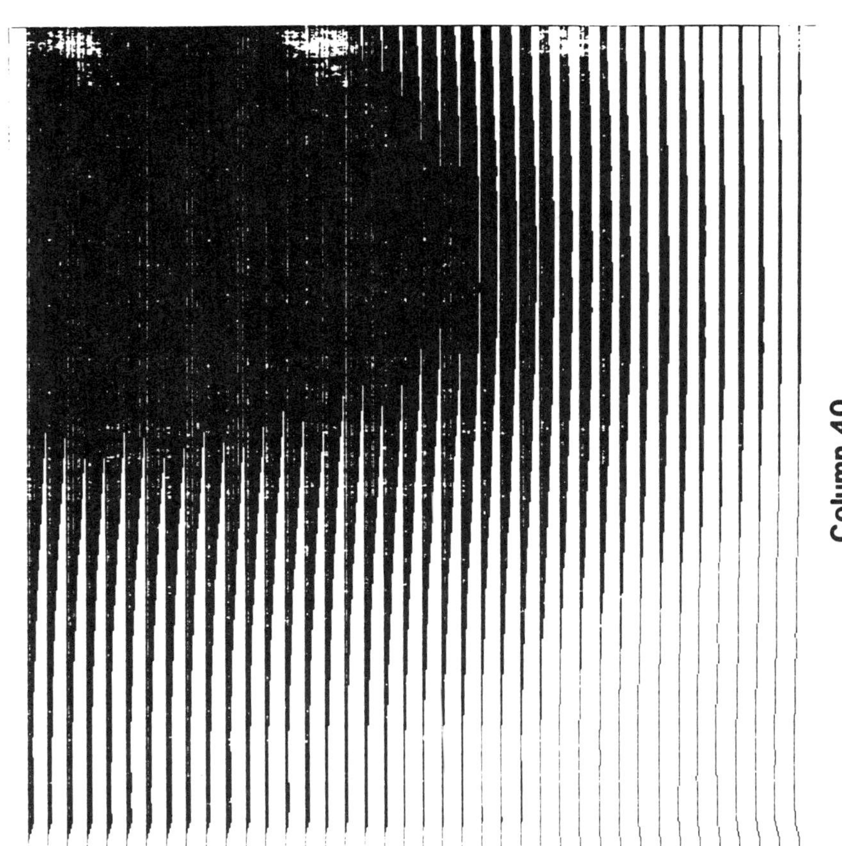

Figure 5.28. *Horizontal depth slice, 9970 ft, reverse polarity.*

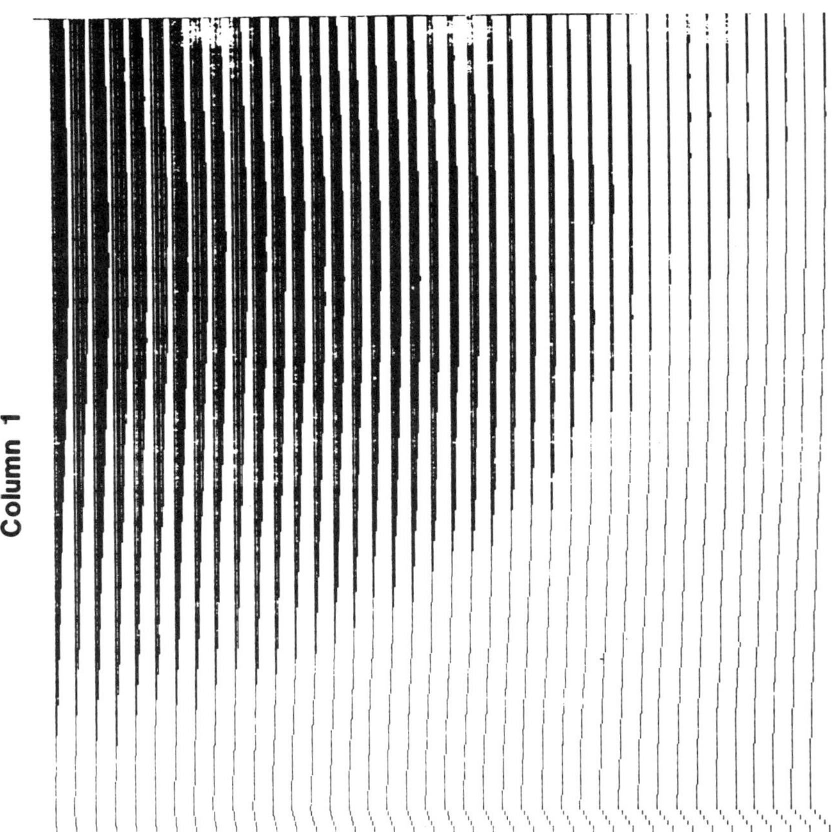

Figure 5.29. *Horizontal depth slice, 9980 ft, reverse polarity.*

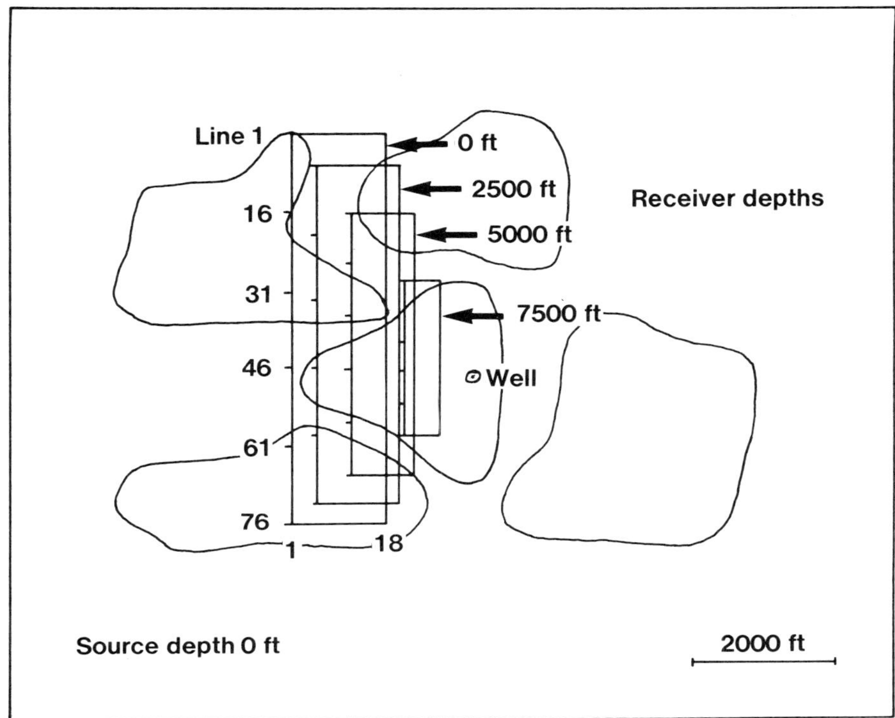

Figure 5.30. *Areal VSP midpoint boxes.*

deviation of an expression relating reflection point to the surface source offset, receiver depth, and orientation of the reflector. It should be noted that, when the receiver is at the surface, the box is identical to the conventional CDP box. As the receiver depth increases, the area of subsurface illumination grows smaller and the reflection angles become wider. With this choice of receiver positions, there is a threefold trace multiplicity in some portions of the survey.

Selected rows of the survey are shown in the next set of figures. In Figures 5.31 and 5.32, the receiver depth was 2500 ft. Along the six lines shown in the two figures, the data have been static-shifted to horizontally align a flat reflector with a surface-to-surface normal incidence time of 1.660 s. The water velocity is assumed to be 11,928 ft/s. (The static shift expression is given in Appendix 5B).

The nearly flat event at 1.5 s on line 46 (Fig. 5.32) is a reflection from the left portion of the center body. A similar event on line 31 (Fig. 5.31) is the reflection from the top of the upper left body. Line 16 starts with a reflection from the edge of the upper right body, and line 1 starts with a diffraction event from the upper left body.

The event between 1.695 and 1.715 s is the reflection from the top of the block. The statics that flatten the reservoir body reflections do not flatten these reflections. The further offset corresponds to the leftmost trace on each line.

Gaps between bodies may be seen on lines 1, 16, and 61. The long train of events below the specular reflections on line 46 is caused by edge diffractions that arise from the entire nose of the left side of the center body.

Portions of the data with the deepest well position are shown in Figures 5.33 and 5.34. The events crossing the tops of the sections are portions of the direct

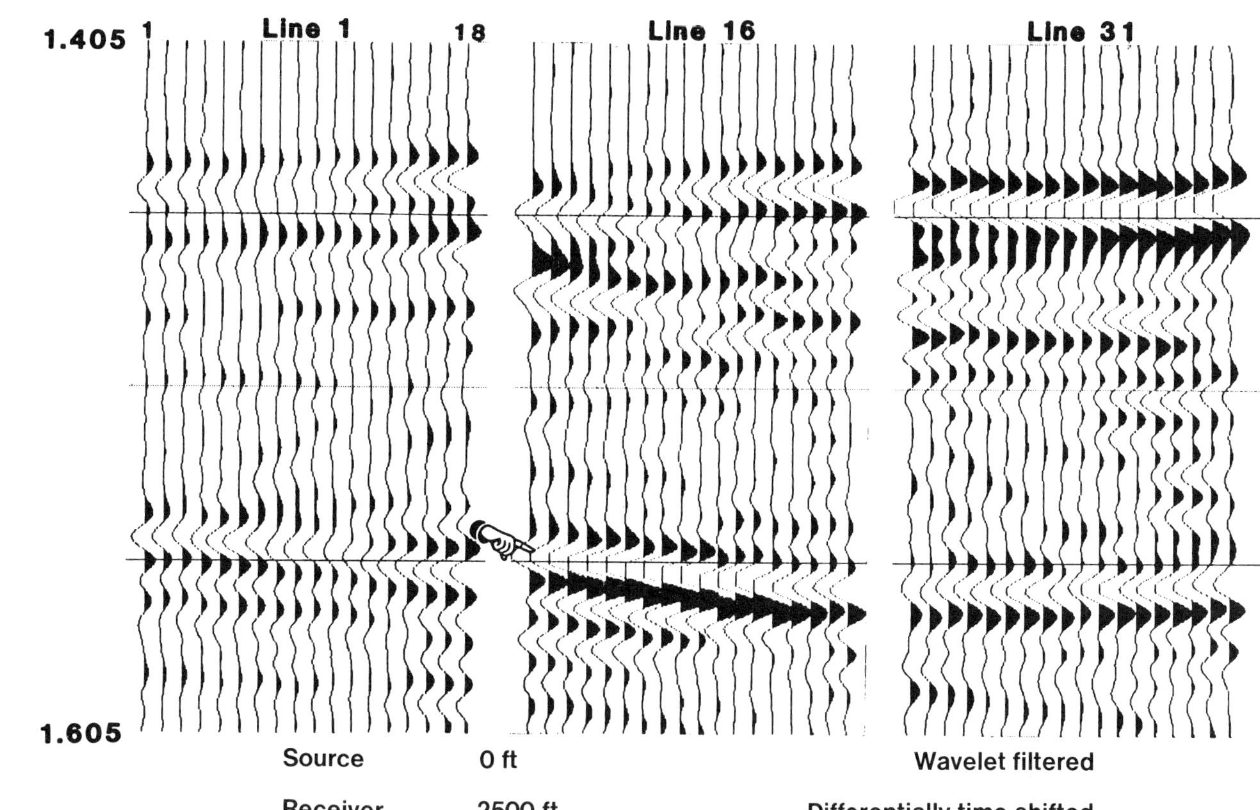

Figure 5.31. Areal VSP lines.

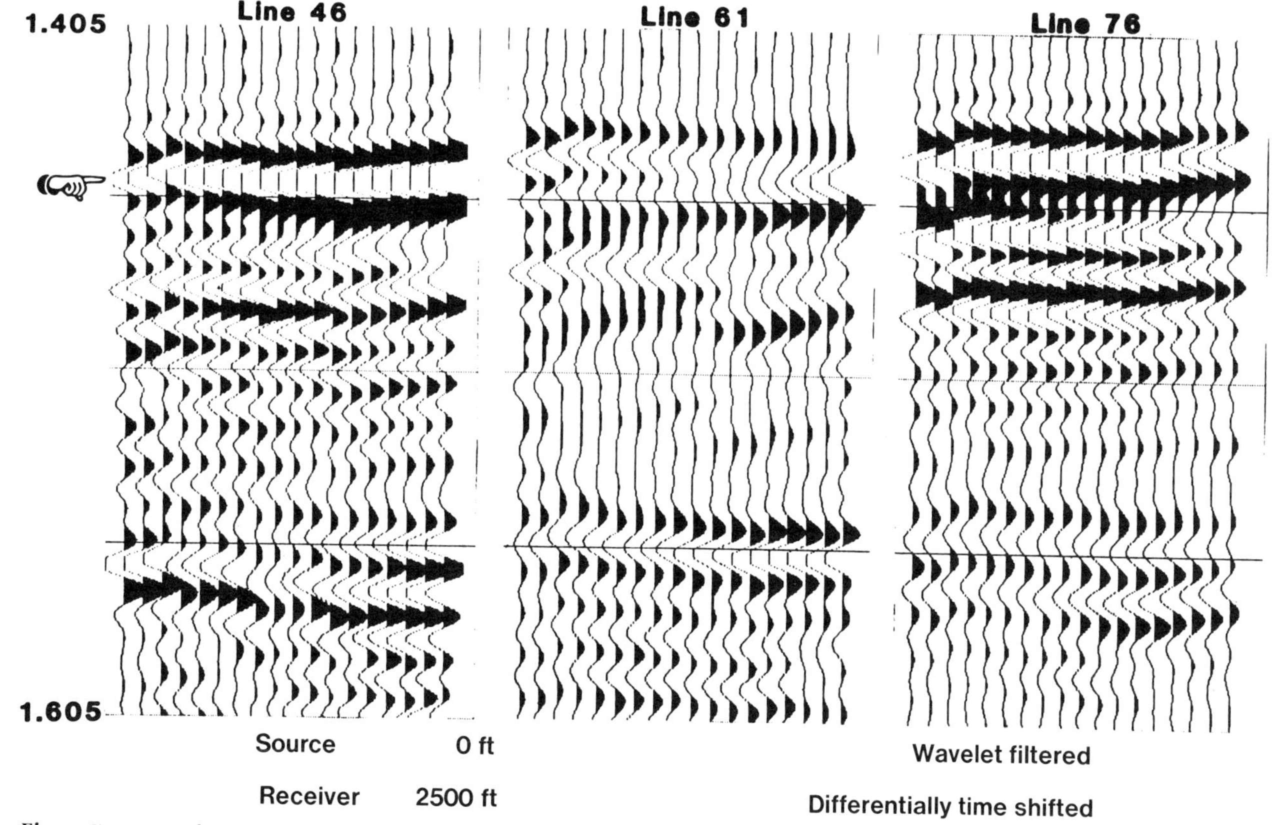

Figure 5.32. Areal VSP lines.

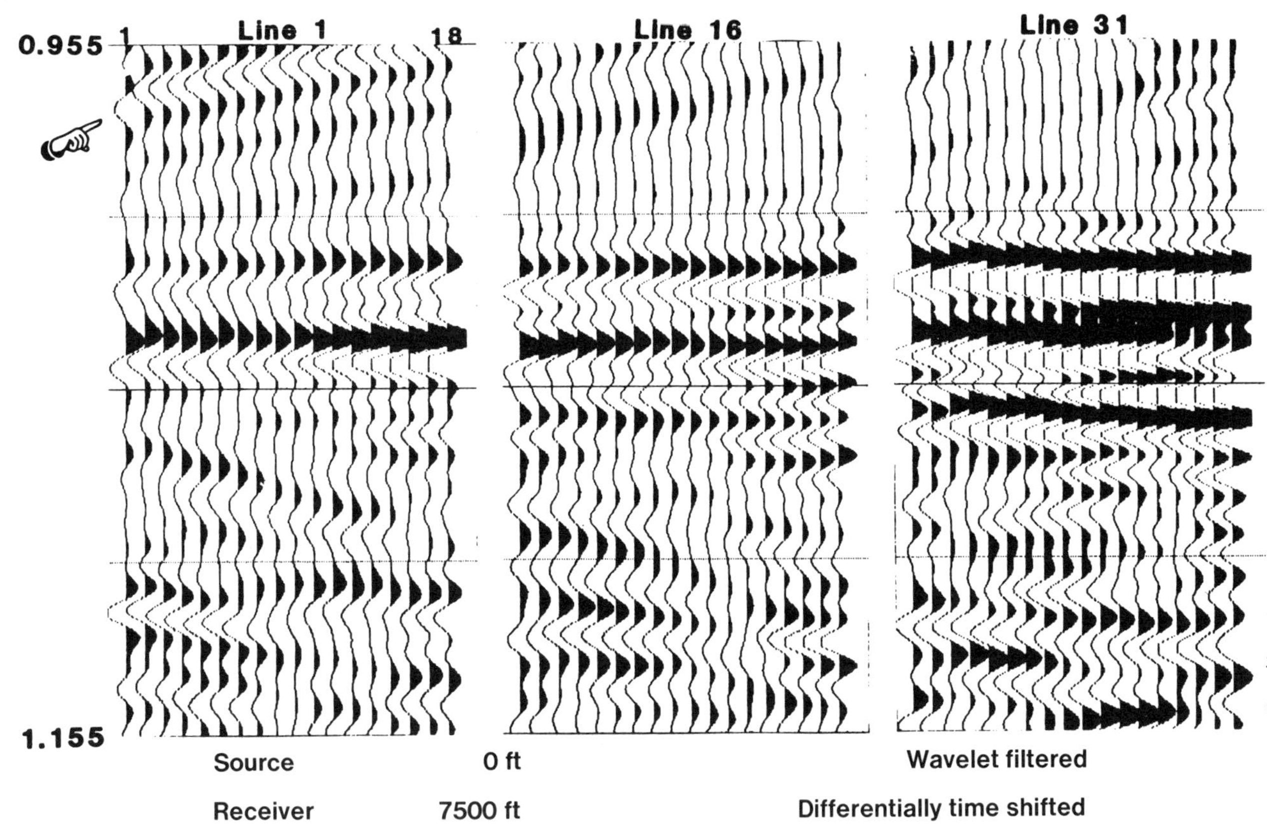

Figure 5.33. *Areal VSP lines.*

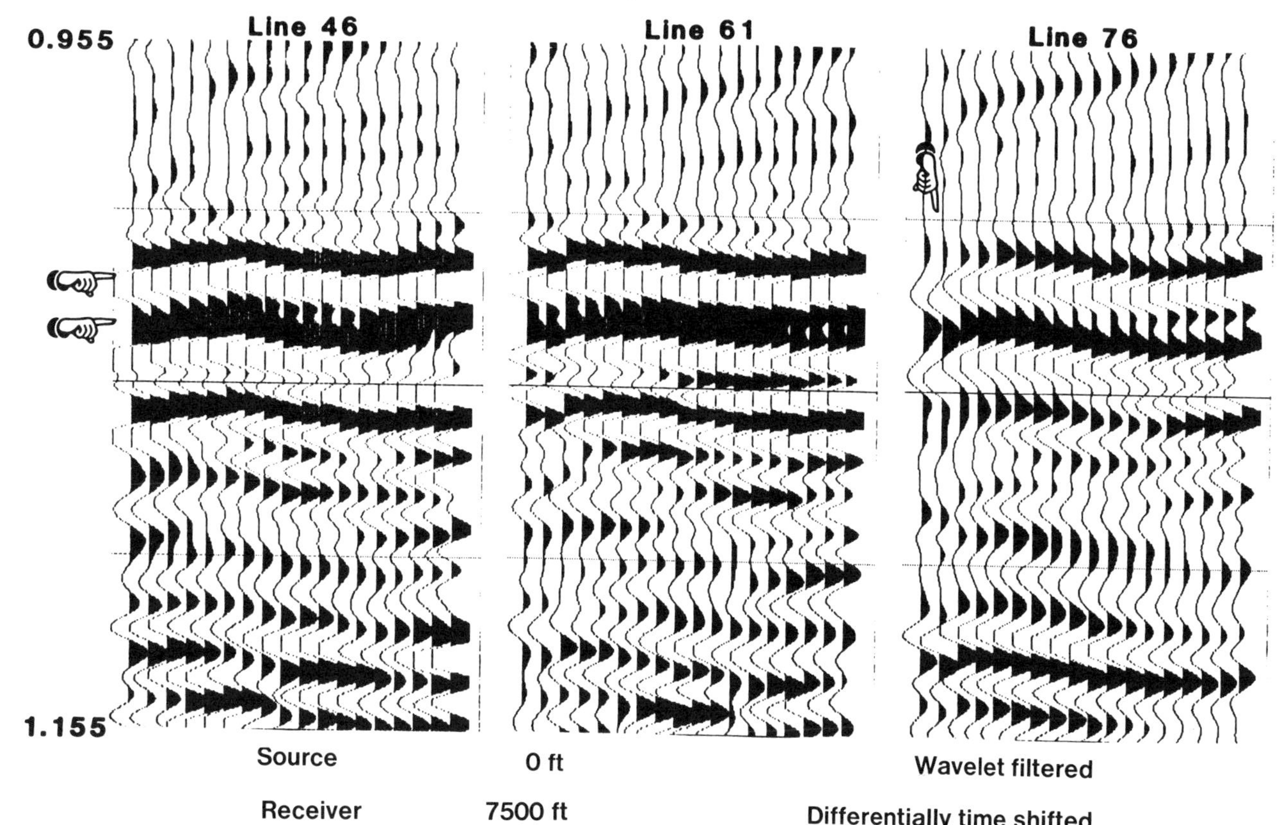

Figure 5.34. Areal VSP lines.

arrival. Lines 31 through 76 have specular reflections from the bodies. The arrivals from the top and bottom of the reservoir body may now be seen as distinct arrivals. The event that follows the top surface reflection by 0.060 s is the previously mentioned source artifact.

It is possible that the events on the two leftmost traces on line 76 are not specular reflections but head waves or channel waves. The waveform shapes are much different from the others, and they have lower frequencies, as head waves have. But, if the Plexiglas velocity is 21,600 ft/s, the critical angle is 33.5° and the critical ray would arrive at a 7500 ft deep reservoir with a horizontal offset of 6900 ft; the actual offset for the leftmost trace is only 6562 ft. It is therefore possible that the lower frequencies in the waves are due to dispersion; in this case, the events are probably channel waves.

CONCLUSIONS. This chapter has presented experimental data of the type that might be recorded over a field of unconnected limestone petroleum reservoirs encased in shale. The results presented are descriptive. Wavelet filtering was used to reshape the long basic waveform to a short, symmetric wavelet. Three-dimensional migration was used to help locate the edges of the reservoirs.

With the near-offset data, convex horizontal edges cause downturning diffractions. An edge with a small radius of curvature gives rise to diffractions that are weaker than those from a large-radius edge; concave edges focus diffractions.

Areal structures cause diffractions to be added to reflections from underlying layers. These structures also cause the reflections to have velocity-related push-down and pull-up, which, together with diffractions from the areal structure, give the appearance of faults. Migration simplifies the picture by collapsing diffractions. However, the push-down, pull-up effects in the time data are carried over into the migrated image. The amplitudes of migrated images of areal structures and of underlying layers are related to the areal distortion of the wavefronts. An areal structure advances or delays a portion of the wavefront, and the reflection amplitude is diminished. The migrated image is correspondingly weakened, and thin-bed effects also carry over. Both of these factors make the extraction of reflection coefficients from migrated data more difficult.

The principles by which diffraction can be understood may be applied to recordings where the sources and receivers are widely separated, as in VSP surveys, and diffractions from areal bodies may be observed.

REFERENCES

Berkhout, A. J., 1980, Seismic migration—Imaging of acoustic energy by wave field extrapolation: New York, American Elsevier Publishers, Inc.

Cerveny, V., and Ravindra, R., 1971, Theory of seismic head waves: Toronto, Univ. of Toronto Press.

Gardner, G. H. F., Gardner, L. W., and Gregory, A. R., 1974, Formation velocity and density—The diagnostic basis for stratigraphic traps: Geophysics, v. 39, p. 770–780.

Keller, J. B., 1962, Geometric theory of diffraction: J. Opt. Soc. Am., v. 52, p. 116–130.

APPENDIX 5A: RELATIONSHIP BETWEEN SOURCE AND REFLECTION POINT OFFSETS WHEN THE RECEIVER IS BURIED AND THE REFLECTOR PLANE IS DIPPING. Assume that a plane reflector with dip angle ζ penetrates a well at depth z (see Fig. 5A.1). Given a receiver A in a well at depth d and a source C on the surface offset a distance x_2 in the downdip direction, find the horizontal offset x of the reflection point B on the reflector.

The basic relationships are

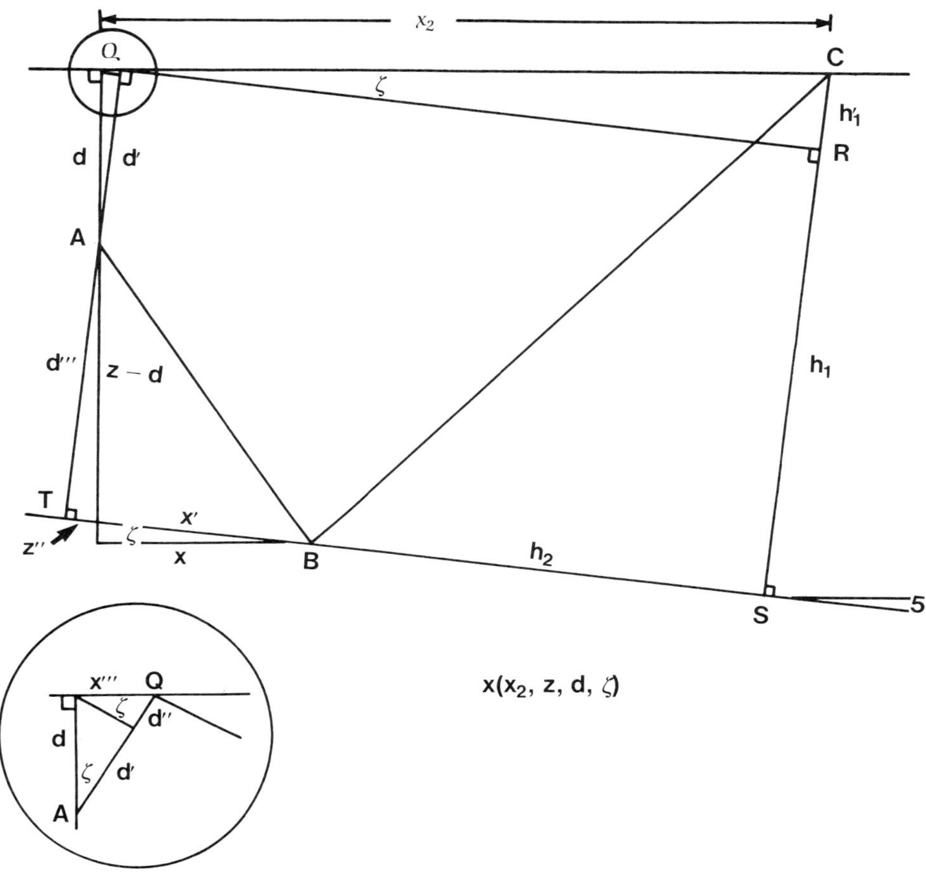

Figure 5A.1. *A reflector (TS) dipping at an angle ζ below a surface QC. A is the receiver for a source at C; CBA is the raypath.*

$$\cos \zeta = x/x' = d'/d = (x'' + x' + h_2)/(x_2 - x'''),$$
$$\sin \zeta = x''/(z - d) = d''/x''' = h'_1/(x_2 - x'''),$$
$$\tan \zeta = x''/d''' = x'''/d.$$

By similar triangles,

$$d'''/(x'' + x') = (h_1 + h'_1)/h_2. \tag{5A.1}$$

Now we have

$$d''' = x''/\tan \zeta = (z - d) \cos \zeta, \tag{5A.2a}$$
$$x'' + x' = (z - d) \sin \zeta + x/\cos \zeta, \tag{5A.2b}$$

$$
\begin{aligned}
h_1 + h_1' &= d' + d'' + d''' + h_1' \\
&= d \cos\zeta + x''' \sin\zeta + x''/\tan\zeta + (x_2 - x''') \sin\zeta \\
&= z \cos\zeta + x_2 \sin\zeta, \qquad (5A.2c) \\
h_2 &= (x_2 - x''') \cos\zeta - x' - x'' \\
&= (x_2 - z \tan\zeta) \cos\zeta - x/\cos\zeta. \qquad (5A.2d)
\end{aligned}
$$

Inserting equations (5A.2) into equation (5A.1) and rearranging, we obtain

$$x = \frac{(z - d)(x_2 \cos 2\zeta - 2z \sin\zeta \cos\zeta)}{2z - d + x_2 \tan\zeta} \qquad (5A.3)$$

Solving for x_2, we obtain

$$x_2 = \frac{xz + (z - d)(x + 2z \sin\zeta \cos\zeta)}{(z - d) \cos 2\zeta - x \tan\zeta} \qquad (5A.4)$$

APPENDIX 5B: VSP STATIC SHIFTS FOR A FLAT REFLECTOR. When the reflector is flat, the expression (5A.3) in Appendix 5A reduces to

$$x = \frac{z - d}{2z - d} x_2.$$

The distance from A to B is

$$\mu_r = \sqrt{x^2 + (z - d)^2}$$

and from B to C is

$$\mu_s = \sqrt{x^2 + z^2}.$$

If the velocity is c, the travel time from C to B to A is given as

$$t = (\mu_r + \mu_s)/c.$$

The normal incidence travel time is

$$t_0 = (2z - d)/c,$$

and the static shift is

$$\Delta t = t - t_0.$$

6. *FINITE ELEMENT AND KIRCHHOFF WAVE THEORY COMPARISONS TO PHYSICAL MODEL DATA*

Lynn L. Chou, Dan D. Kosloff, Edip Baysal, and Fred J. Hilterman

INTRODUCTION. This chapter examines a pinchout structure and compares time sections from physical models with synthetic sections obtained through numerical modeling. In addition to the inherent interest in this model, it has features that serve as severe tests of the numerical modeling algorithms.

The numerical calculations were performed in two ways: by the three-dimensional (3-D) Kirchhoff summation scheme of Hilterman (1981) and by the two-dimensional (2-D) forward modeling algorithm of Kosloff and Baysal (1981). This example tests the capability of the summation scheme in handling a multilayer structure using a vertical replacement velocity scheme. For the Fourier method, this problem shows the difficulties in approximating dipping lines on a numerical mesh and in handling thin layers with small velocity contrasts. We will show that both modeling methods handle these problems well.

COMPARISON BETWEEN ZERO OFFSET PHYSICAL MODEL DATA AND NUMERICAL CALCULATIONS. Figure 6.1 shows the configuration of the model. The layers were composed of different types of RTV silicone rubber, with scaled velocities of 7920 ft/s for L1, 7066 ft/s for L2, 8640 ft/s for L3, and 7476 ft/s for L4. The water velocity was 11800 ft/s.

A zero-offset profile was run along a dip line at a scaled distance of 3500 ft above the top of the model. The physical data time section is shown in Figure 6.2. The reflections from the material boundaries of this structure are predictable and easily distinguished.

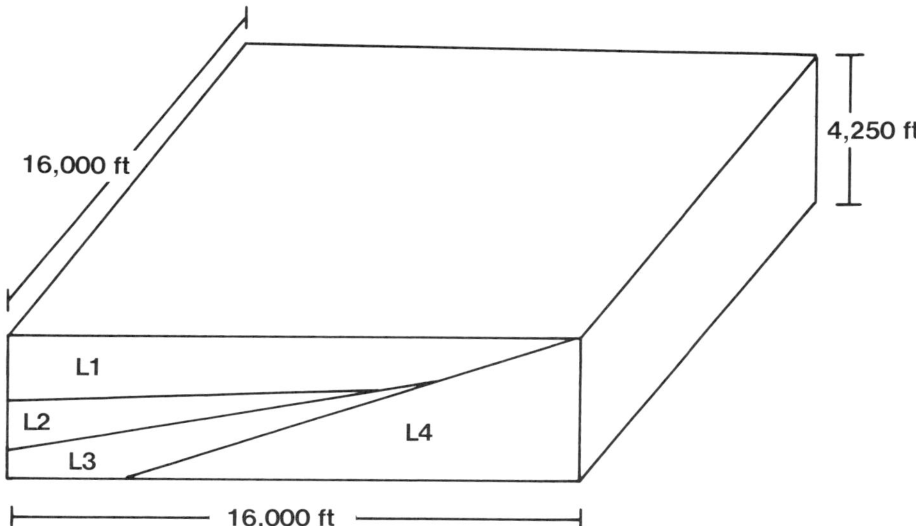

Figure 6.1. *Configuration of the physical model.*

Figure 6.2. *Physical data time section.*

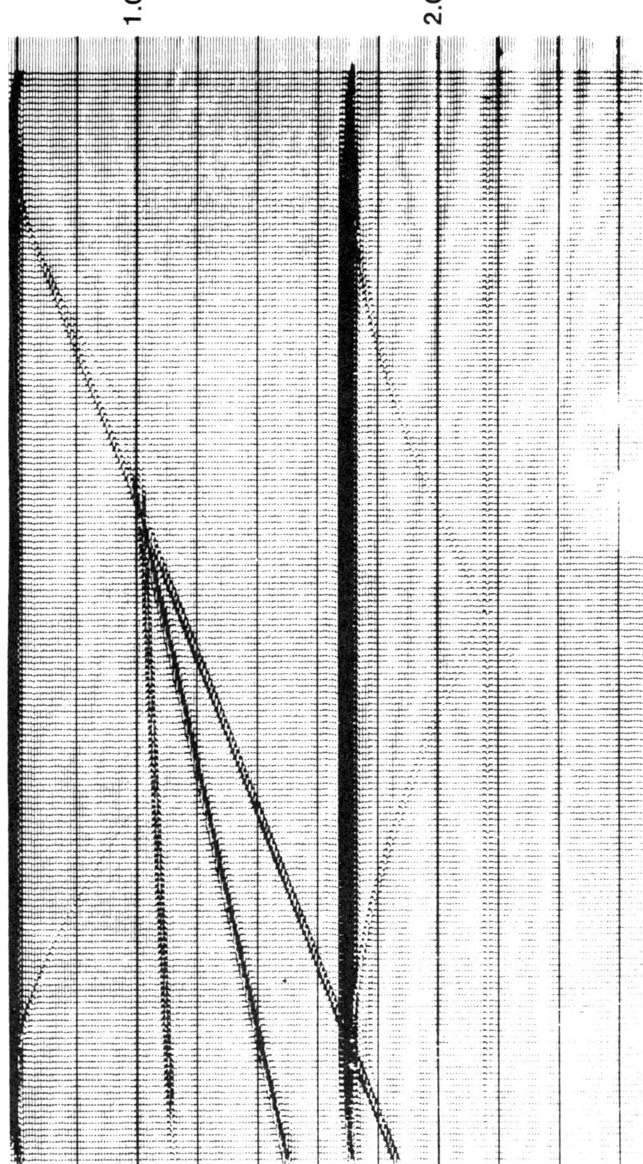

Figure 6.3. *Synthetic data time section.*

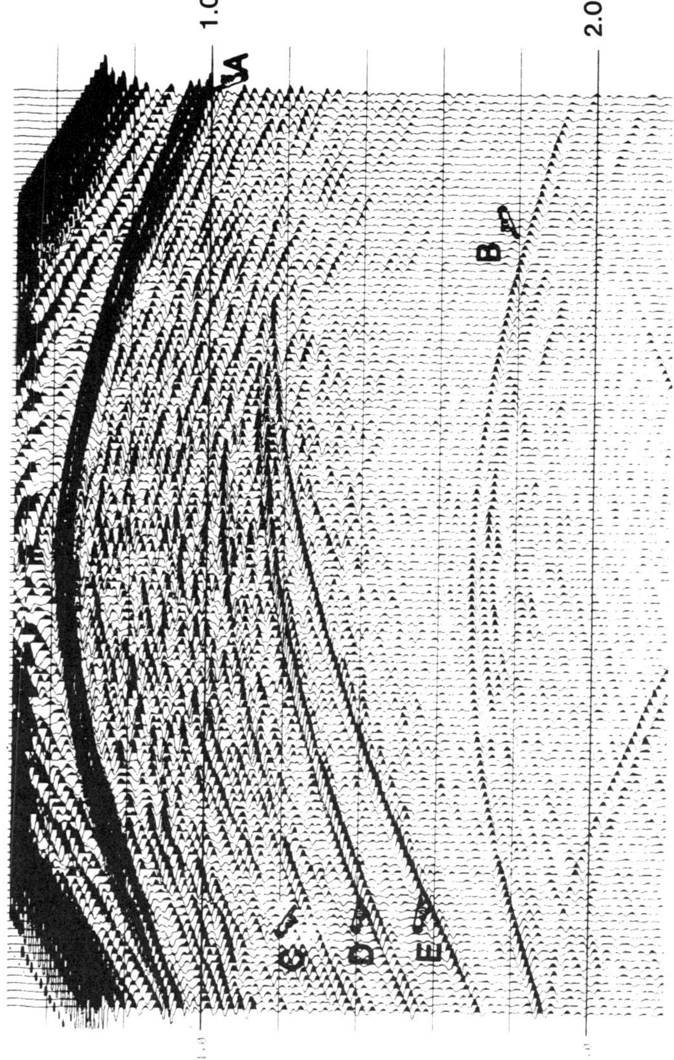

Figure 6.4. *Time section from common shot gather along a dip line.*

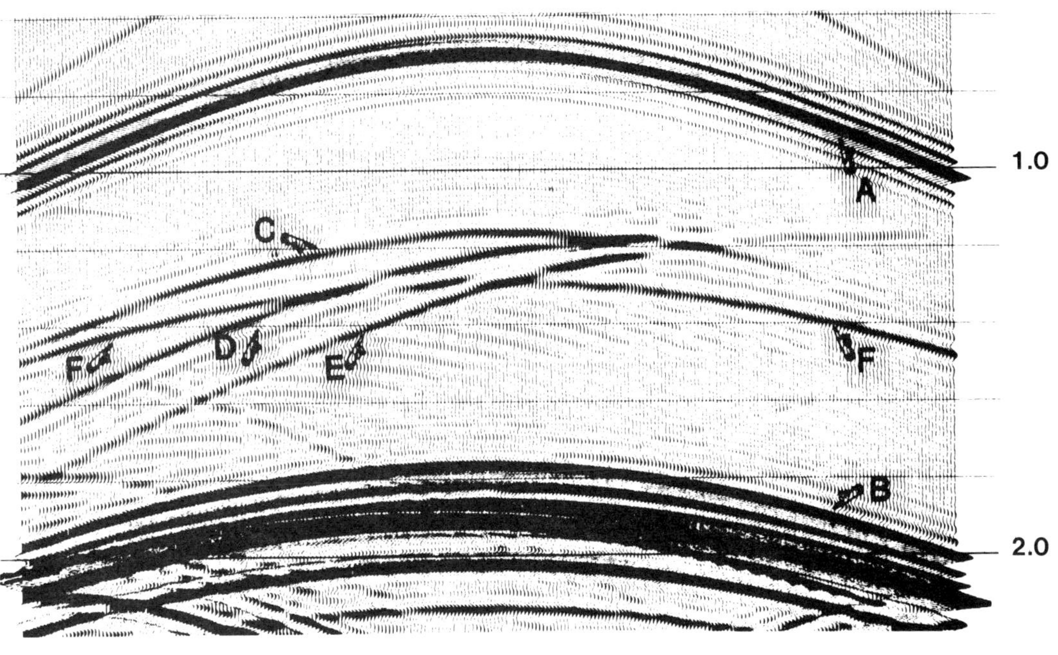

Figure 6.5. *Time section from forward modeling calculations, using the Fourier method algorithm.*

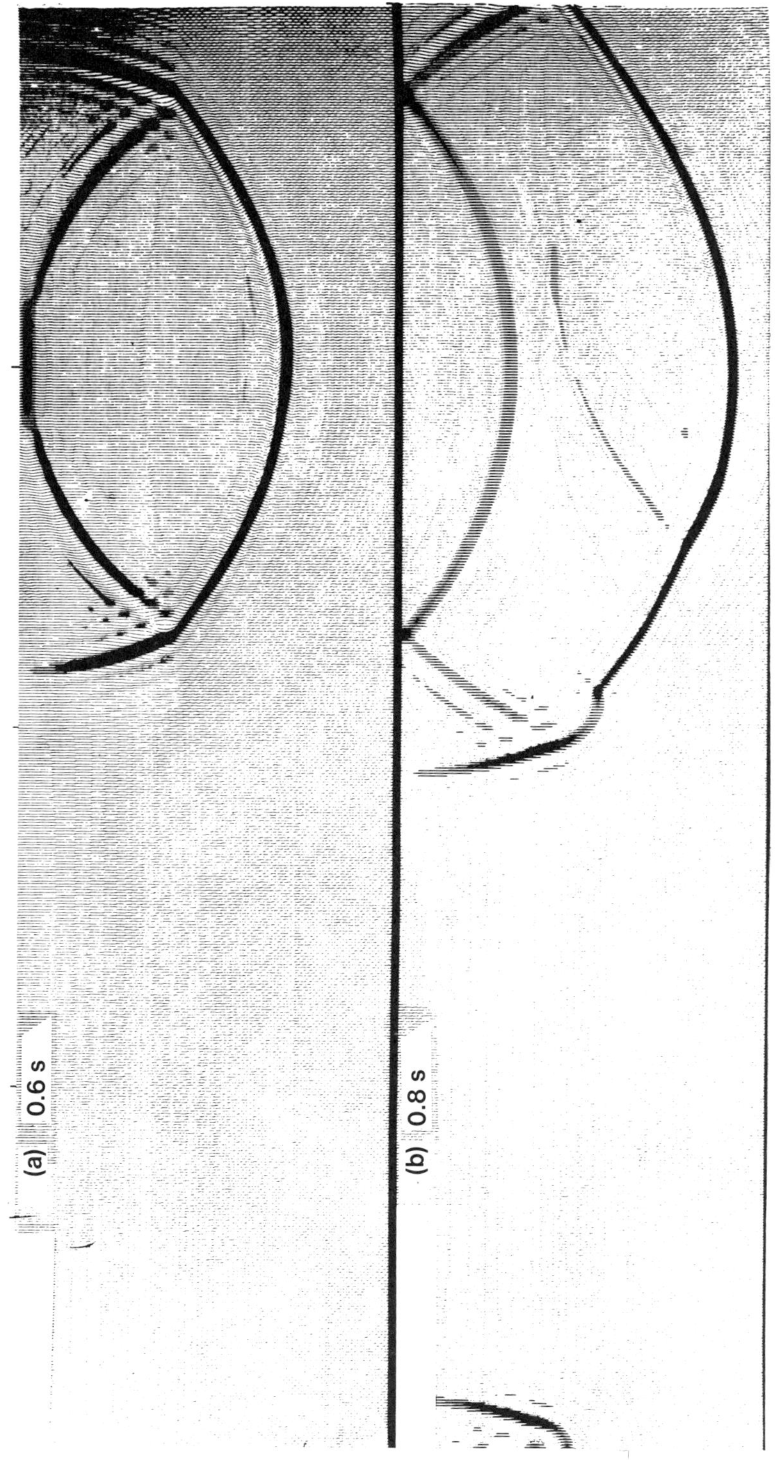
(a) 0.6 s
(b) 0.8 s

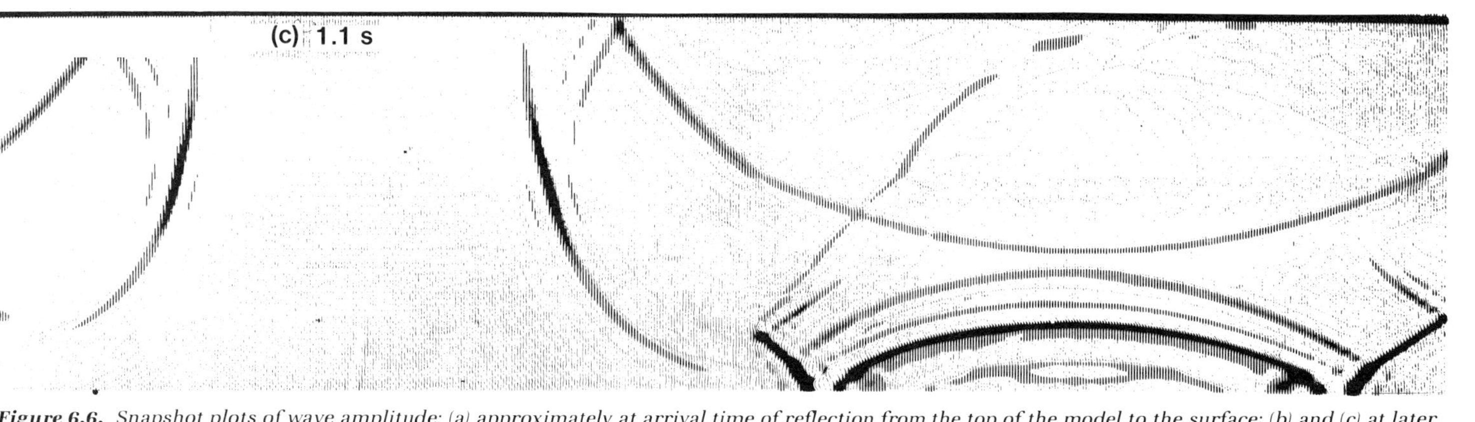

Figure 6.6. *Snapshot plots of wave amplitude: (a) approximately at arrival time of reflection from the top of the model to the surface; (b) and (c) at later times.*

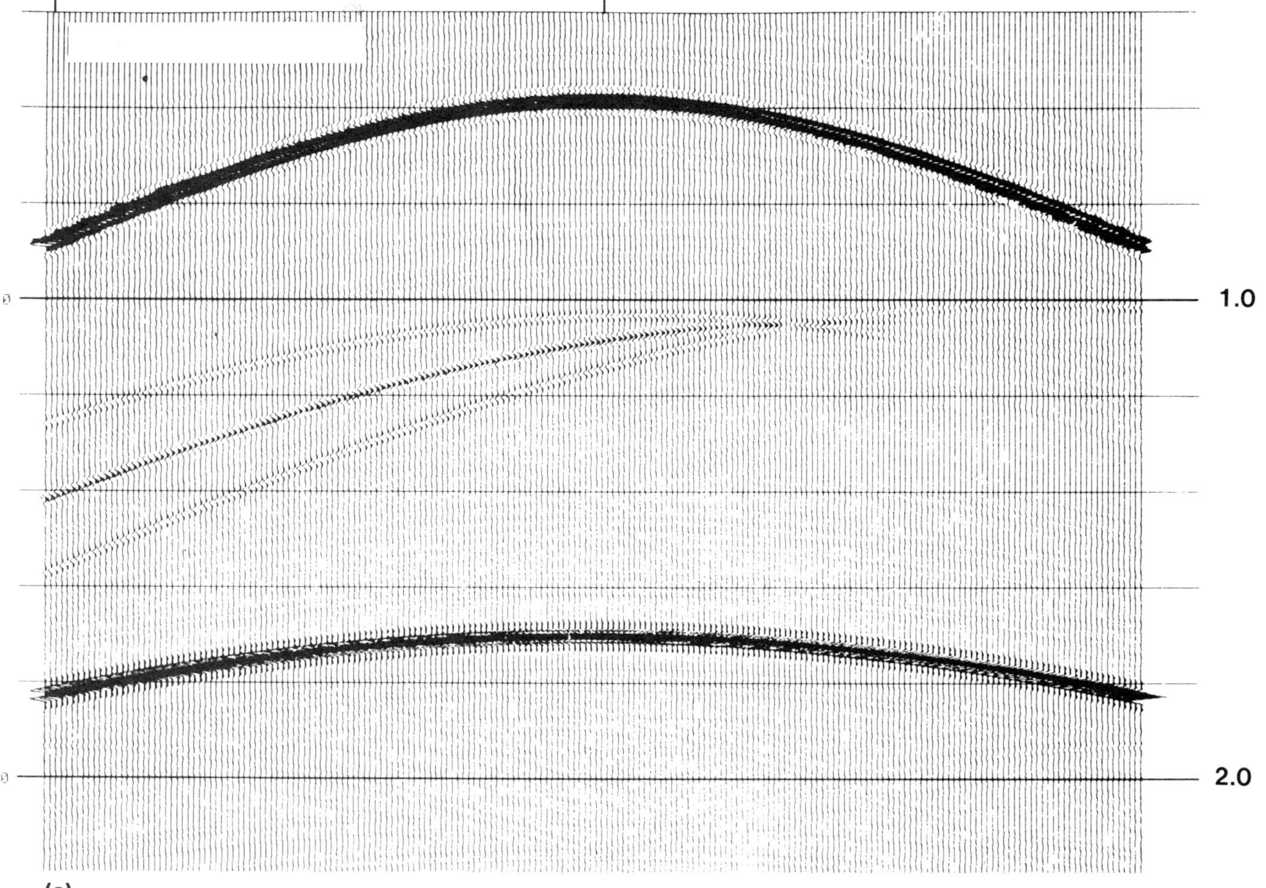

(a)

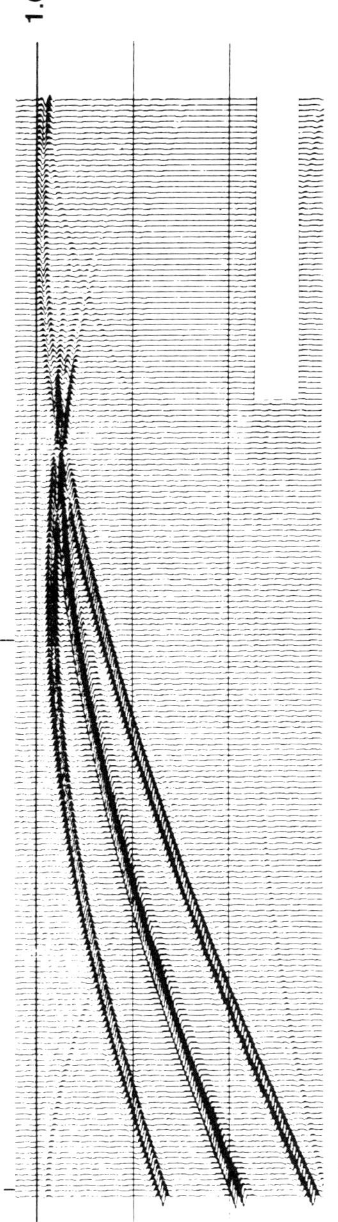

Figure 6.7. *Time section from the Kirchhoff algorithm.*

Figure 6.3 is a synthetic time section obtained from the 3-D Kirchhoff algorithm. The shot configuration was the same as that for the physical model, and a velocity replacement ratio of 12,000/7,600 was used in generating the input model. The main events in this figure agree well with those in Figure 6.2 from the physical model.

COMMON SHOT GATHER RESULTS. This case considers a common shot gather along a dip line, with a shot location above the center of the model at the same height above the model as in the previous example. The resulting time section is shown in Figure 6.4. The reflections from the top and bottom of the model (A and B) are very prominent in this section. The reflections from the interval boundaries (C, D, and E) are also present but are weaker.

Figure 6.5 is the time section from forward modeling calculations using the Fourier method algorithm. Events A through E correspond very closely to the events on the physical model time section (Figure 6.4). However, there is an additional event, F, which did not appear in Figure 6.4. We interpret this event as a surface multiple that cannot appear in the physical model because of the absence of a surface.

The snapshot plots enabled us to verify the interpretation. Figure 6.6a shows a wave approximately at the time of arrival of the reflection from the top of the model to the surface. Figures 6.6b and 6.6c present amplitudes at later times. The downward propagation and subsequent reflection and arrival of the surface multiple can be seen. Reflections from the bottom of the model are also present.

Finally, Figure 6.7a and 6.7b depict a time section from the Kirchhoff algorithm. Once again, the events closely correspond to the events on the physical model section in Figure 6.4.

CONCLUSIONS. Comparisons of Kirchhoff and finite element time sections with physical model time sections are good. Each theoretical algorithm offers some advantages over the other, but, in general, either algorithm will definitely benefit the interpreters.

REFERENCES

Hilterman, F. J., 1981, Forward 3D modeling programs: SAL Progress Review, v. 7.

Kosloff, D. D., and Baysal, E., 1981, Forward modeling by the Fourier method: SAL Progress Review, v. 7.

7. VELOCITY ESTIMATION FROM THREE-DIMENSIONAL DATA

J. K. Owusu and G. H. F. Gardner

INTRODUCTION. The effectiveness of the two-step approximation to three-dimensional migration and its use in the determination of velocity estimates from areal seismic data has been demonstrated (Owusu and Gardner, 1980). The velocity estimates obtained were found to be less dependent on the reflector geometry than those derived from conventional common-depth-point (CDP) data. The contaminating cross-dip and cross-curvature effects were removed, but the in-line dip and curvature effects still remained. With complete three-dimensional (3-D) seismic data coverage, and using a complete 3-D migration technique without any approximations, much better velocity information might be obtained.

Three-dimensional, multifold seismic surveys generate notably large amounts of data for the analysis of subsurface problems. A typical marine survey, for example, may have source points 100 ft apart along parallel lines that are 200 ft apart. Thus, recording with a 48-channel streamer, each square mile surveyed will contain more than 65,000 traces. A typical 3-D velocity analysis will use at least 10,000 traces. The large number of traces greatly improves the signal-to-noise ratio and, because of the use of migration techniques, correspondingly more meaningful velocity estimates can be derived.

Notwithstanding the improved results obtained, these methods have remained unattractive because they require a great deal of computer time and storage capacity. To produce a velocity analysis display using conventional philosophy, as discussed by Sattlegger (1975, 1980), would require that each trace be migrated separately for each velocity in a suite of constant velocities. Each trace of N time samples migrated for M different velocities would thus require about $N \times M$ normal moveout calculations. This is a time-consuming process and, for the large quantities of data from a 3-D seismic survey, would require many computer hours, if not days.

This chapter explores a new velocity analysis computer algorithm, based on a complete 3-D Kirchhoff integral migration before stack. The algorithm reduces the number of computations from about $N \times M$, as required before, to about N. The output can be regarded as migration of the data for all possible velocities. Rather than the usual plot of semblance as a function of velocity V and time T, the display is a plot of instantaneous power as a function of depth Z and time T. The maxima in this plot are picked and the velocity obtained from the relation $V = 2Z/T$.

The reduction from $N \times M$ computations to N constitutes a great saving in the computer time. However, further time savings in data reading and writing overhead and in data storage can be achieved by data compression. The amplitude recovered by correlating a Vibroseis sign bit recording with a sign bit representation of the sweep signal is remarkably similar to that obtained using a 16- or 32-bit digitization for both the recording and the sweep signal. This property suggests that, for some data-processing routines, there will be little loss in accuracy using a sign bit digitization, but there will be a great gain in data compression. Velocity analysis is one such routine. Velocity determination primarily depends on arrival times and, hence, the phase of events rather than their amplitudes. Cochran (1973) demonstrated that comparably good velocity information can be obtained from CDP data by semblance computation using sign bit data. The sign bit data will provide several advantages. First, data storage

This chapter was also presented as a paper at the SEG convention in Houston, November 1980.

requirements will be reduced 16 times compared to integer data storage. Second, because of the data compression, large blocks of data can be read and written at one time. Third, reducing the data into sign bits has the added advantage of equalizing the variations in the power between traces. This should prove very valuable, especially for land data, which typically is very noisy.

This velocity analysis is intended primarily to help in the detection of overpressured subsurface zones in sand-shale sequences before a drilling rig is moved to a well site. Conventional drillability profiles are based on velocity measurements, amplitudes, and other attributes obtained from CDP gathers at depth points that sometimes are not related to the points along the axis of the proposed well. The algorithm discussion here provides drillability profiles that relate to the axis of the proposed well.

The algorithm has been implemented and tested using a number of synthetically generated model data as well as data acquired over a physical model in a tank.

THEORY AND DEFINITIONS. The implementation of 3-D migration can be divided into a series of operations. First, each input trace is differentiated with respect to time. The purpose of this step is to retain the shape of the wavelet through subsequent operations. This can be checked by migrating data generated by a single uniform horizontal interface plane in the subsurface. The pulse resulting from the migration should be similar to the pulse in each input trace. If the differentiation step is omitted, the migrated pulse will be the integration of the input pulse. Next, each sample of the input data is time-shifted, weighted, and added to all the output traces.

Any time T_0 on the output trace is associated with a time T on the input trace by the traveltime formula

$$T = [(T_0/2)^2 + SP^2/V^2]^{1/2} + [(T_0/2)^2 + RP^2/V^2]^{1/2}, \tag{7.1}$$

where RP = distance from the receiver station R to the output location P,

SP = distance from the source station S to the output location P,

V = the root-mean-square (rms) velocity corresponding to time T_0 on the output trace at P.

The time shift ΔT for the sample at time T is given by

$$\Delta T = T - T_0. \tag{7.2}$$

This requires that a new shift ΔT be calculated for each sample of the input data trace. For velocity analysis, which requires that the migration be done with many different velocities, this process would be extremely time-consuming. The computations can be arranged so that each data trace is added to each output trace with the same shift for all samples. To do this, we replace V by

$$V = 2Z/\Delta T \tag{7.3}$$

in equation (7.1) and transform the data into a logarithmic time scale. This is accomplished by rearranging equation (7.1) and taking the natural logarithm to obtain

$$\log T = \log T_0 + \log \left[\frac{(1 + SP^2/Z^2)^{1/2}}{2} + \frac{(1 + RP^2/Z^2)^{1/2}}{2} \right]. \tag{7.4}$$

Defining the logarithmic time scale by

$$U = CK * \log T - CL, \tag{7.5}$$

where CK and CL are suitable scale factors, we obtain

$$U = U_0 + \Delta U, \tag{7.6}$$

where the shift in U is given by

$$\Delta U = CK * \log \left[\frac{(1 + RP^2/Z^2)^{1/2}}{2} + \frac{(1 + SP^2/Z^2)^{1/2}}{2} \right]. \tag{7.7}$$

Thus, in the transform space, the shift ΔU for any fixed value of Z is the same for all samples of the trace. This is particularly advantageous and is the key to the time-economic viability of the algorithm.

To construct the velocity analysis display, a range of Z values is selected. For convenience, it is selected on a logarithmic scale. A depth sample S is defined by

$$S = DK * \log Z - DL, \tag{7.8}$$

where the constants DK and DL are chosen such that the depth samples $S = 1, NS$ span the appropriate range of Z values.

LOGARITHMIC SAMPLING OF DATA. Equal sampling intervals in U and S imply unequal sampling intervals in T and Z, respectively. To illustrate the logarithmic transformation, consider the following case:

$$U = 500 * \log T - 3106, \tag{7.9}$$

where constants CK and CL have been chosen to be 500 and 3106, respectively, so that T ranges from 0.5 to 2.0 s when U ranges from 1 to 655.

Table 7.1 shows the relationship between the index U, the time T in milliseconds on the trace, and the nearest time sample number N, at 4 ms sampling rate. The following points may be noted:

(1) Between U samples 1 and 2, the time interval is 1 ms.
(2) Between U samples 347 and 348, the time interval is 2 ms.
(3) Between U samples 550 and 551, the time interval is 3 ms.
(4) Between U samples 654 and 655, the time interval is 4 ms.

Thus, equal sampling of the U variable corresponds to sampling in time with an interval that is proportional to time. The number of time samples for the range of 0.5 to 2.0 s is 377, where the corresponding number of U samples is 655.

An uneven sampling of the T scale is thus implicit in the transformation from T to U. If 4 ms is the largest time interval desired on the T scale, then choosing the constant CK so as to obtain 4 ms sampling at 2 s will require oversampling at earlier times. From a practical standpoint, this might be a useful way to sample seismic data that contain high frequencies at earlier times and therefore need more frequent sampling then than at later times, when the high frequencies have been attenuated.

Figure 7.1 shows a 10 Hz input sine wave generated at a 4 ms sampling rate; Figure 7.2 shows the differentiated trace transformed to the U domain, using equation (7.9); and Figure 7.3 shows the sine wave resampled on a logarithmic scale.

Table 7.1. *Relationships between the index U, the time T, and the nearest time sample N*

U	T (ms)	N
1	499.7	125
2	500.7	125
.	.	.
.	.	.
.	.	.
347	998.25	250
348	1000.25	250
.	.	.
.	.	.
.	.	.
550	1498.17	375
551	1501.17	375
.	.	.
.	.	.
.	.	.
654	1998.20	500
655	2002.20	501

APERTURE SIZE. Summing a large volume of data (as from 3-D surveys) in the migration procedure has several benefits, which have been discussed earlier. These benefits are offset, however, by the increased addition of random noise. For this reason, it is desirable to use only those data that fall within the portion of the migration surface that contains the main contribution to the image. To find that portion of the migration surface requires a coherence criterion.

Gardner, French, and Matzuk (1974) established such a criterion for a horizontal reflector. The equation for the migration surface in a medium of velocity V can be approximated for a midpoint spacing r, which is small compared to depth Z, by

$$t = \frac{2Z}{V} + r^2/ZV. \tag{7.10}$$

Coherence occurs if the increase in t does not exceed half a period for the pulse. Hence, the distance r for which coherence can occur is given by

$$\frac{r^2}{ZV} = \frac{T}{2}, \quad r = \left(\frac{ZVT}{2}\right)^{1/2}, \tag{7.11}$$

where T is the period of the pulse. This criterion is used in the selection of the aperture size. As Z increases, the radius of the aperture is increased.

APERTURE WEIGHTING. Kuhn and Alhilahi (1977) pointed out the importance of a directivity function as a component of the weighting factors for migration. This directivity function simply beam-steers the energy toward its origin. Thus, in order to image a point P on a dipping reflector, the data aperture must include the normal incidence point.

In constructing drillability profiles, the data used should include the normal incidence point. The point is determined by starting at P in the direction of the normal to the reflector and tracing a ray to the surface. With zero-offset data, this

Figure 7.1. *Example of a sine wave.*

requires that the data used must have source S and receiver R close to the normal incidence point. The location of the source and receiver relative to the center of the aperture determines the weight applied. For nonzero-offset data, a modified procedure can be followed. Snell's law requires that the normal incidence ray bisect the angle between the rays to S and R. Hence, given S and R, a ray to the surface can be traced that initially bisects the S and R rays. The surface location is known as the bisector point. Data used must be those for which the bisector point is close to the normal incidence point.

A simple formula can be obtained that locates the bisector point with acceptable accuracy. Define a point B on SR such that B divides SR in the ratio of the traveltimes for S and R (just as the bisector of the vertical angle of a triangle divides the base in the ratio of the sides). Hence, defining $R = (1 + RP^2/Z^2)^{1/2}$ and $S = (1 + SP^2/Z^2)^{1/2}$, the coordinates (XB, YB) of the bisector point are given by

$$XB = (XS \times R + XR \times S)/(S + R),$$

$$YB = (YS \times R + YR \times S)/(S + R), \tag{7.12}$$

Figure 7.2. *Differentiated version of a sine wave.*

where (XS, YS) and (XR, YR) are the source and receiver coordinates, respectively.

The location of the bisector point relative to the center of the aperture is used to determine the weight W for a given trace, given the Z-level into which it is to be added. In this algorithm, the aperture is defined at a given Z-level as an ellipse with a semimajor axis A, a semiminor axis B, and the angle θ between A and the x-axis. The center of the aperture is given as (XC, YC). The equation for the ellipse is given by

$$CXX \times X^2 + CXY \times x \times y + CYY \times y^2 = 1, \tag{7.13}$$

where $CXX = (\cos\theta/A)^2 + (\sin\theta/B)^2$,

$CXY = 2\cos\theta\sin\theta\,(1/A^2 - 1/B^2)$,

$CYY = (\sin\theta/A)^2 + (\cos\theta/B)^2$,

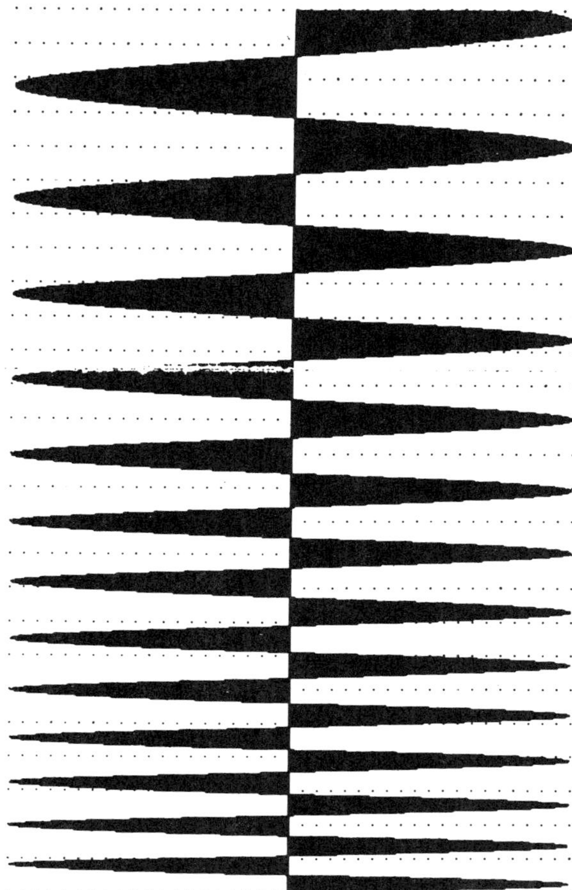

Figure 7.3. *Sine wave resampled on the logarithmic scale.*

$$x = XC - XB,$$

$$y = YC - YB.$$

The weight is then defined as

$$W = 1 - CXX \times x^2 - CXY \times x \times y - CYY \times y^2. \tag{7.14}$$

From equation (7.14), the weight W is 1 when the bisector point is at the center of the aperture, zero when it is on the boundary, and negative when it is outside the aperture. When W is negative, the trace is not added to this Z-level.

SUMMARY OF THE ALGORITHM. The algorithm is executed in two phases: (1) the data-preparation phase and (2) construction of velocity analysis or drillability profile display.

The operations in the data preparation phase can be summarized as follows:

(1) The input traces are sorted, using specified data windows to assign data to corresponding disk files.

(2) The input traces are differentiated.
(3) The differentiated trace is then Hilbert-transformed to form an analytical signal.
(4) The analytical signal is then resampled with a logarithmic scale.
(5) The resampled analytical signal can optionally be converted into sign bit.
(6) The disk files of analytical signals are optionally saved on tape for later use, or the data are immediated used in the drillability profile display phase.

The drillability profile display phase can be summarized as follows:

(1) The input parameters required are:
 (a) A description of the axis of the well, as specified by the coordinates at the surface and at a number of breakpoints along the well axis;
 (b) At each breakpoint, a description of the aperture, as specified by the coordinates of its center, its major and minor axes, and its orientation relative to the x-axis;
 (c) The logarithmic scale conversion constants *DK* and *DL*, which are also the number of *S* samples necessary to select the *Z*-levels for the output.
(2) By linear interpolation, the coefficients *CXX*, *CXY*, and *CYY* in equation (7.14) are calculated for each selected *Z*-level along the well axis. Also, the coordinates of the aperture center are calculated for each *Z*-level.
(3) The resampled analytical signals prepared in the data preparation phase are ready for further processing. For each selected signal, the only information needed from its header are the coordinates of the shot and receiver locations.
(4) For a given analytical signal, the coordinates of the bisector point are computed, using equations (7.12), for each selected *Z*-level. Combining these coordinates with the coordinates of the center of the aperture, the weights to apply to the entire analytical signal for all the selected *Z*-levels are calculated using equation (7.14).
(5) The shift for each *Z*-level is calculated, using equation (7.7), for the given analytical signal.
(6) Using the appropriate shift and weight values, the signal is weighted and shifted for each *Z*-level and added into the buffer for the level.
(7) The envelopes of the accumulated analytical signals are then formed, having cycled through all the data and all the depths.

APPLICATION OF THE ALGORITHM TO MODEL DATA. The algorithm was tested using data from several computer-simulated models and physical model data collected in the tank.

Horizontal Layer Model. The model consisted of a horizontal layer of 5000 ft thickness and a medium velocity of 12,000 ft/s.

A computer-simulated sixfold CDP areal survey was generated along parallel straight lines, with 100-ft spacing between lines and 100-ft spacing between CDP points. A square area of data, 2000 ft by 2000 ft, was used to construct the velocity diagram for an output location at the center of the square; thus, 2646 traces formed the data base. A CDP gather was generated with the near source-receiver offset distance chosen as 1500 ft and the sources and receivers incremented by 300 ft. The data were then input to the data preparation phase and resampled, using the constants $CK = 500$ and $CL = 3106$ in equation (7.5). The index U ranges from 1 to 655, so that T ranges from 0.5 to 1.848 s. Twenty depth levels were chosen—that is, the index S ranges from 1 to 20—using the constants $DK = 10.50$ and $DL = 77.00$ in equation (7.8). Thus, the depths selected ranged in value from 1683 to 10,281 ft.

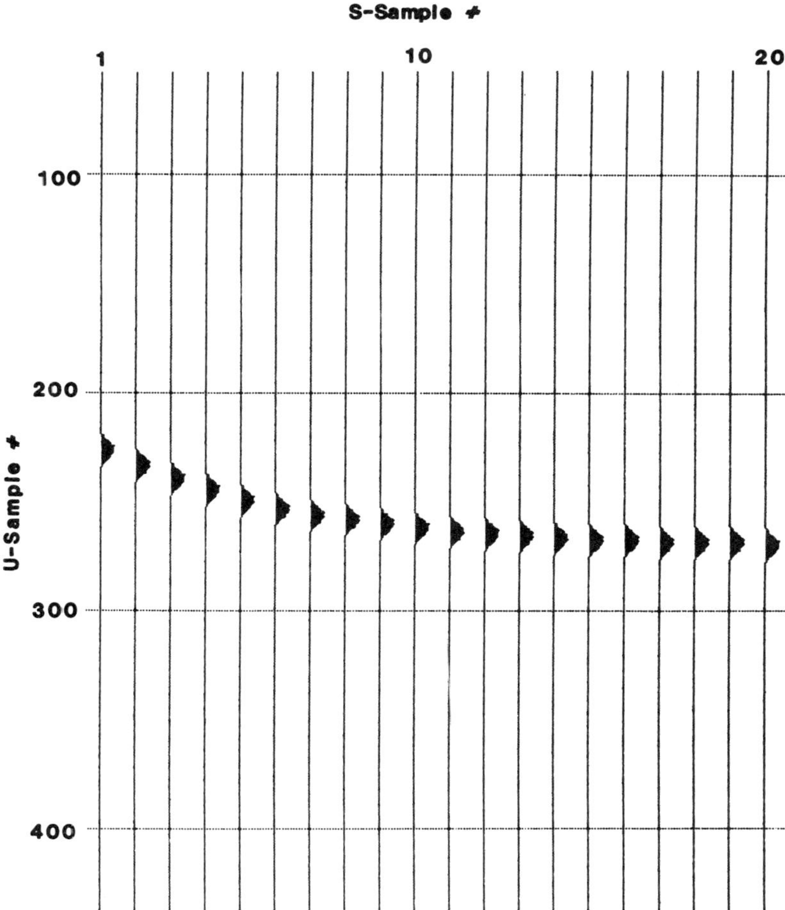

Figure 7.4. *Velocity diagram when only the near trace of the sixfold gather over a 12,000 ft/s horizontal layer has been migrated and added. The envelope of the analytical signal was plotted.*

Figure 7.4 shows the velocity diagram when only the near trace, whose source-receiver midpoint occurs at the output location, was migrated and added to it. The envelope of the analytical signal was plotted. The figure shows a typical example of how the shift applied to the trace varied with depth. The shift values applied to the trace make the wavelet follow a curve whose curvature varies with the source and receiver separation. Figure 7.5 shows the case in which only the far offset trace, from the chosen CDP gather, was migrated and added to the velocity diagram. The intermediate source-receiver offset traces displayed shift curves whose curvatures ranged between those displayed by the two extreme offset traces.

Figure 7.6 shows the accumulation of all six migrated analytical signals. The envelopes of the accumulated signals were plotted. By adding the traces, they

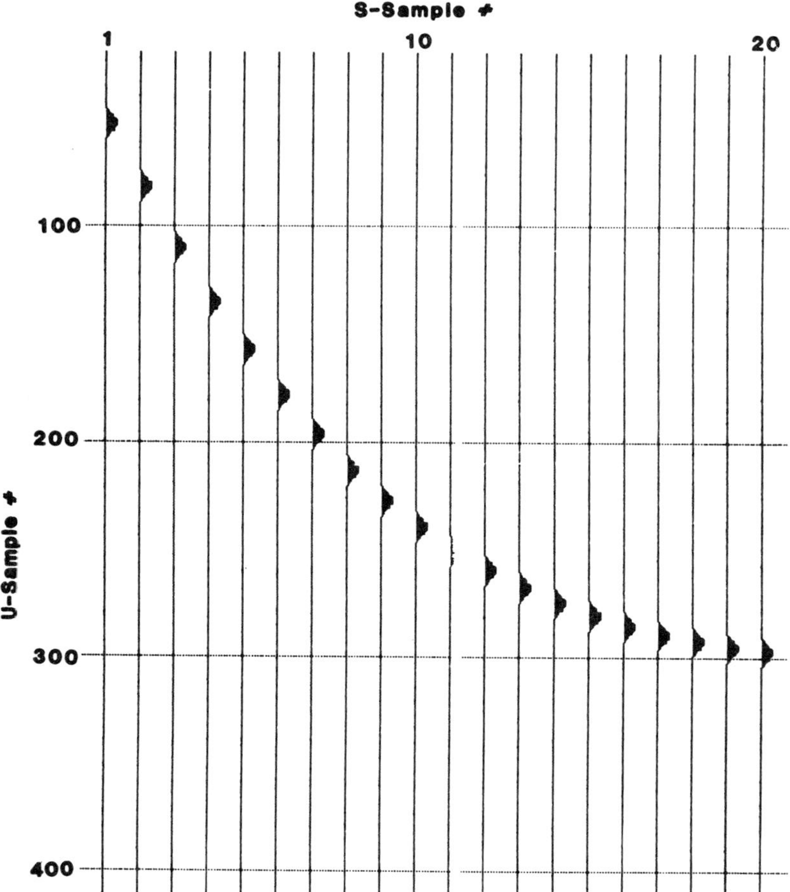

Figure 7.5. Velocity diagram when only one far trace has been migrated and added.

interfere constructively where there was coherence in the input data and destructively where there was no coherence.

To interpret the diagram, the largest peaked event, a result of constructive interference, is generally picked to give the apparent depth of the subsurface point scattering the energy and also the corresponding time. The velocity is related to the depth Z and the traveltime T by the equation $V = 2Z/T$. On the *U-S* diagram, a constant velocity function is a slanted line. Lines corresponding to constant velocity functions 11,000, 12,000, and 13,000 ft/s have been superimposed on the diagram. From the diagram, two peaks occur on *S* samples 12 and 13 that are larger than all others. The most constructive interference can be perceived to occur between the two largest peaks. The event was therefore picked between *S* samples 12 and 13 at a *U* sample of about 260. This gives a velocity of 12,000 ft/s at a time of about 0.838 s and an apparent depth of 5,000 ft.

Figure 7.7 shows the real part of the accumulated analytical signals whose envelope was shown in Figure 7.6. The isolated events of lower amplitude may be removed by using more data to cause destructive interference.

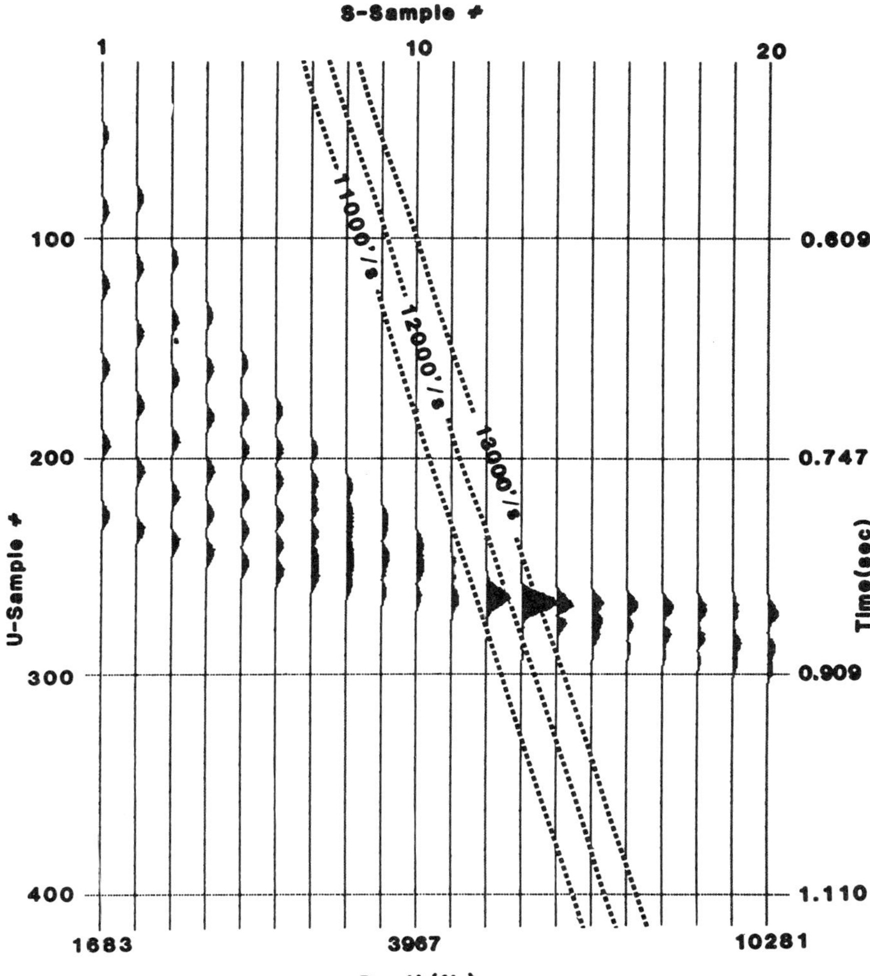

Figure 7.6. *Velocity diagram when all six traces have been migrated and accumulated.*

The aperture for the migration was then chosen to include all 2646 traces. Figure 7.8 shows the velocity diagram obtained. The largest peaks occur on *S* samples 12 and 13. For the same reason as explained earlier, the event is picked between the two peaks. The pick occurs at a velocity of about 12,000 ft/s. Notice how the isolated, uninterfered peaks occurring at shallower depths indicated by *S* samples 1 through 10 on Figure 7.7 have become very small. For the deeper depths, indicated by *S* samples 14 through 20, there are still some partly interfered peaks, which tend to become more distinct and separated. These peaks, considered to be noise with respect to the image, can be reduced in size by using a larger range of offsets and a smaller increment between offsets. Also, at greater depths, a larger aperture for the migration would be required. If necessary, the area over which the data were collected should be increased.

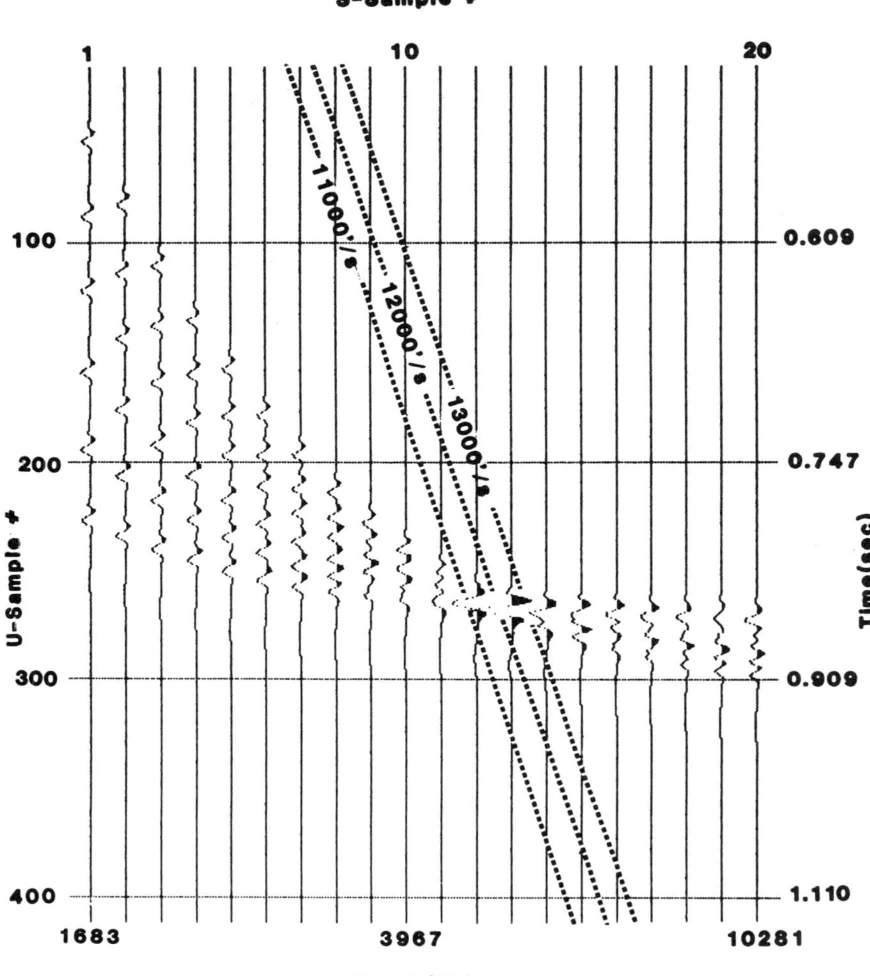

Figure 7.7. *The real part of the analytical signal.*

Dipping Reflector Model. The model used consisted of a single dipping reflector of 30° dip. A desired vertical well was located such that it intersected the reflector at a depth of 5000 ft. The velocity of the medium above the reflector was 12,000 ft/s.

A computer-simulated sixfold CDP areal survey was generated over the model along parallel straight lines, with the sources and receivers oriented along the lines in the dip direction of the reflector. The line spacing was 100 ft and the CDP spacing was 100 ft. A square area of data, 2000 ft by 2000 ft, was used to construct the velocity diagram at the well location, shown in Figure 7.9. The center of the data area was chosen to be at a distance of 5000 ft (tan 30°) away from the well location. This is the point at which a normal to the reflector, at the intersection of the well and the reflector, intersects the horizontal surface. A total of 2646 traces were used to construct the velocity diagram. Each trace was obtained by

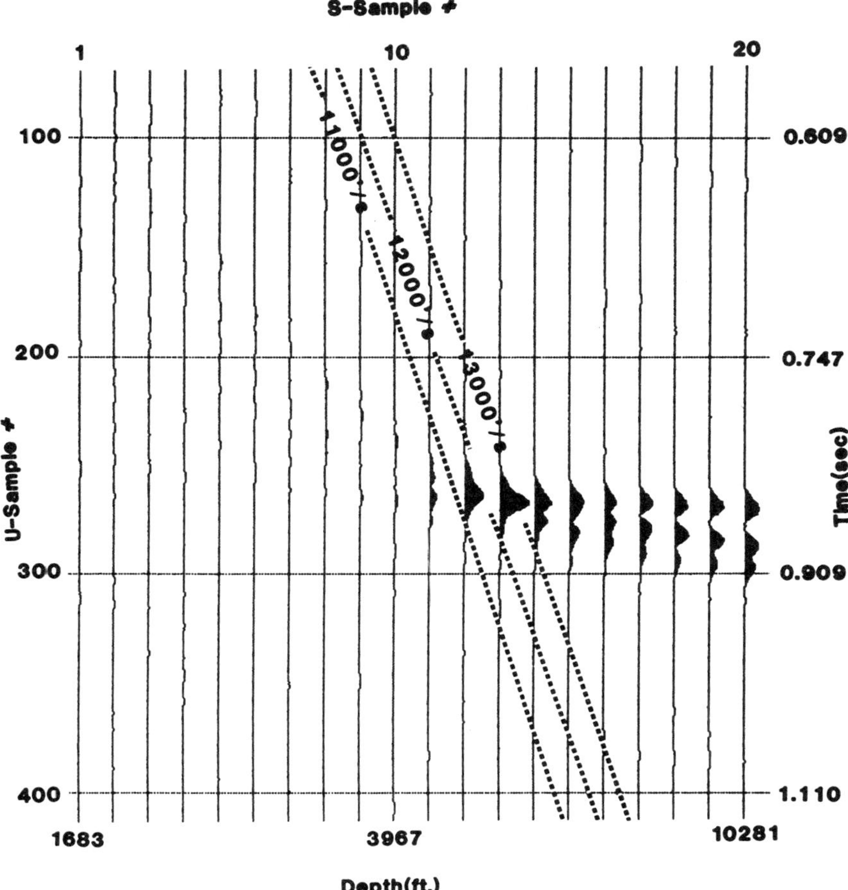

Figure 7.8. Velocity diagram when all 2646 traces simulated over the horizontal layer model were migrated and added.

calculating the reflection point for the given source and receiver location (Dunkin and Levin, 1971) and then calculating the traveltime from the source to the reflection point and back to the receiver.

Figure 7.10 shows the velocity diagram obtained. It shows two main peaks of almost identical size but separated by about ten *U* samples, which is equivalent to about 23 ms. The largest offset used was 4200 ft. The time diffusion might be eliminated if larger offset values including 5000 ft were used. The event picked occurred halfway between *S* samples 12 and 13 at a *U* sample of 262. The corresponding velocity pick fell on the 12,000 ft/s line.

In conventional velocity analysis, CDP data at or near the well location would have been selected. Levin's (1971) well-known structural effect would have affected the velocity obtained, giving a higher velocity of 12,000/(cos 30°) $\simeq$ 13,856 ft/s. In the algorithm presented here, by properly selecting the data aperture to include the normal incidence point, the velocity obtained is not affected by structure.

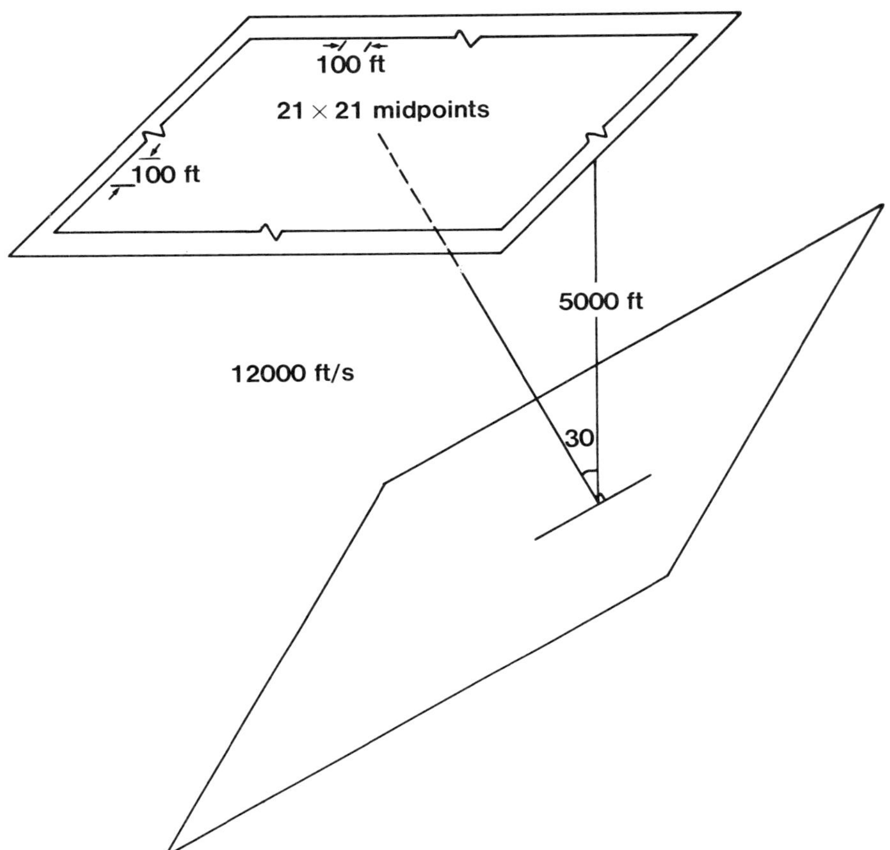

Figure 7.9. Areal data simulation over a dipping interface. Center of square area is the normal incidence point.

Point Scatterer Model. The model used consisted of a point scatterer at a depth of 5000 ft in a medium of velocity 12,000 ft/s.

The computer-simulated sixfold CDP areal survey was generated over the model along parallel straight lines that were separated by 100 ft and at CDP spacing of 100 ft. A square area of data, 2000 ft by 2000 ft, was used to construct the velocity diagram. The data base consisted of 2646 traces. The traces were obtained by calculating the traveltime from each source location to the point and back to its corresponding receiver. Figure 7.11 shows the data aperture over the point scatterer. The well axis for the velocity diagram was chosen to be the vertical line passing through the point.

Figure 7.12 shows the velocity diagram obtained. Apart from what can be considered the resultant background noise of the operation, two main events of identical size and character occur at S samples 12 and 13. The event picked between the two occurs at U sample 262, corresponding to a time of 0.852 and a velocity of 12,000 ft/s.

The background noise probably results from an insufficient number of offsets and the offset range used. A larger fold and larger offset range, with small increments in offsets, would decrease the amount of the background noise.

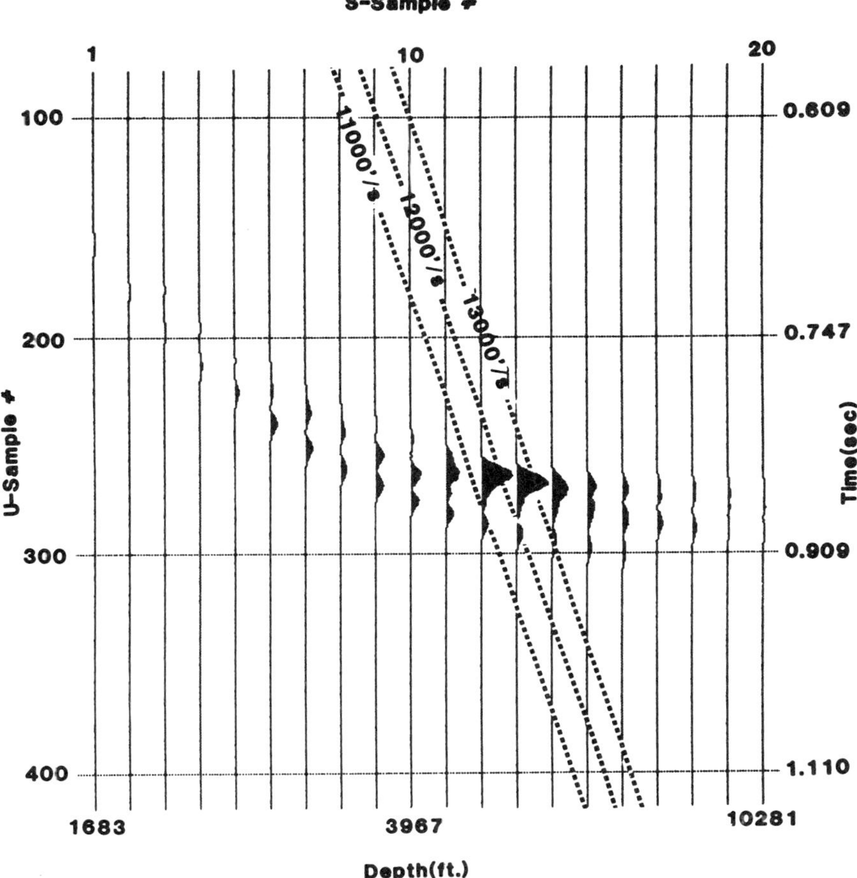

Figure 7.10. *Velocity diagram constructed using 2646 traces simulated over the dipping interface.*

Multilayered Media Model. Thus far, the alogorithm has been tested successfully using single-layer models. Generally, the sedimentary basins that are the target of exploration can be considered layered, so the algorithm was tested using a layered media model. Also, in the previous model studies, no noise was added to the data. In this study, the effect of random noise was examined.

The model used is illustrated in Figure 7.13. It is the same model used by Taner and Koehler (1969) to illustrate CDP velocity spectra.

A computer-simulated 12-fold CDP areal survey was generated over the model along parallel straight lines, with line spacing of 100 ft and CDP spacing of 100 ft. A square area of data, 2000 ft by 2000 ft, was used to construct velocity diagrams for an output location at the center of the square. Thus, 5292 traces formed the data base. Each trace for a given source and receiver offset was obtained by ray-tracing through the model, using equations given by Slotnick (1959). The reflection spikes obtained, whose amplitudes were proportional to the reflection coefficients, were then convolved with a zero-phase wavelet with a central frequency of 30 Hz. The wavelet is shown in Figure 7.14. A time shift was applied after convolution so that

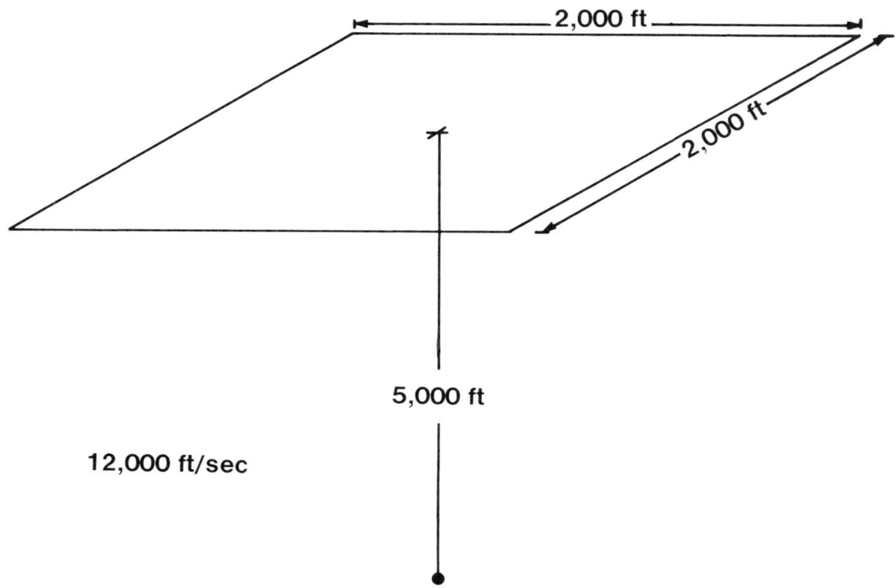

Figure 7.11. *Data aperture over the point scatterer.*

the reflection arrival times coincided with the peak amplitude of the wavelet. A 12-fold CDP gather for the data is shown in Figure 7.15. The near source-receiver offset distance was 1500 ft, and the far offset distance was 4800 ft.

Random noise with a uniform distribution of amplitude was added to the traces with a signal-to-noise ratio of about 2:1. An example of the resampled analytical signal formed from the differentiated data is shown in Figure 7.16. Figure 7.17 shows the trace in Figure 7.16 reduced to sign bit. The data were resampled with constants $CK = 217.16$ and $CL = 999.04$ in equation (7.5). The index U ranged from 1 to 700, corresponding to a range of T from 0.1 to 2.5 s.

The aperture used for the migration was selected, using equation (7.11), to increase linearly to a radius of 1100 ft at a depth of 9000 ft. The aperture is smaller than that required for a complete migration. It was used to show how this affects the character of the migrated pulse.

Figure 7.18 shows the velocity diagram when one data trace (without any noise) has been added to it. Twenty depth levels were selected, using the constants $DK = 5.59$ and $DL = 30.86$ in equation (7.8). The S samples ranged from 1 to 20, with a corresponding depth range of 298 to 9841 ft. The envelope of the migrated analytical signal was plotted, and the shift applied to the trace was different for each depth. Other traces will have different shift curves that will interfere with one another. The accumulation of all the traces will show the events for which there is maximum coherence in the input data.

Figure 7.19 shows the accumulation of all 5292 analytical signals when the input traces were 16-bit, with random noise added, as in Figure 7.16. The envelope of the analytical signal was plotted, and the discrete events corresponding to the reflectors are apparent. Straight lines have also been added to indicate the locus of events with constant velocity. It can be seen that the magnitude of the peaks correspond very well with the magnitude of the reflection coefficients.

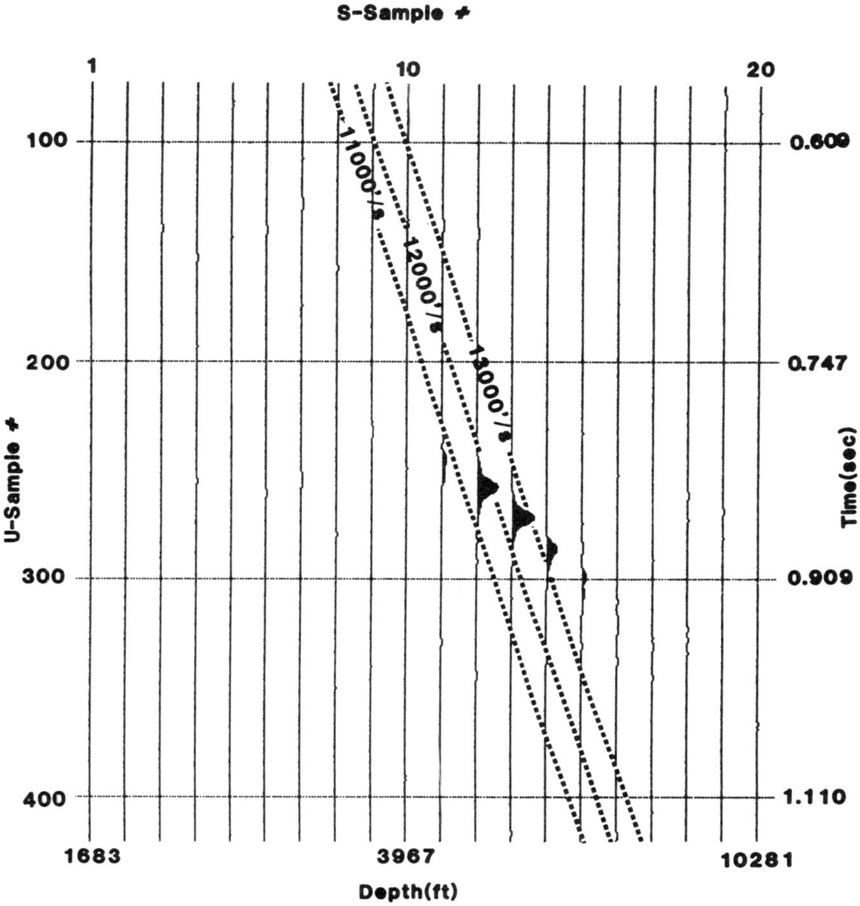

Figure 7.12. Velocity diagram constructed using 2646 traces simulated over the point scatterer.

The result of reducing the input data to sign bit and constructing the velocity diagram, using the same parameters, is shown in Figure 7.20. The velocity inferred from the location of the peaks is identical for either 16-bit or 1-bit representation of the data. The amplitudes of the peaks, from the sign bit representation, no longer correspond so closely to the magnitude of the reflection coefficients.

A close-up view of Figure 7.20 around the peak corresponding to the third reflector at a depth of 3400 ft is shown in Figures 7.21, 7.22, and 7.23. Here the depth and time scales have been expanded by migrating the data with $DK = 66.05$ and $DL = 527.78$ in equation (7.8), with the S samples ranging from 1 to 20. The corresponding depth values ranged from 2998 to 3998 ft. The velocity lines are superimposed on the diagrams.

Figure 7.21 shows the envelope function. The peak is very well defined in both time and depth, although the envelope appears to have a flattened top. This may

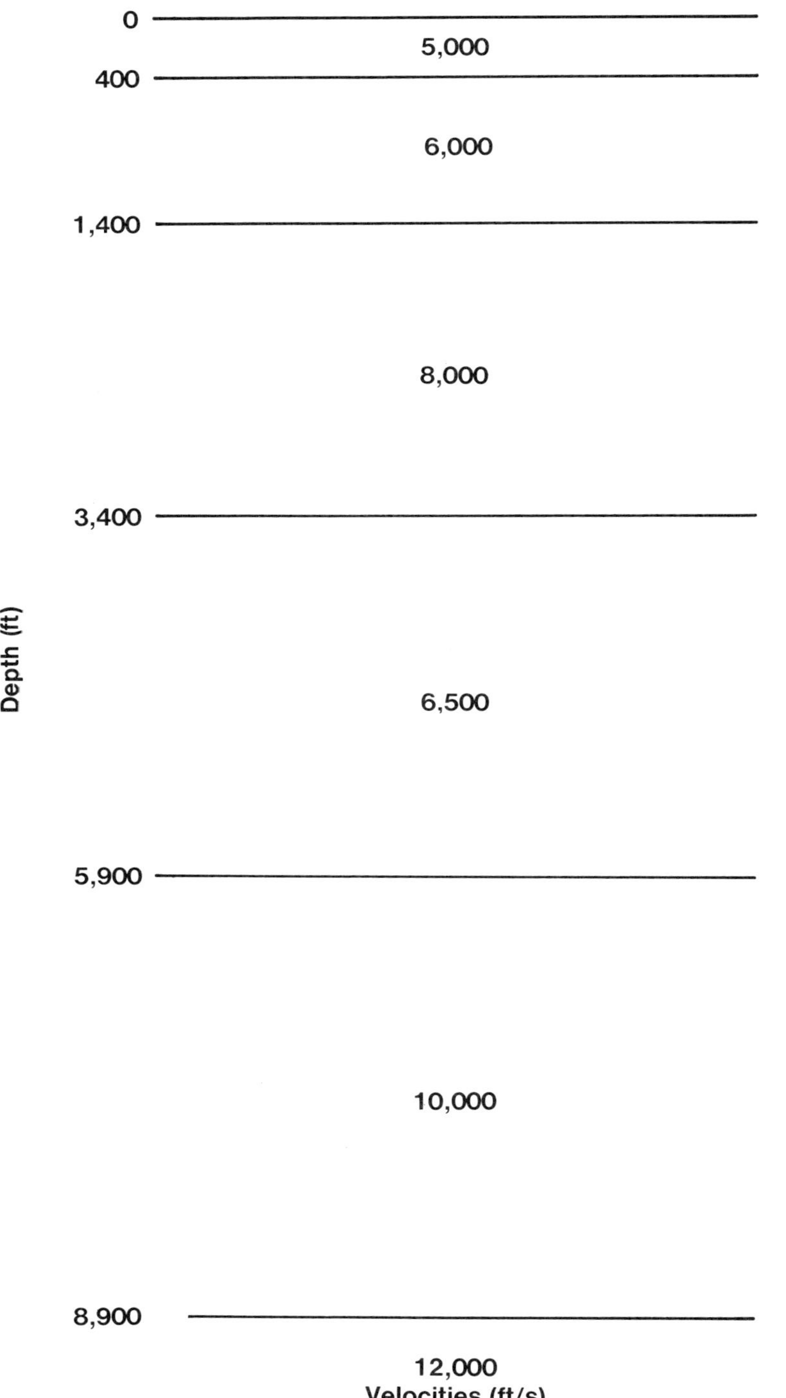

Figure 7.13. *Multilayered model used to generate 5292 three-dimensional synthetic seismic traces.*

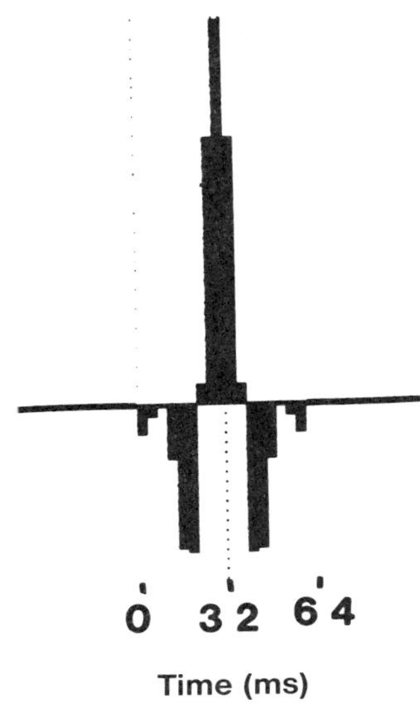

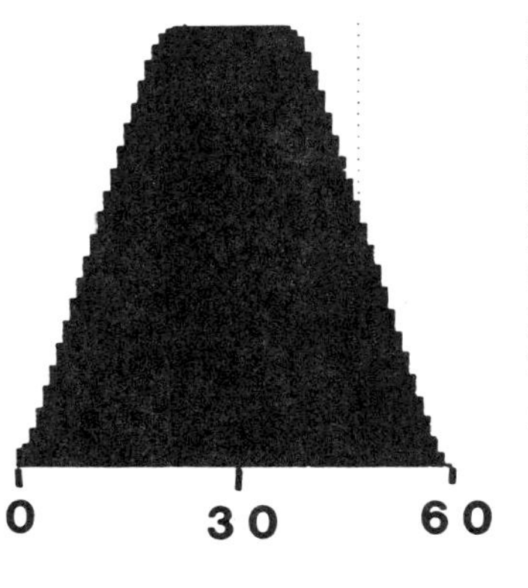

Figure 7.14. *The zero-phase wavelet that was convolved with the reflection coefficients of the model, and its amplitude spectrum.*

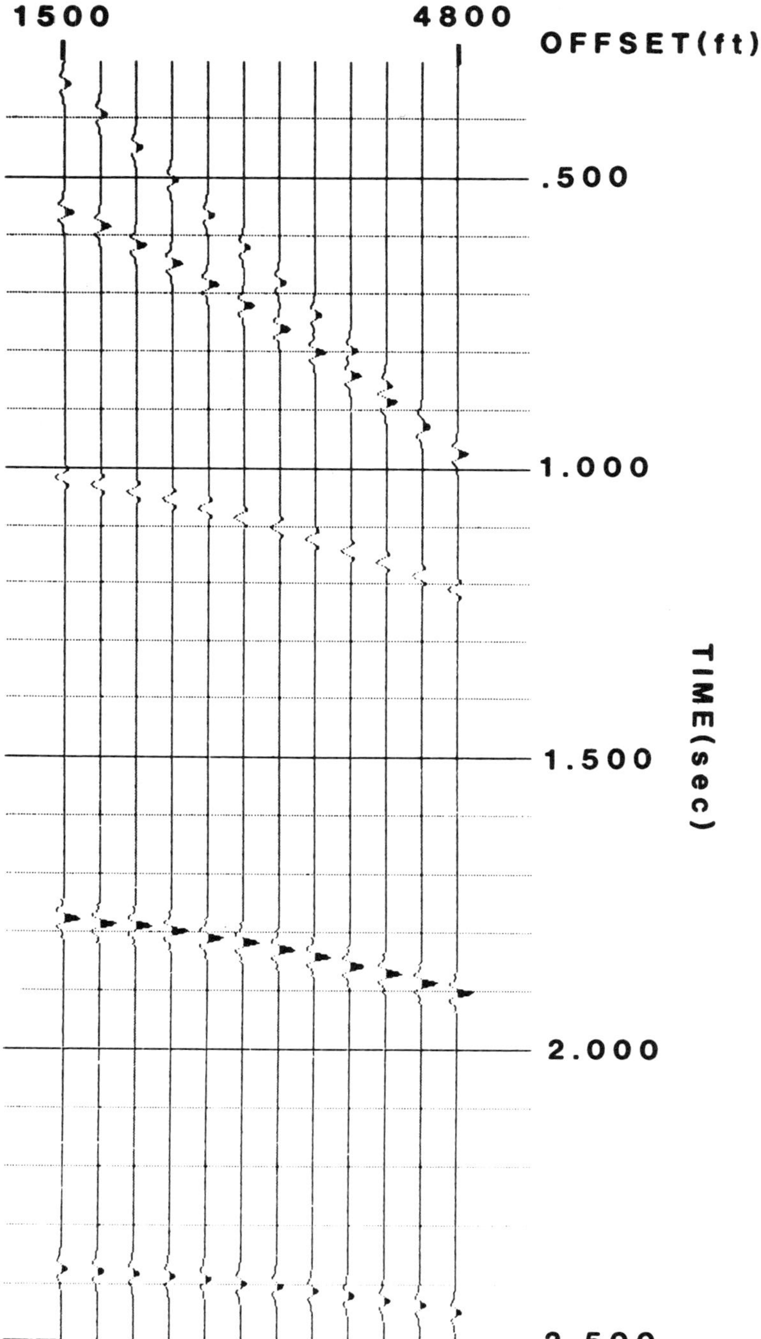

Figure 7.15. A 12-fold CDP gather from the areal data.

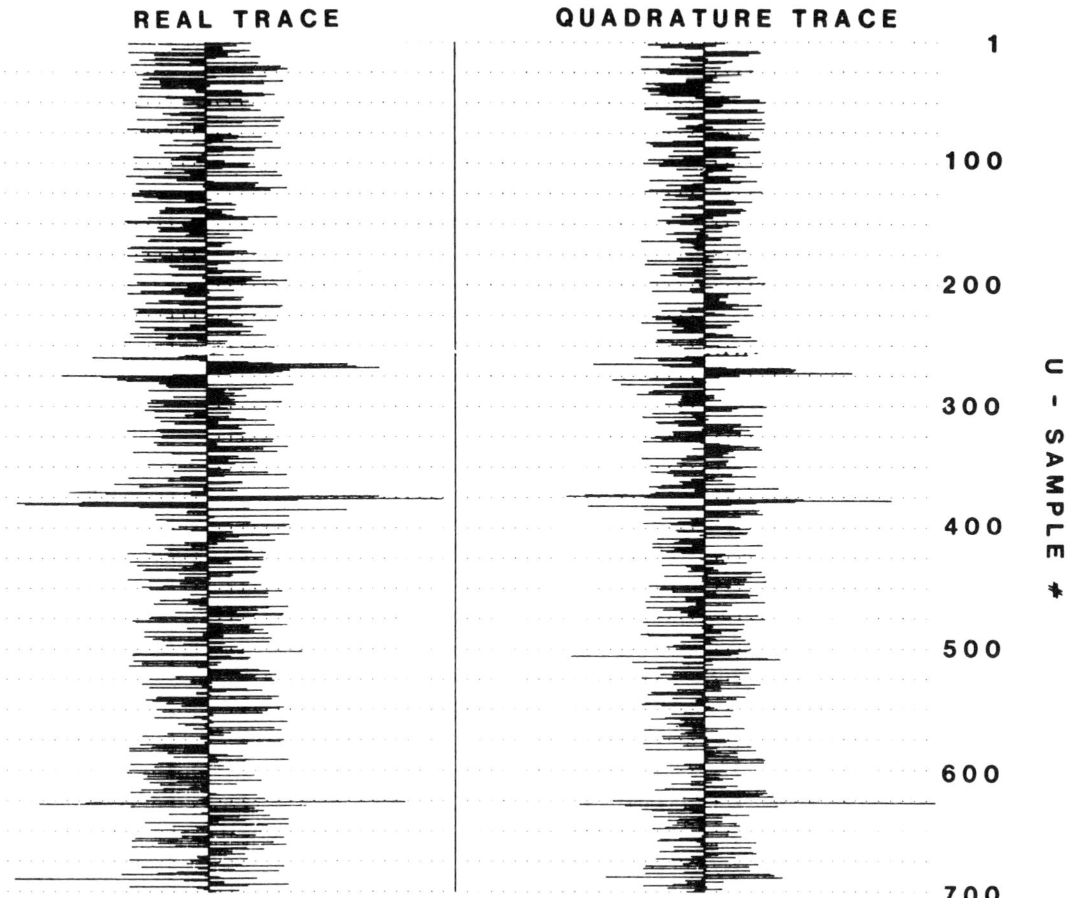

Figure 7.16. *Example of the analytical signal after differentiation and resampling, with noise added.*

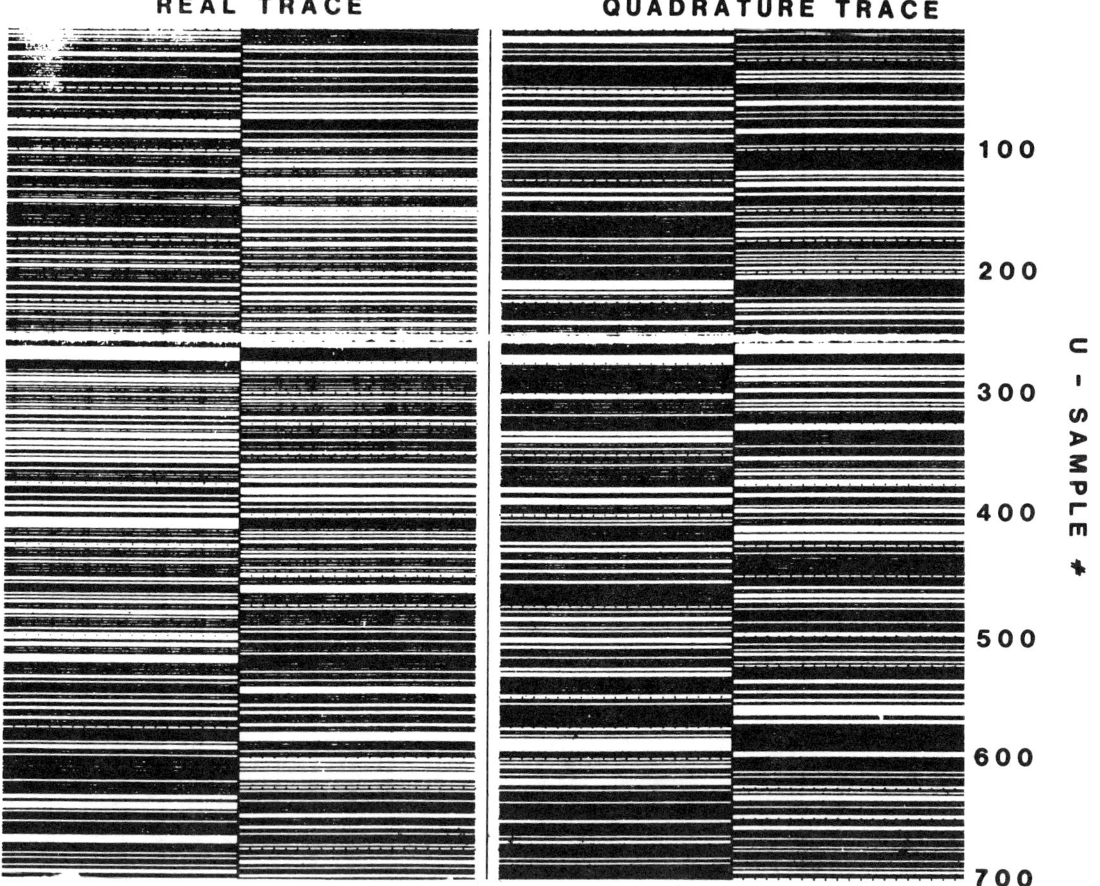

Figure 7.17. *The same signal in Figure 7.16 reduced to sign bits.*

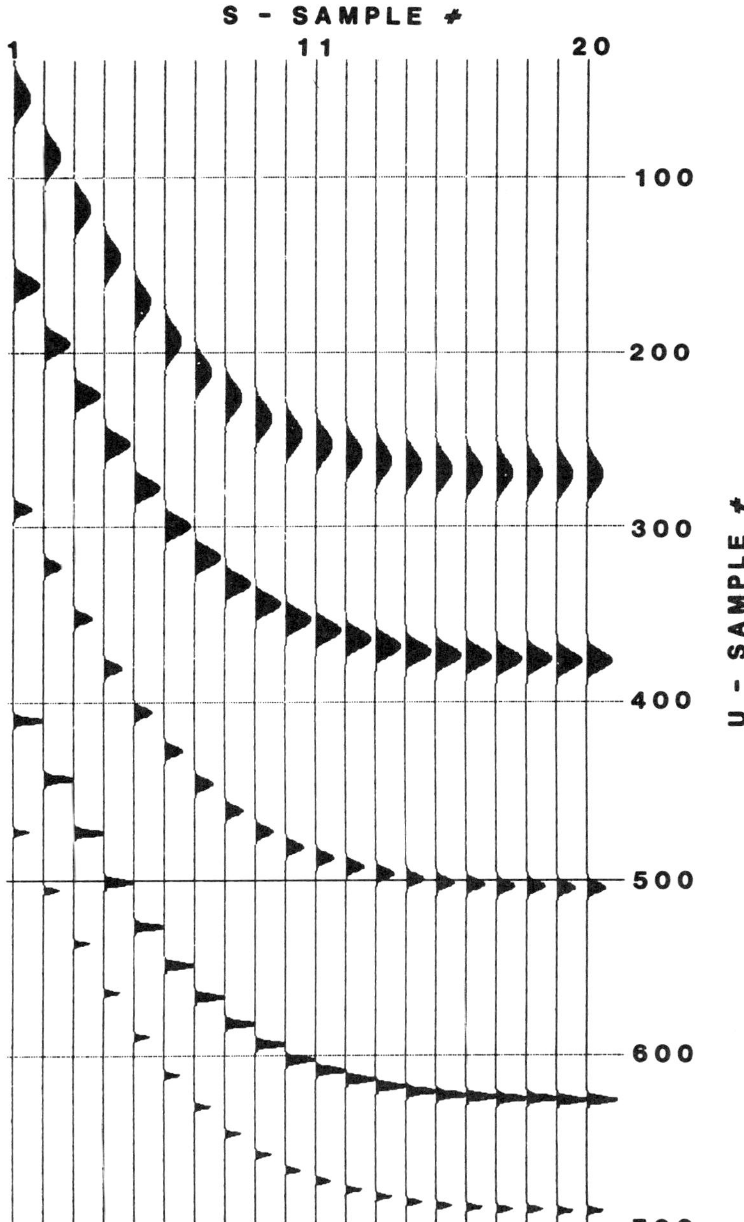

Figure 7.18. Velocity diagram when one 16-bit analytical signal has been added. The envelope of the signal is plotted. No noise has been added.

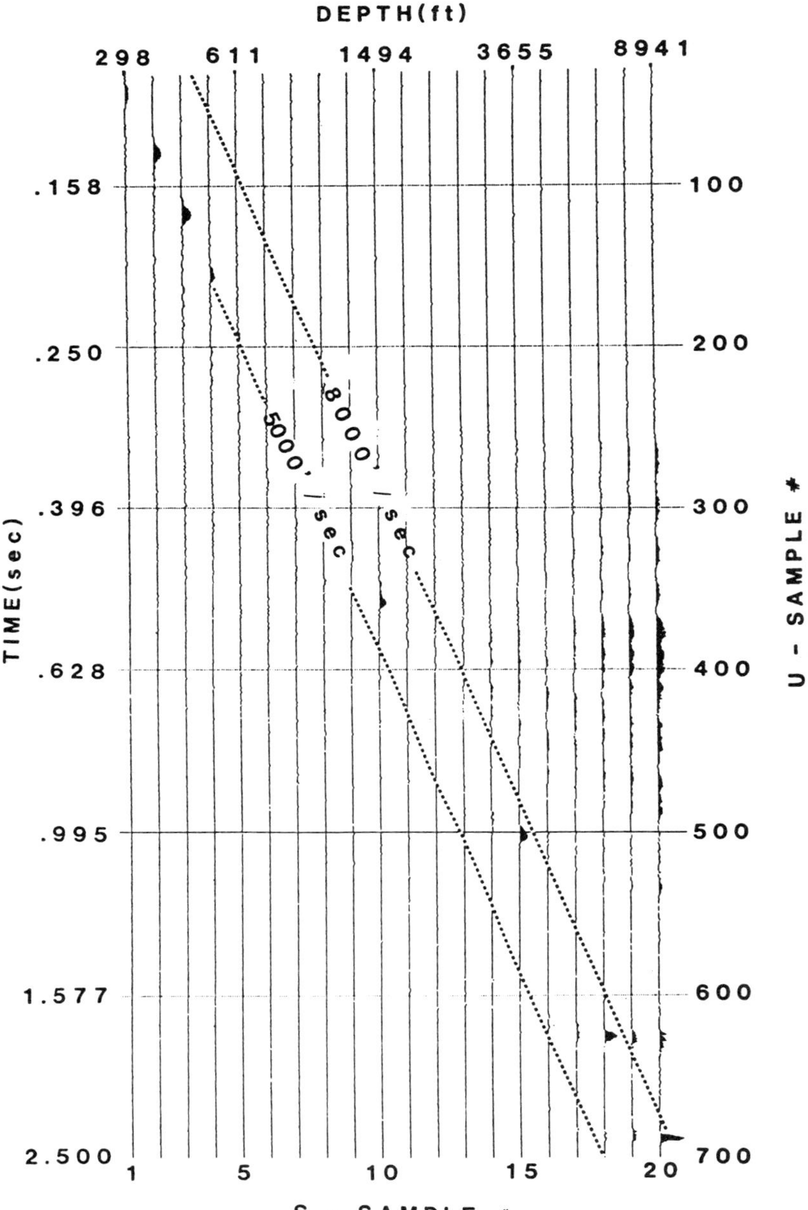

Figure 7.19. Velocity diagram when 5292 16-bit signals have been added. Random noise has been added to the signals.

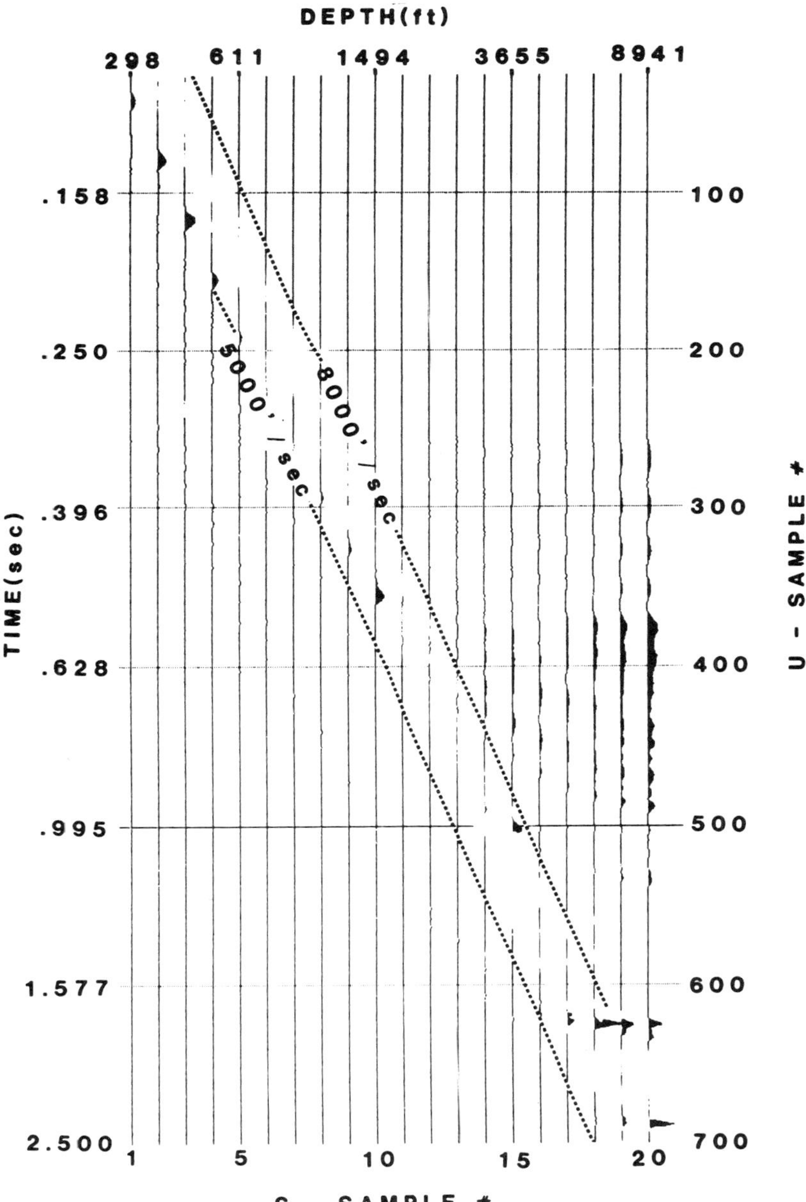

Figure 7.20. *Velocity diagram when the same data used to construct Figure 7.19 are reduced to sign bits.*

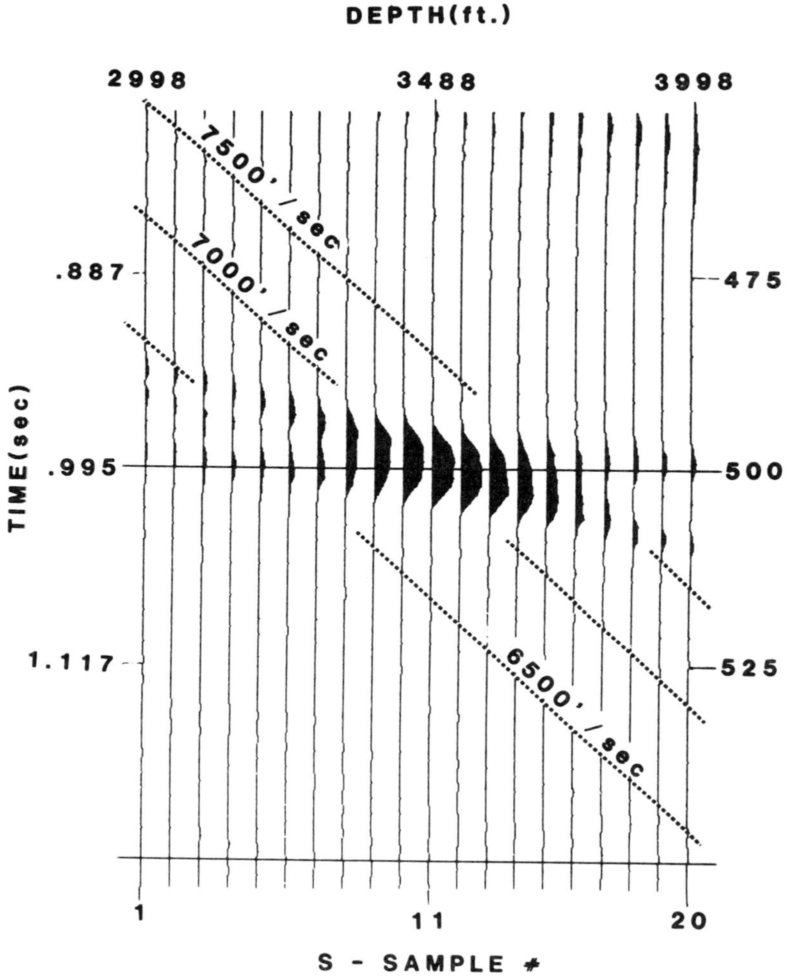

Figure 7.21. *Enlargement of Figure 7.20, showing the envelope.*

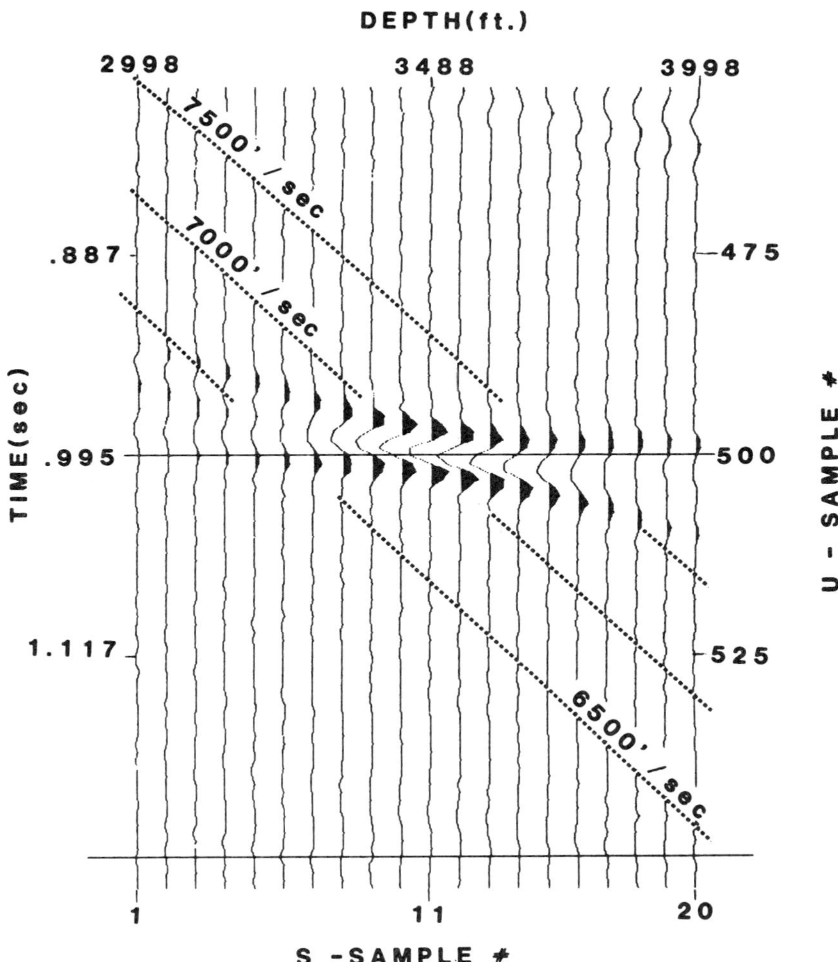

Figure 7.22. The same enlargement as Figure 7.21, showing the real part of the analytical signal.

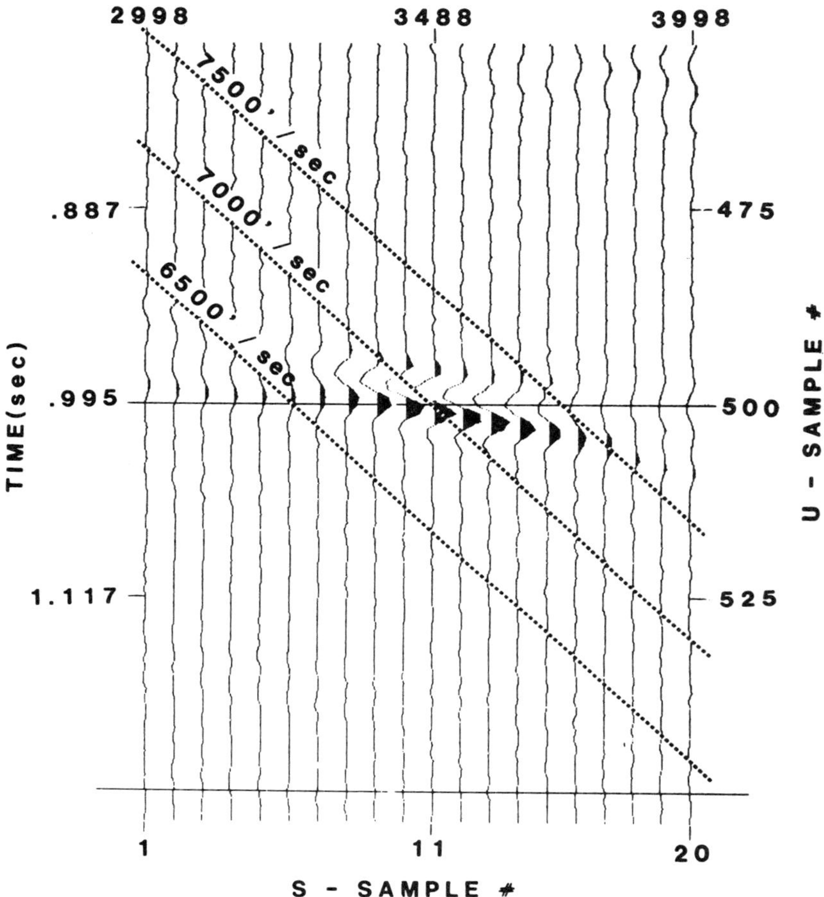

Figure 7.23. *The same enlargement as Figure 7.22, showing the imaginary part of the analytical signal.*

result from the low level of noise present with the signal. As O'Brien, Kamp, and Hoover (1979) have shown, clipping of the signal tends to occur when the signal is greater than the noise and the noise is uniformly distributed.

Figures 7.22 and 7.23 show the real and imaginary parts of the migrated analytical signal constructed from the sign bit data.

It may be noted that the shallowest reflector at 400 ft depth is not well focused (Fig. 7.19). It appears as about four pronounced peaks, the largest of them appearing at *S* samples 2 and 3, between which the right depth occurs. The poor image at that depth may be attributed to the fact that the least source-receiver offset used was 3.75 times the depth. Offsets of about one-tenth the depth must be used in order to focus the shallow reflectors.

It may also be noted that, at large depths and high velocities, the display has substantial background noise. This may be caused by two factors. The first factor is the number and spacing of the offsets in the data-acquisition specifications. A

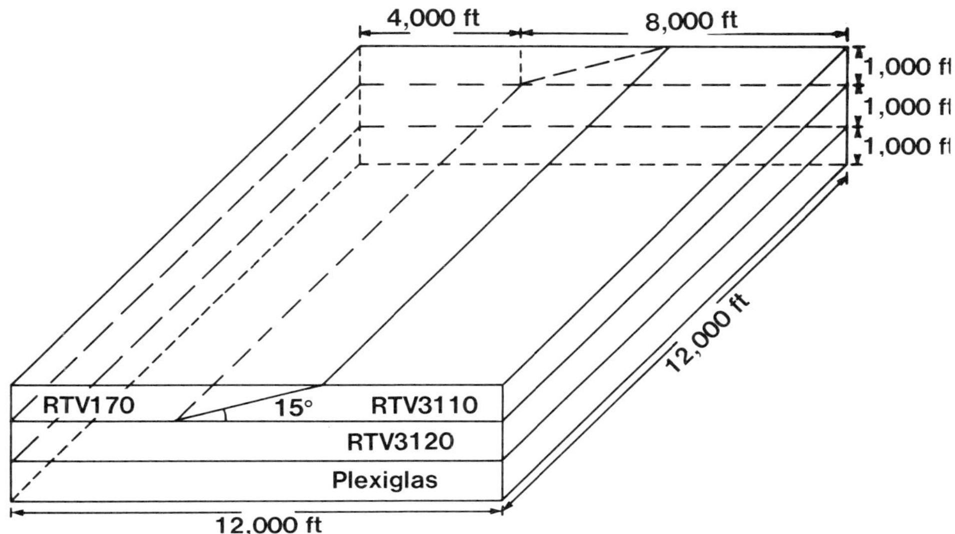

Figure 7.24. Faulted, multilayered physical model over which an eightfold, cross-spread array of seismic data was collected in a water tank. A total of 6720 traces were collected.

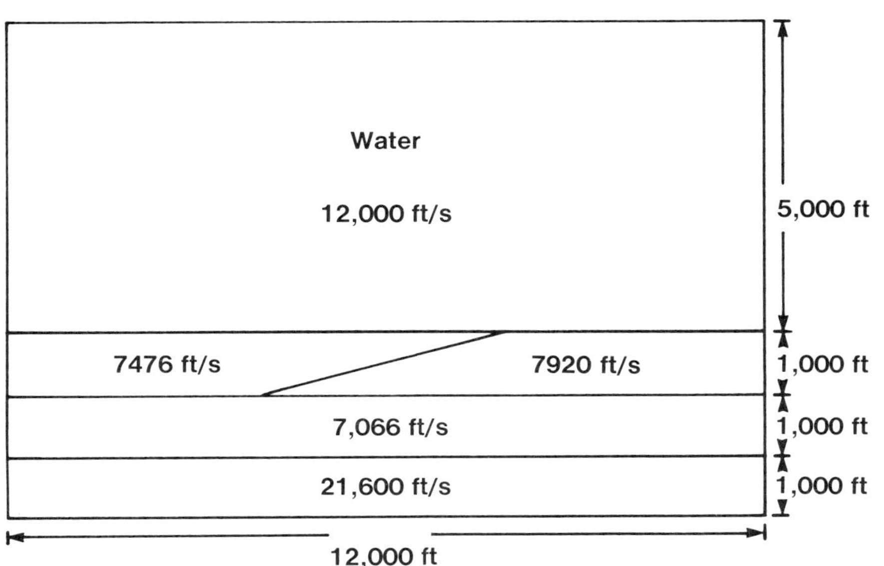

Figure 7.25. A cross section of the physical model.

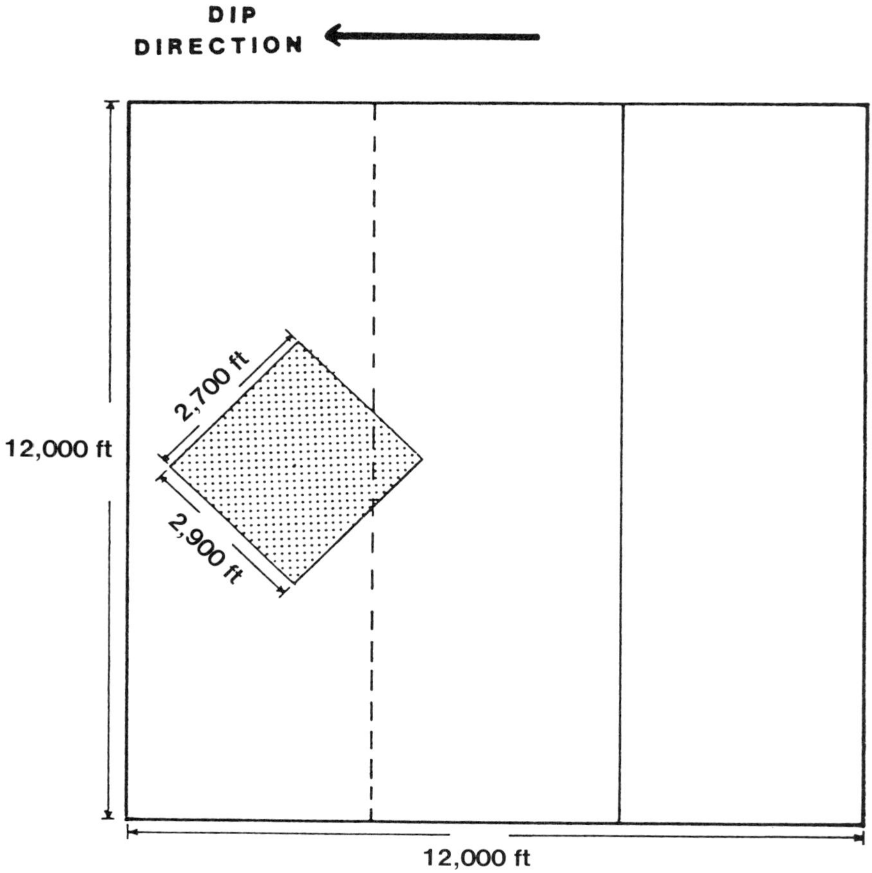

Figure 7.26. Plan view of the model, showing the location of data coverage.

larger range of offsets and a smaller increment between offsets would suppress the noise. The second factor is the data aperture used for the migration. At greater depths and greater velocities, a larger aperture and an increase in the area over which data were collected may be required.

The effect of migration aperture can also be seen in the waveforms in Figures 7.22 and 7.23. The real component should be zero-phase, since the original wavelet was zero-phase. This is not the case, however, so a larger aperture may be necessary to correct the phase.

The velocity derived from Figure 7.21 is about 7050 ft/s. The rms velocity for this reflector is 6950 ft/s—an error of about 1.4 percent. A difference of this order is to be expected, because the rms velocity gives the traveltime for small offsets, and the velocity derived from the data is an average over the actual offset range.

Faulted Layered Model. After testing the algorithm using computer-simulated model data, it was considered worthwhile to test it on physical model data because of the unavailability of three-dimensional field data.

Model Description. Figure 7.24 shows a schematic diagram of the physical model used. It was essentially a two-dimensional, layered model. The top layer consisted of two different silicone rubber materials, RTV 170 and RTV 3110, separated by a fault of 15° dip. The RTV 170 was on the downdip side of the fault. The second layer consisted of Plexiglas. The model was placed in the water tank, with the source and receiver transducers positioned at an equivalent depth of 5000 ft. Each model layer had an equivalent thickness of 1000 ft. The properties of the model materials are summarized in Chapter 1. Figure 7.25 shows a cross section of the model as it was positioned in the water.

Data Acquisition. An eightfold, cross-spread array was shot over the model, with the arrays at an angle of 45° to the dip direction of the fault. The source-receiver midpoints covered a square area of 2700 ft by 2900 ft and contained 6720 traces, which formed the data base. The center of the square was chosen to be about 1000 ft from the projection of the base of the fault. It was also chosen to contain the location of the intersection of a normal ray from the fault plane and the horizontal plane of data collection. Figure 7.26 shows the plan view of the model, with a dotted square area showing the location of the data coverage. The midpoint spacing was 100 ft. Each midpoint had eight traces, whose source-receiver offset distance ranged between 2100 ft and 9400 ft. Each data trace had the source and receiver oriented at a different azimuth.

Figure 7.27 shows a typical record obtained for a fixed receiver location, with the source transducer occupying 15 stations on either side of the receiver, along the source line. The station spacing was 200 ft. Figure 7.28 shows the same record with uniformly distributed, unfiltered random noise added to it. The signal-to-noise ratio was about 2:3 at the strongest reflection and about 0.0033 at the weakest.

The data were then differentiated and resampled, using the constants $CK = 762.86$ and $CL = 5098.41$ in equation (7.5). The resampled analytical signal had noise added, and the resampled noisy analytical signal was also converted into sign bit data.

Velocity Estimation. Figure 7.29 shows the locations of two vertical wells A and B on the map of the model. Well A passes through the center of the data window, and well B passes through the center of the fault plane. Velocity diagrams were constructed for well A to study the effects of adding noise to the data and also using sign bit resampled data. Velocity diagrams were constructed for well B to study the effect of using different aperture sizes and orientations on the velocity estimation over a fault.

The data were migrated using the constants $DK = 29.87$ and $DL = 250.30$ in equation (7.8). The index S ranged from 1 to 20, with corresponding depths ranging from 4506 to 8512 ft.

Well A: The data were migrated using a circular aperture whose radius varied linearly from 200 ft at zero depth to a radius of 1300 ft at a depth of 9000 ft.

Figure 7.30 shows the velocity diagram constructed along the well axis, using the data without any noise added to it. Plotted here are the envelope of the analytical signal accumulated by migrating the input data within the prescribed time-varying aperture. The amplitudes of the peaks correspond to the magnitudes of the reflection coefficients. The reflection event from the interface of RTV 170 and RTV 3120 (reflection coefficient = 0.0033) was so weak that it barely showed on Figure 7.30. In Figure 7.31, the same velocity diagram displayed at 5dB higher gain shows the weak reflection event.

Figure 7.32 is a display of the real part of the analytical signal constructed from the input data. The second reflecting event shows very lightly. The input wavelet character appears to have been well preserved.

All four interfaces can be identified by their peaks on the velocity diagram. The approximate velocity and time picks have been summarized in Table 7.2. The estimated values are not the exact velocity and time values expected. The velocity estimates were found to be lower by about 3 percent. The deviations could be reduced by selecting a depth-time window around the picks and choosing the

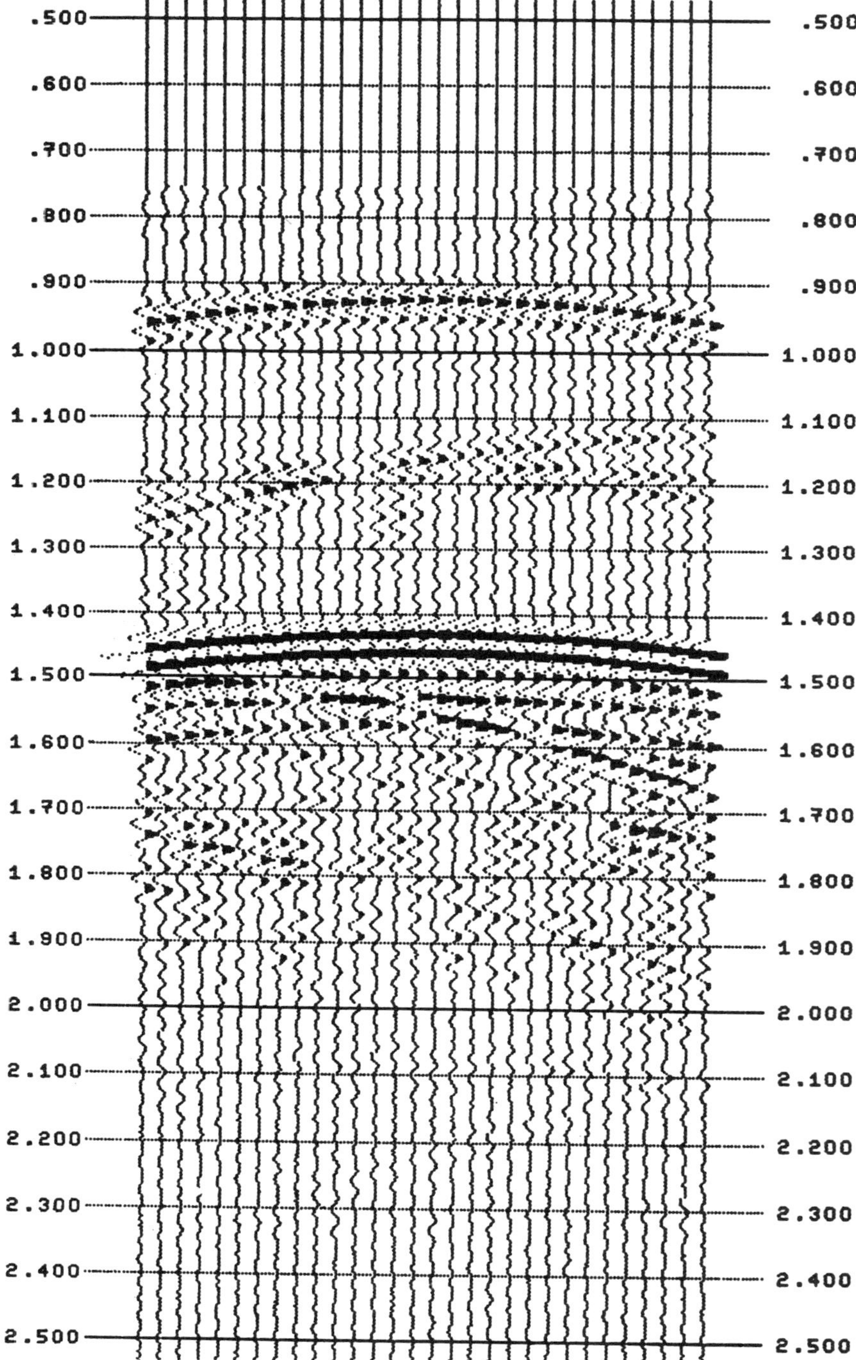

Figure 7.27. *A typical fixed receiver record.*

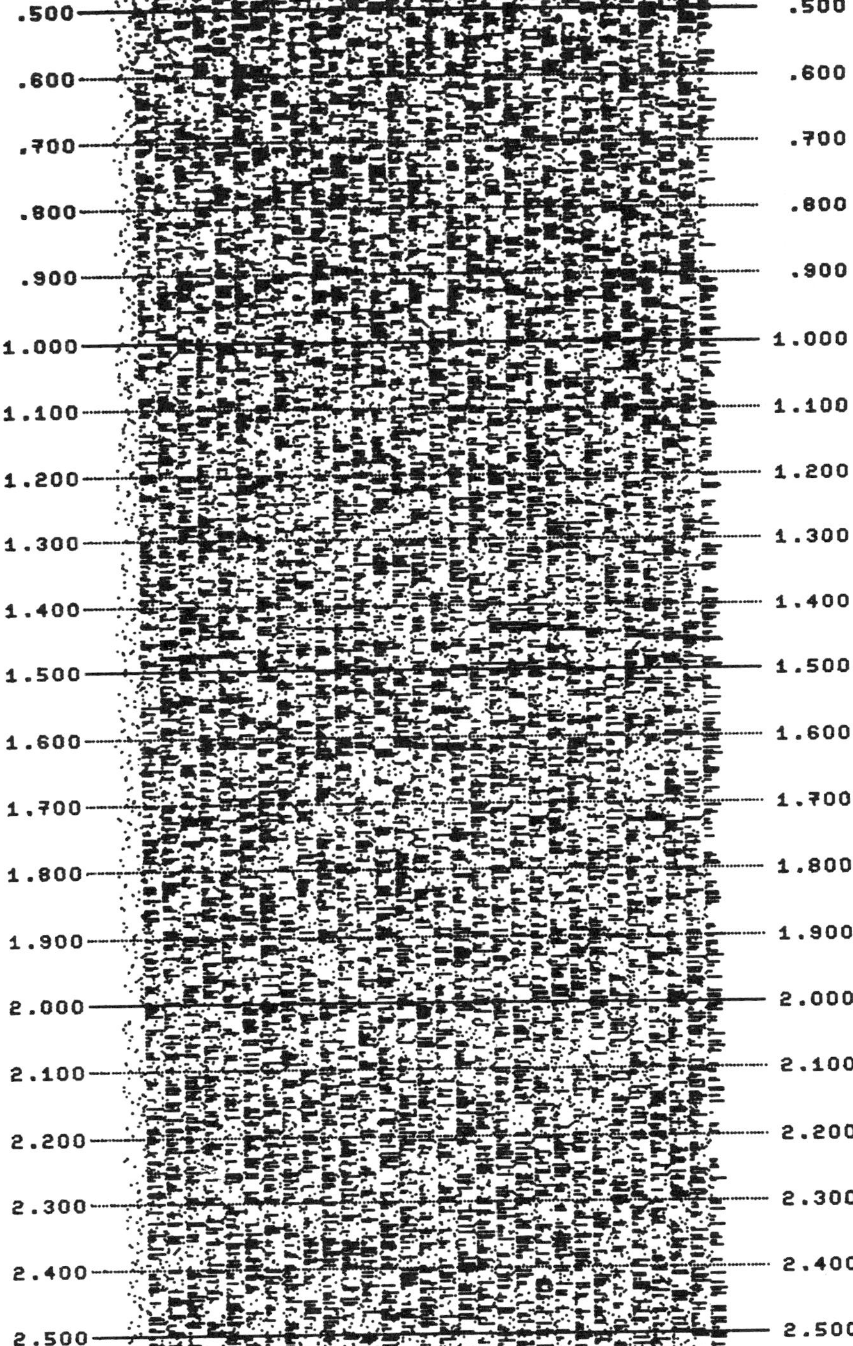

Figure 7.28. *The same record as in Figure 7.27, with random noise added. The signal-to-noise ratio is about 2:3 for the strongest reflection and about 1:300 for the weakest.*

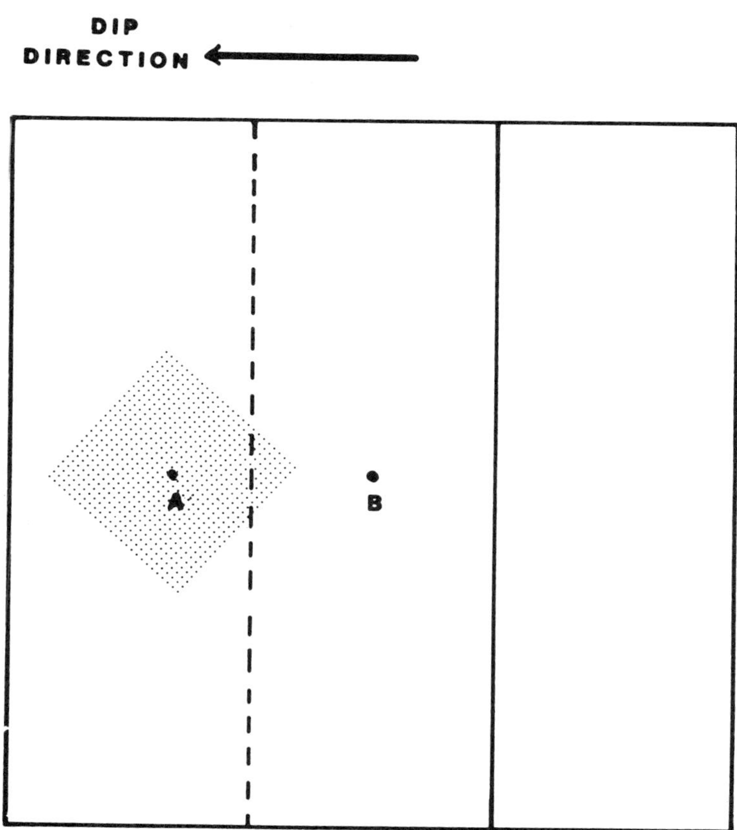

Figure 7.29. *Map of the physical model, with well locations A and B indicated.*

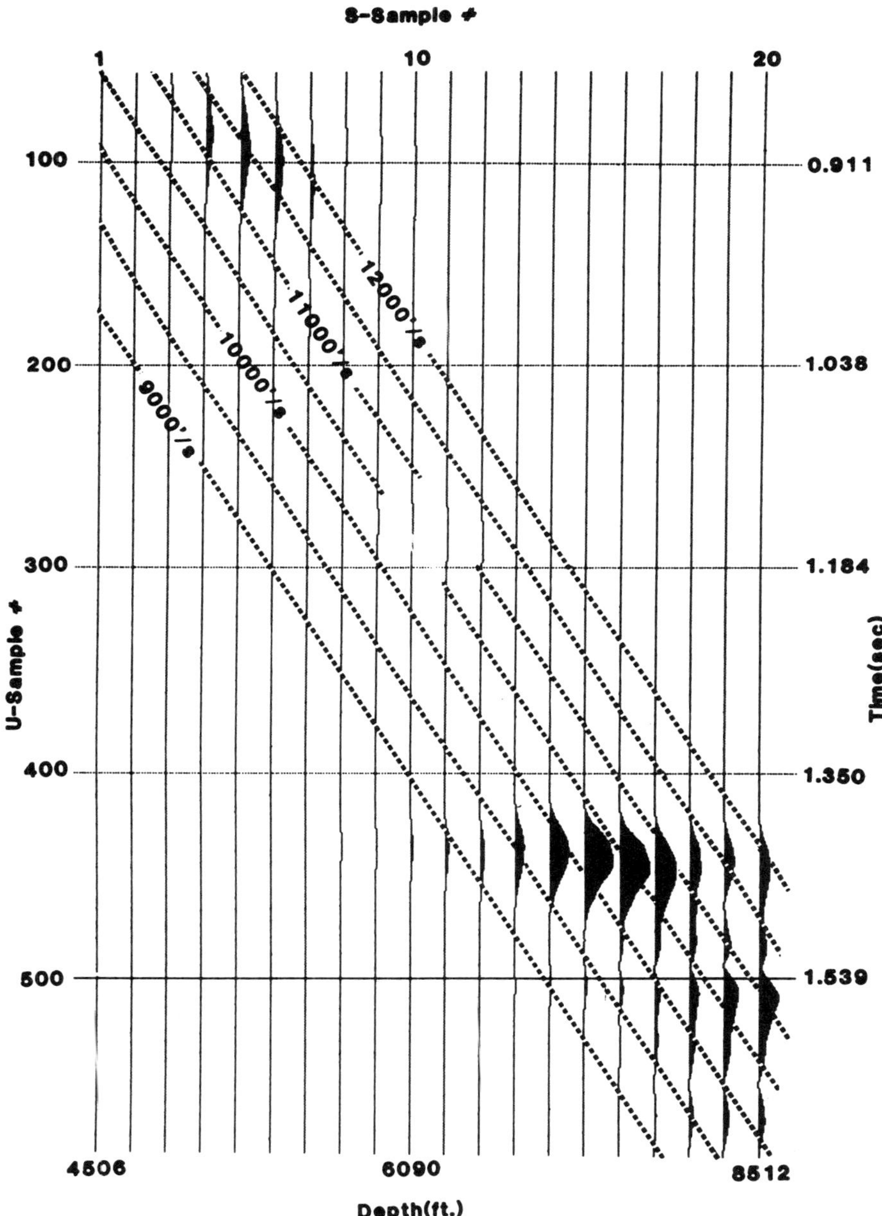

Figure 7.30. *Velocity diagram for well A, constructed using 6720 analytical signals, without any noise.*

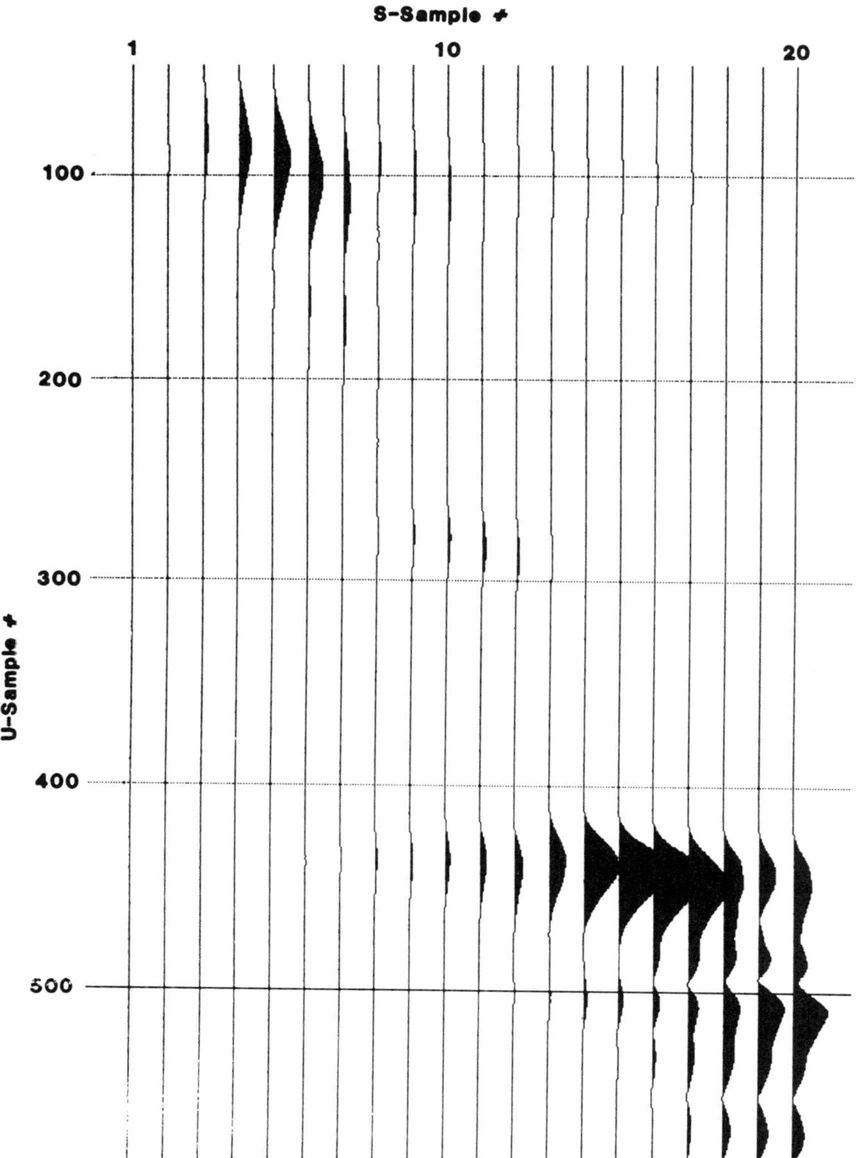

Figure 7.31. Velocity diagram in Figure 7.30 displayed at a higher gain to show the weak reflection event.

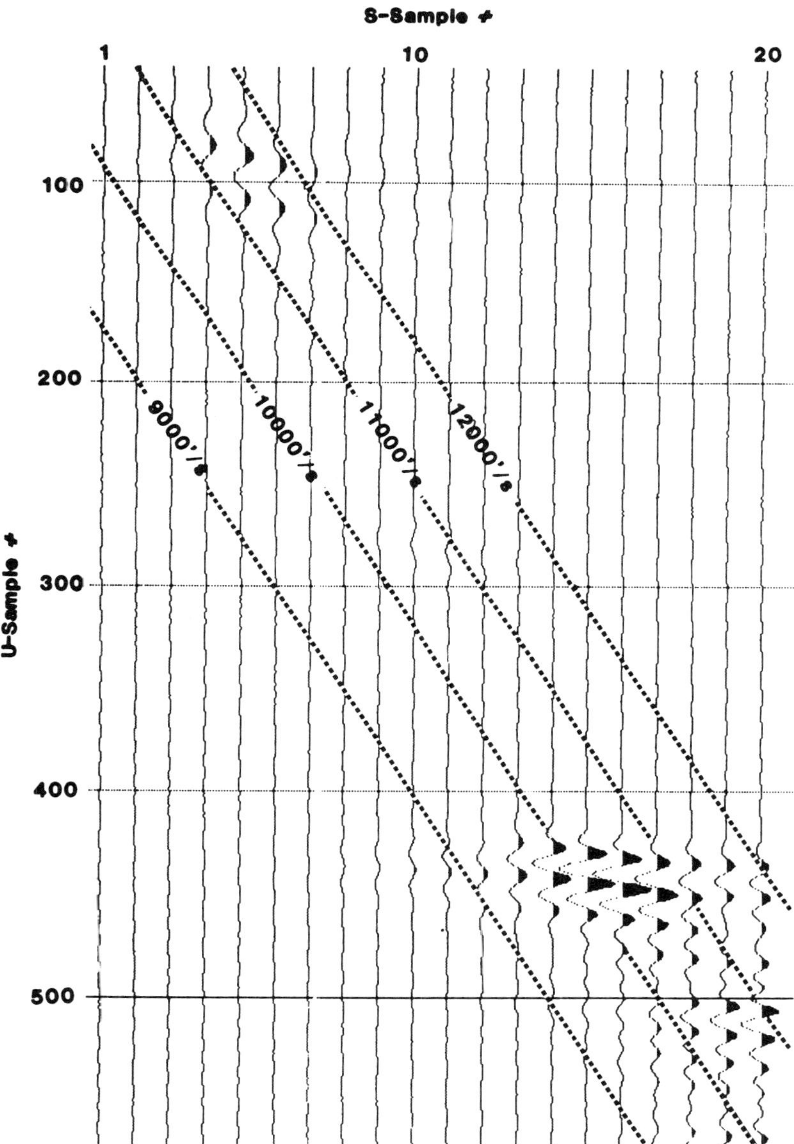

Figure 7.32. *Velocity diagram in Figures 7.30 and 7.31, showing the real part of the analytical signal.*

Table 7.2. *Approximate velocity and time picks, well A*

Event No.	*S* Sample No.	Apparent Depth (ft)	*U* Sample No.	Estimated Traveltime(s)	Expected Traveltime(s)	Estimated Velocity (ft/s)	Expected Velocity (ft/s)
1	5	5152	92	0.901	0.833	11500	12000
2	11	6298	285	1.161	1.101	10750	11072
3	15.5	7323	442	1.426	1.384	10250	10379
4	19.5	8371.7	508	1.554	1.477	11000	11412

appropriate constants *CK, CL, DK,* and *DL* for a closer and detailed look at them.

The velocity diagram was also constructed using the input data with random noise added. This was repeated after the resampled analytical signal was reduced to sign bit. The same parameters were used as were used before.

Figure 7.33 shows the velocity diagram constructed using the noisy input data; the envelope of the analytical signal is plotted. A comparison with Figure 7.30 shows that, although the noise is not completely removed, the signal-to-noise ratio has improved and the same events picked before are apparent. Here, again, the second event, occurring at about *U* sample 285 and *S* sample 11, is not easily identifiable. Figure 7.34 shows the real part of the analytical signal. A close examination of the diagram shows that the second event is detectable. The signal-to-noise ratio appears to have improved, and all other events appear very prominently. The character of the wavelet appears to have been preserved.

Figure 7.35 shows the velocity diagram constructed after reducing the resampled analytical signal of the noisy data to sign bits. The three events that were prominent before are again prominent here. The amplitudes of the peaks still bear some resemblance to the reflection coefficients. The second event is still not apparent. Figure 7.36 shows the corresponding real part of the analytical signal. Again the wavelet is recovered for the three prominent events. The second event once again shows an indication of its presence. Figure 7.37 shows the velocity diagram filtered, using a 101-point low pass, with frequency cut at 60-70 Hz. Although the output is ringy, all four events are clearly present at the expected locations.

Well B: Without any random noise added, the data were migrated to construct velocity diagrams for points along well B, which was sited over the fault. The objective was to make a comparative study of how the velocity diagrams would be affected by using migration apertures of different sizes, shapes, and orientation.

The square of data midpoints generated during data collection does not include the well location. The center of the data window was selected to include the normal incidence point of the fault. The fault plane was expected to be well imaged, whereas the underlying reflectors were not expected to be well imaged.

Figure 7.38 shows a velocity diagram constructed for a circular aperture of constant radius 200 ft for all depths considered. This essentially picks a CDP gather from the data. The envelope of the analytical signal is plotted. The aperture size chosen contained only eight midpoints, or 64 traces. The diagram shows events such as the top reflector, the fault plane, and, very strongly, the Plexiglas/RTV interface. These events are not focused, however, and velocity interpretation from the diagram is difficult.

Figure 7.39 shows the velocity diagram constructed for a circular aperture that increases monotonically with depth, to a radius of 1300 ft at a depth of 9000 ft. As expected, the fault plane event appears to be well imaged, as shown by the most prominent peaks occurring around *U* sample 200. The largest peak appeared to be between *S* samples 8 and 9 and for an approximate *U* sample of 190. The corresponding velocity pick was about 11300 feet per second for an apparent depth of about 5793 feet and a travel time of about 1.025 seconds. Four more possibly identifiable events appear on the diagram but with extremely low amplitudes due to possible destructive interference during the migration.

Figure 7.40 shows the velocity diagram constructed for an elliptical aperture whose major axis was oriented in the dip direction of the fault. The minor radius was constrained to be 200 ft for all depths considered, whereas the major radius was allowed to increase monotonically with depth, as before. Once again, the fault plane reflection appeared to be the most prominent event. Figure 7.41 shows the real part of the analytical signal. The plot shows the preservation of the

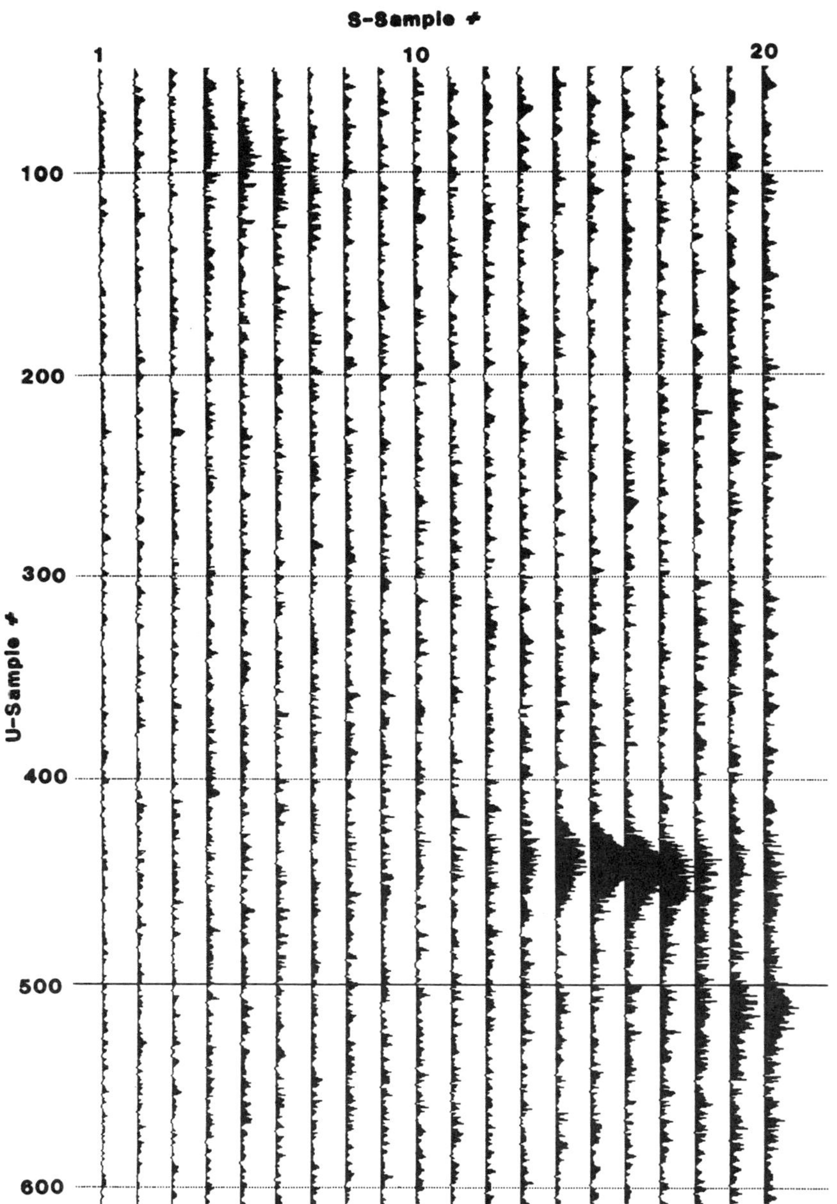

Figure 7.33. *Velocity diagram constructed with the noisy data. Both the envelope and the real part of the analytical signal are displayed.*

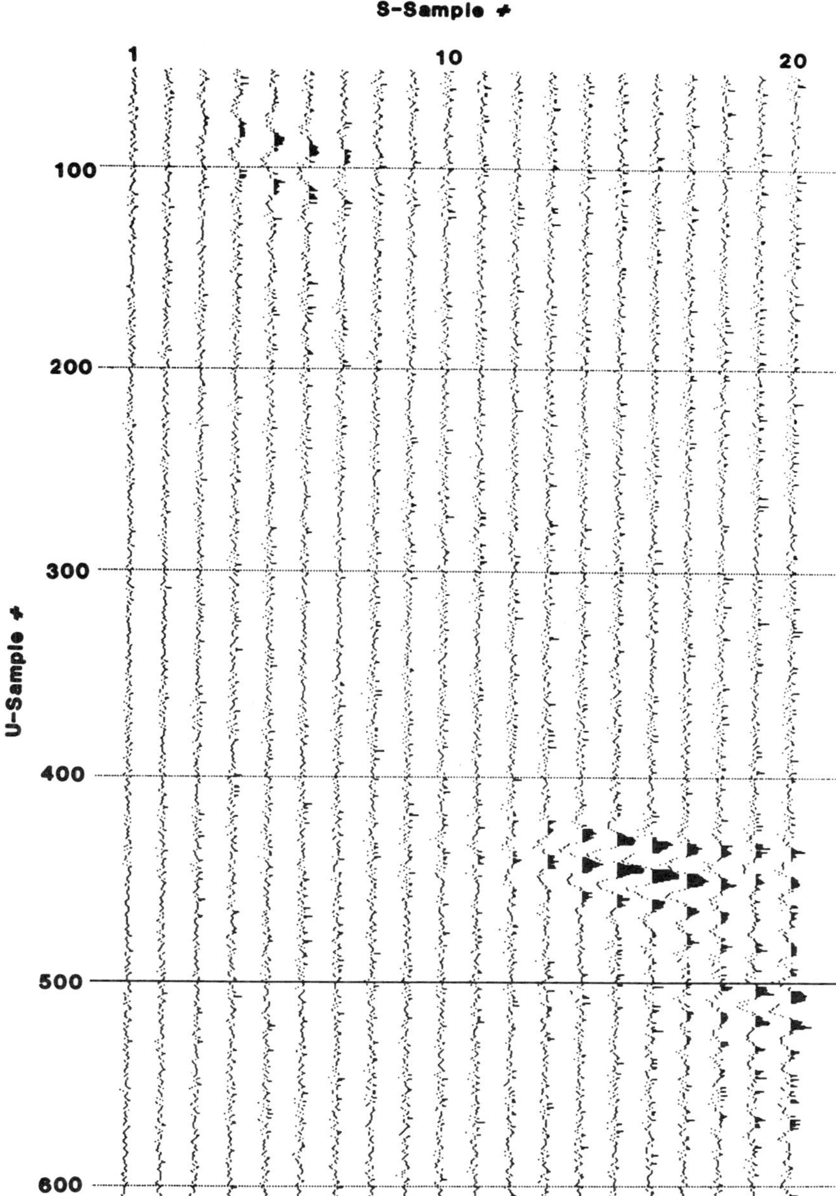

Figure 7.34. Velocity diagram showing real part of analytical signal for noisy data.

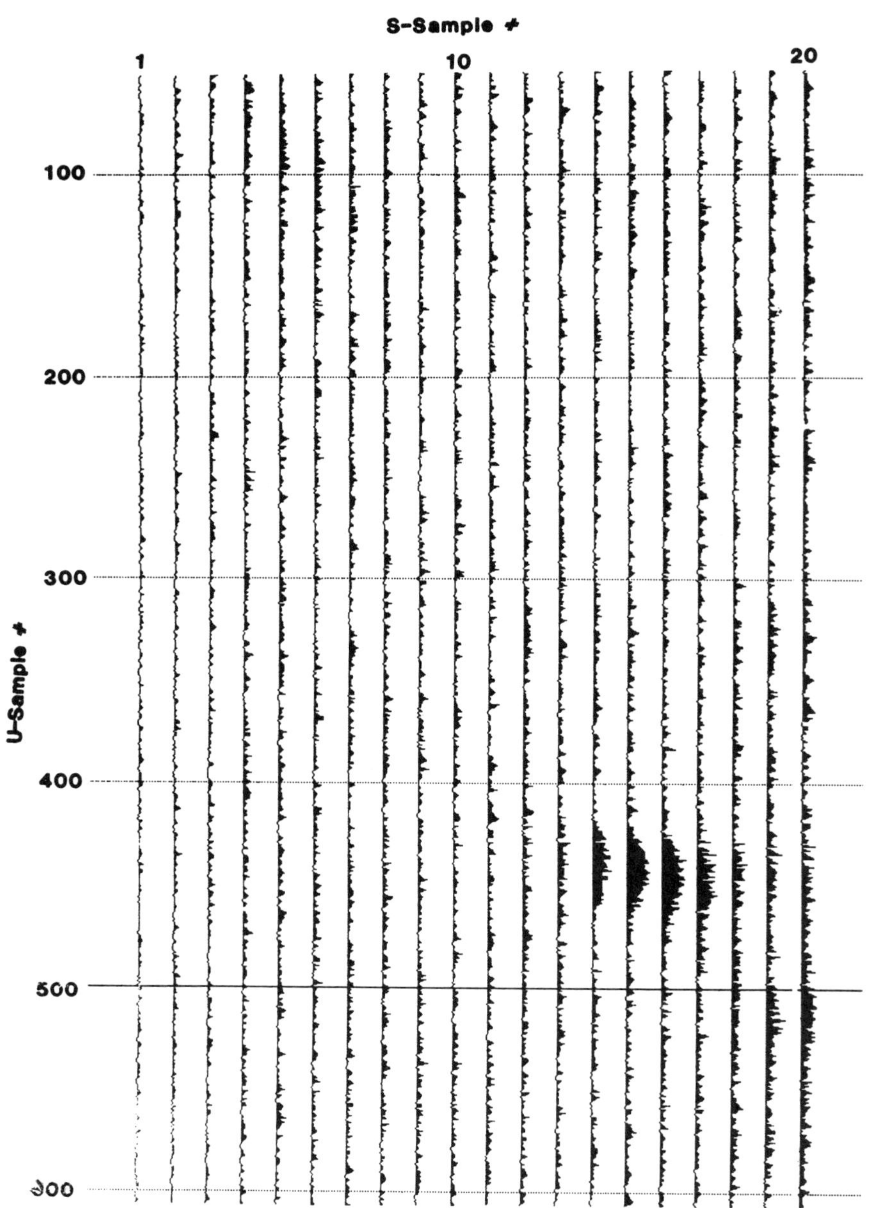

Figure 7.35. Velocity diagram constructed when the data used to construct Figure 7.33 are reduced to sign bit. The envelope and real parts are shown with and without filtering.

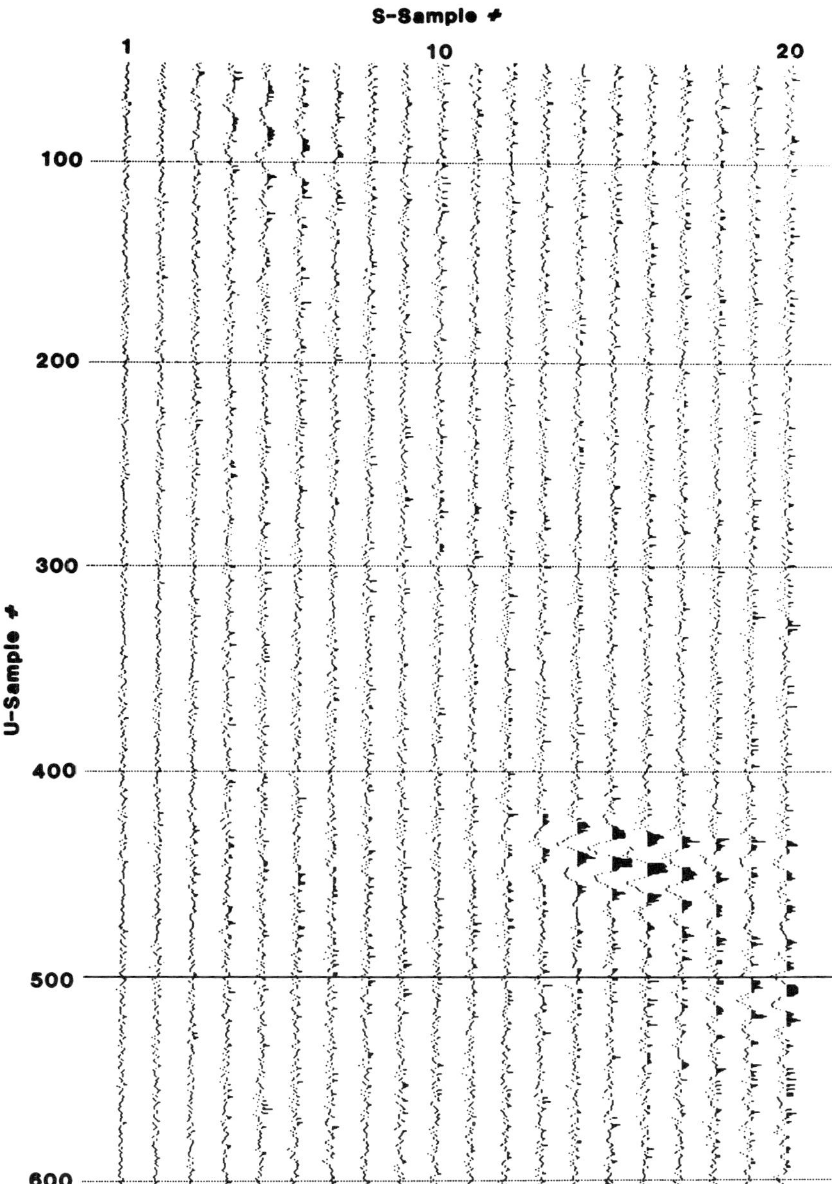

Figure 7.36. *Velocity diagram obtained using sign bit data. Real part of analytical signal is displayed.*

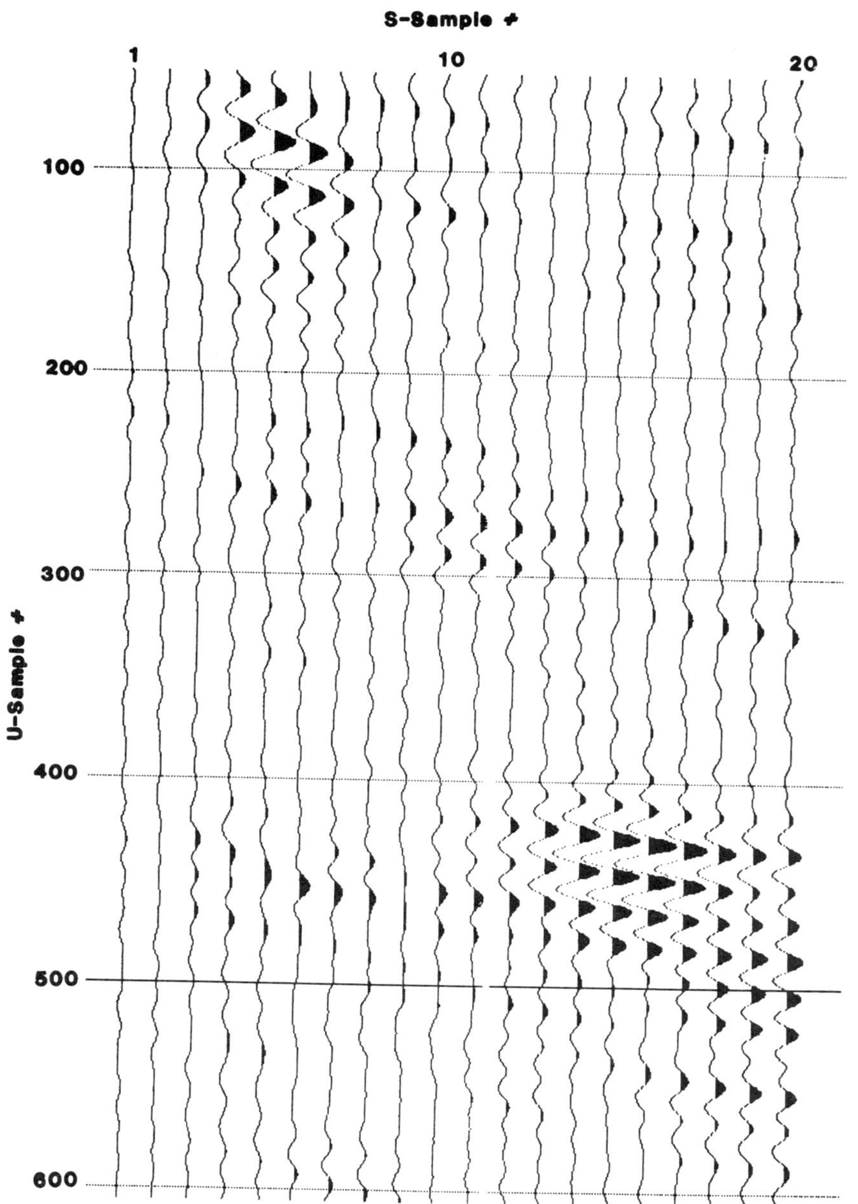

Figure 7.37. Filtered (sign bit) velocity diagram. The real part of the analytical signal is displayed.

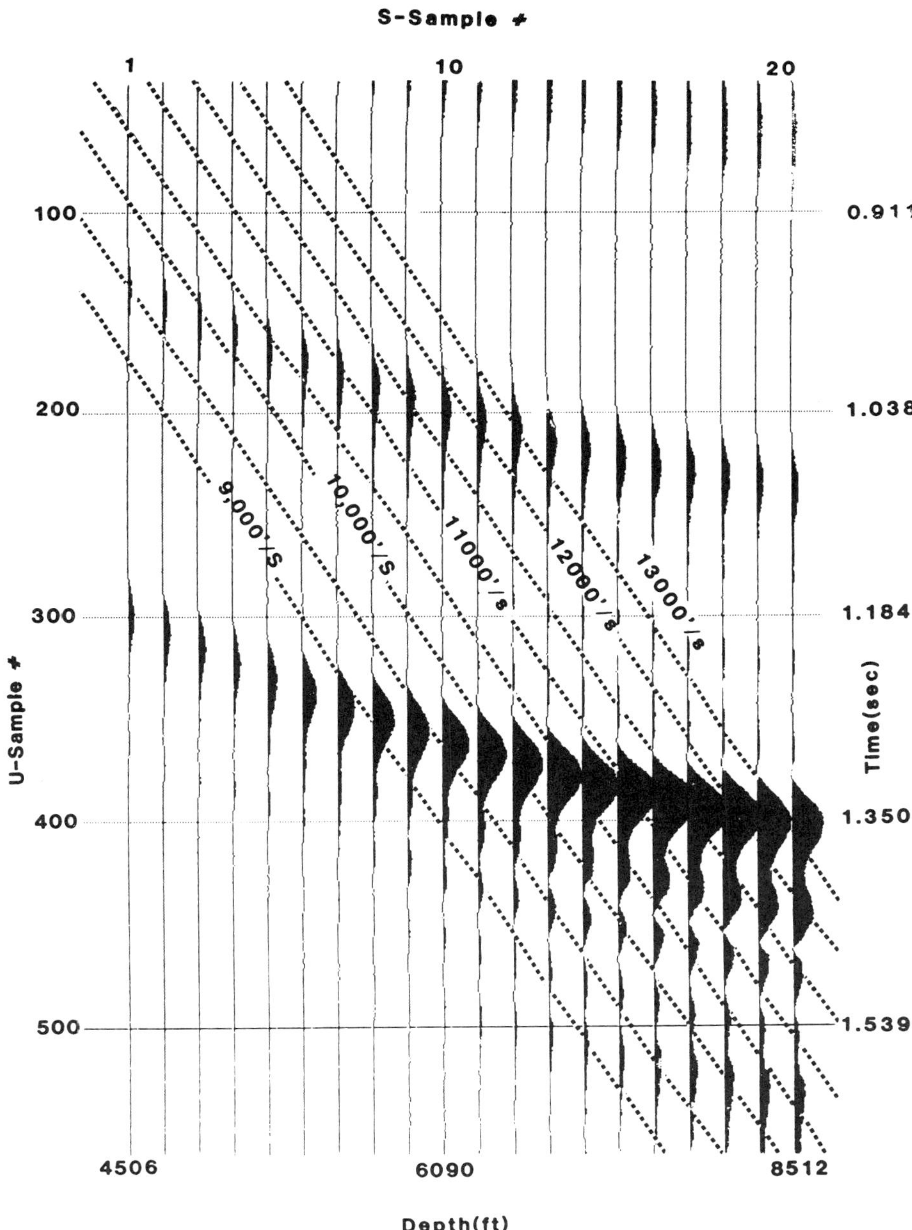

Figure 7.38. *Velocity diagram constructed for well B, over the fault, using a circular aperture of constant radius 200 ft with depth.*

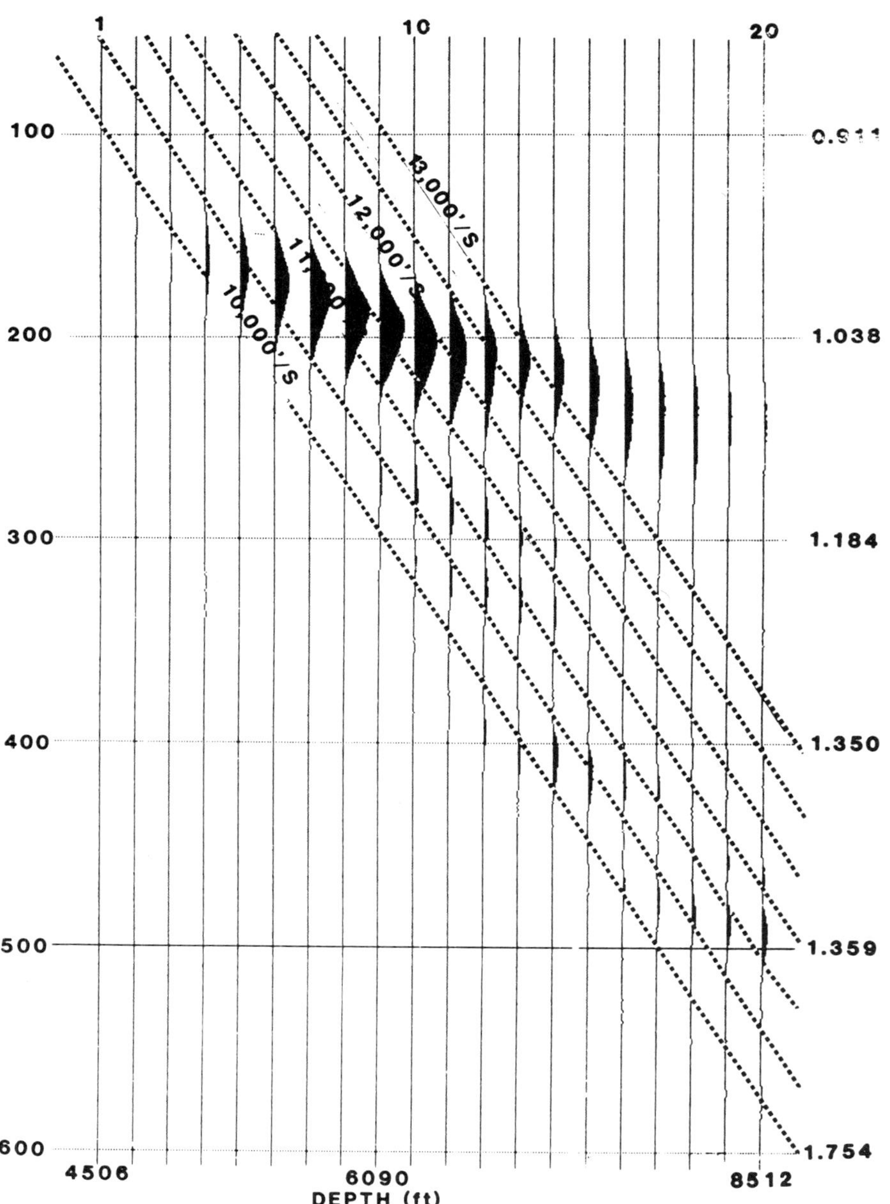

Figure 7.39. *Velocity diagram constructed for well B, using a circular aperture that increases monotonically with depth.*

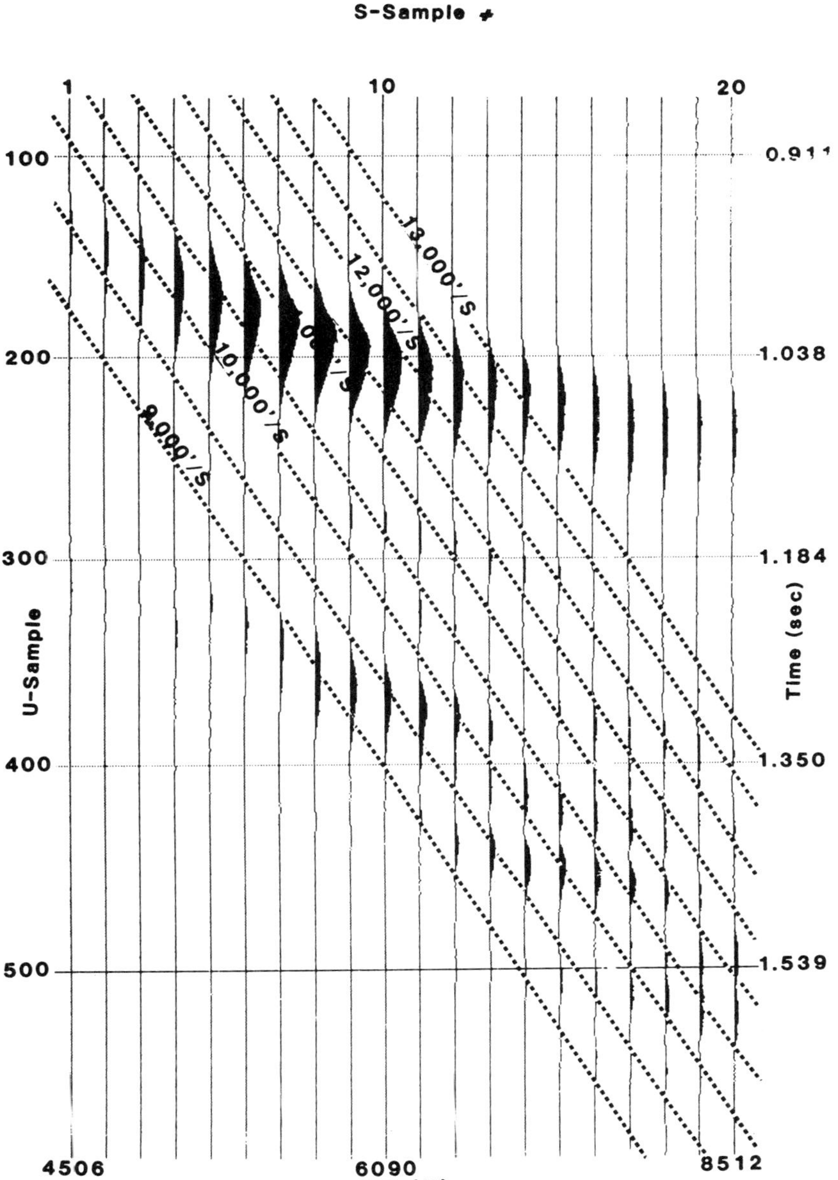

Figure 7.40. *Velocity diagram constructed for well B, using an elliptical aperture whose major axis increased monotonically with depth and was oriented in the dip direction.*

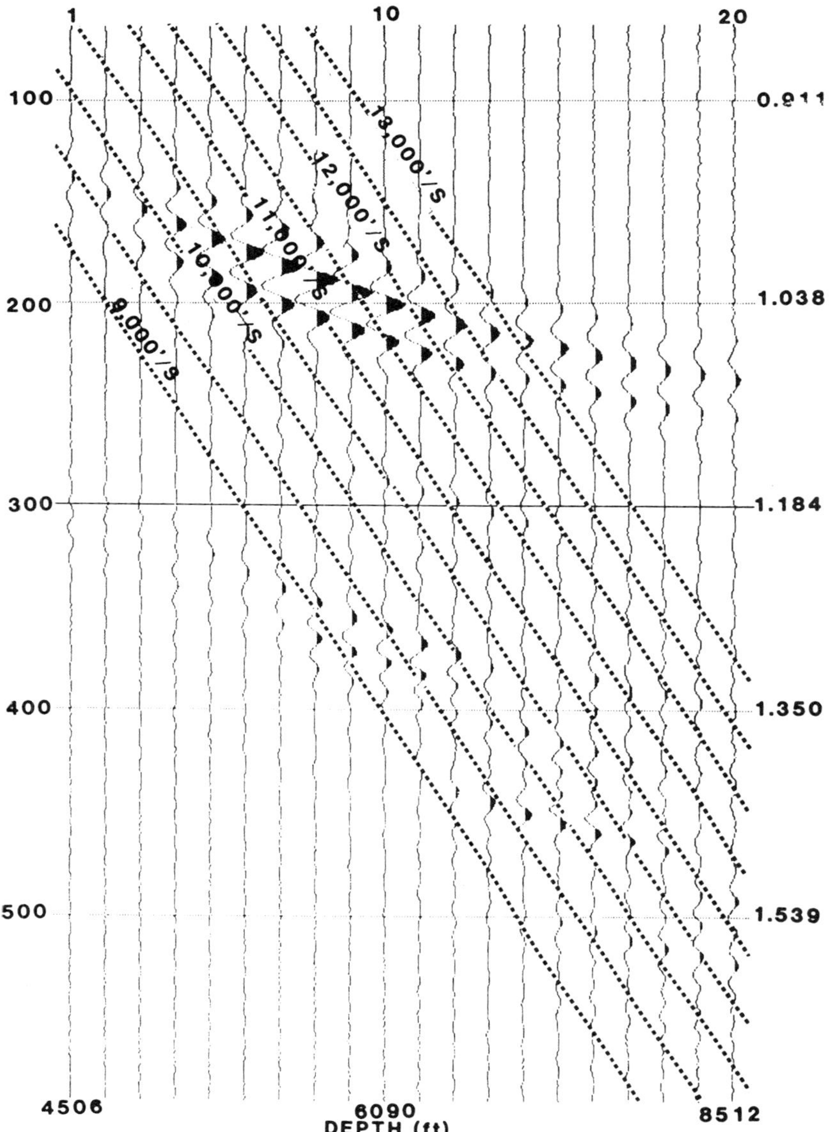

Figure 7.41. *Velocity diagram showing the real part of the analytical signal. The major axis of the elliptical aperture is oriented in the dip direction.*

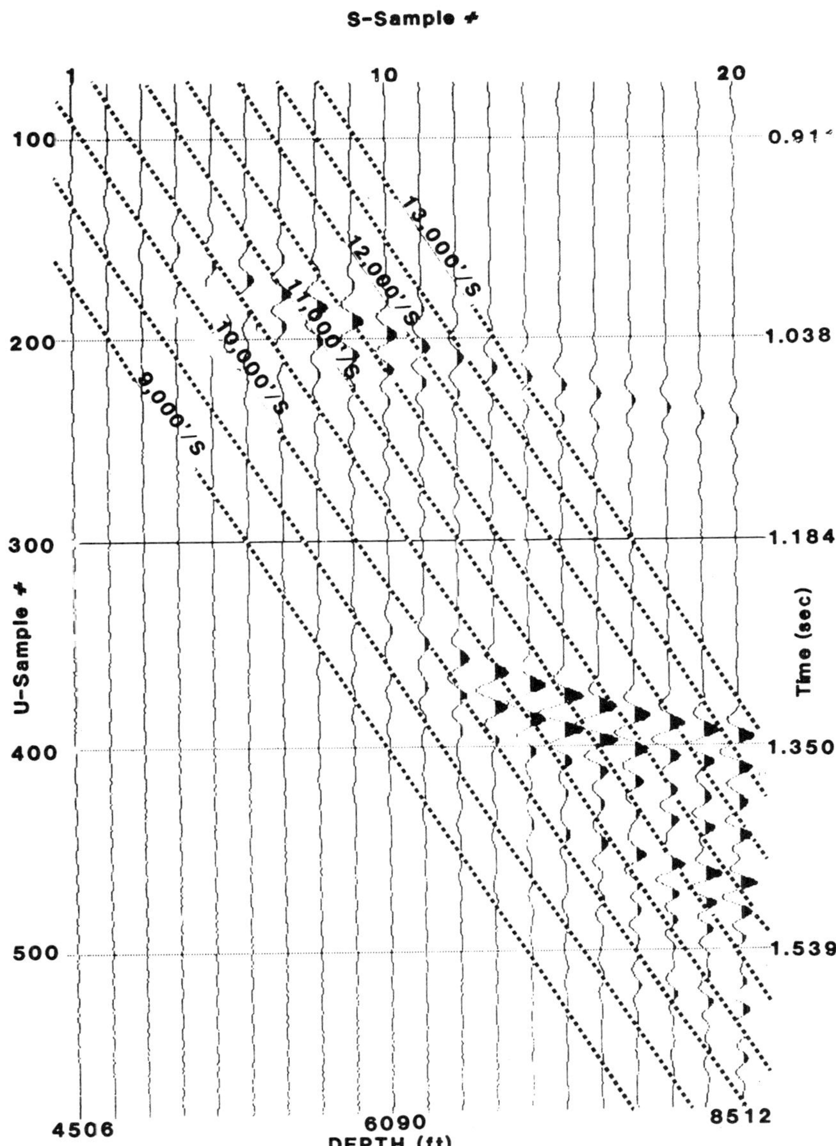

Figure 7.42. *Velocity diagram showing the real part of the analytical signal, constructed using an elliptical aperture, with the major axis oriented in the strike direction.*

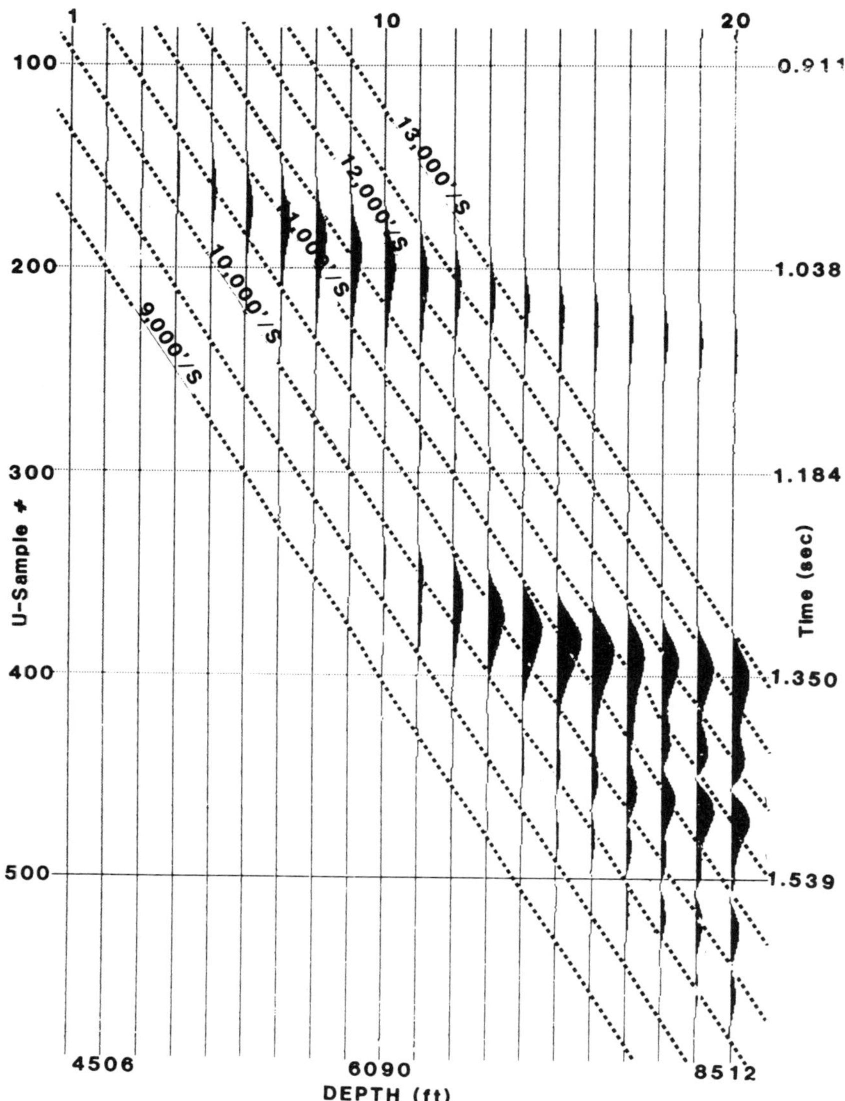

Figure 7.43. *Velocity diagram constructed using an elliptical aperture, with the major axis oriented in the strike direction, showing the envelope function of the analytic signal.*

typical wavelet in the signal. Using Figures 7.40 and 7.41, the largest peak was selected between *S* samples 8 and 9 at a *U* sample of about 192. The corresponding velocity pick was about 11,300 ft/s for an apparent depth of about 5793 ft and a traveltime of about 1.025 s. Four more possibly identifiable events appear on the diagram, but with small amplitudes due to possible destructive interference during the migration.

Figure 7.42 shows the velocity diagram constructed for an elliptical aperture whose major axis was oriented in the strike direction of the fault. Again, the fault plane reflection appears well imaged. This time, however, it is not the most prominent event. The underlying events have a great deal of gain to their energy content. Figure 7.43 shows the real part of the analytical signal. Measuring peak to trough on the well-preserved wavelet, and combining this with the envelope plots, the largest peak was selected to be between S samples 8 and 9 at a U sample of about 188. The corresponding velocity pick was about 11,334 ft/s at an apparent depth of 5792 ft and a traveltime of 1.022 s.

The velocities determined for the fault reflection are about the same for the elliptical apertures oriented in the strike and the dip directions of the fault and also for the depth-increasing circular aperture. They averaged about 11,300 ft/s. Although they differed by some finite amounts, the differences can be attributed to the picking of the events. A close-up of the velocity diagram for a small window around the pick would very likely correct the small differences.

CONCLUSION. The algorithm outlined in this chapter can be a very fast algorithm, which, with the large amount of data used, can yield very stable velocity estimates. Because it is based on three-dimensional migration before stacking, the shooting geometry is unimportant; the only information needed is the shot and receiver coordinates. Its advantage over other velocity analysis algorithms is that it can be used to obtain velocity estimates over any structure over which conventional velocity analysis programs cannot provide useful results (Blackburn, 1980). Thus, the velocity information obtained is structure-independent.

The algorithm has also been shown to work for sign-bit-only data, preserving the wavelet of the signal. This is particularly advantageous because it reduces computer storage requirements, cuts down the input-output computer overhead, and works well with the noisy data typical of land surveys.

REFERENCES

Blackburn, G., 1980, Errors in stacking velocity—True velocity conversion over complex geologic situations: Geophysics, v. 45, p. 1465–1488.

Cochran, M. D., 1973, Seismic signal detection using sign bits: Geophysics, v. 38, p. 1042–1052.

Dunkin, J. W., and Levin, F. K., 1971, Isochrons for a three-dimensional seismic system: Geophysics, v. 36, p. 1099–1137.

Gardner, G. H. F., French, W. S., and Matzuk, T., 1974, Elements of migration and velocity analysis: Geophysics, v. 39, p. 811–825.

Kuhn, M. J., and Alhilahi, K. A., 1977, Weighting factors in the construction and reconstruction of acoustical wavefields: Geophysics, v. 42, p. 1183–1198.

Levin, F., 1971, Apparent velocity from dipping interface reflections: Geophysics, v. 36, p. 510–516.

O'Brien, J. T., Kamp, W. P., and Hoover, G. M., 1979, Theory of amplitude recovery from sign-bit recorded data with applications to seismic problems—I & II: Presented at the 49th Annual Meeting of the SEG, New Orleans.

Owusu, J. K., and Gardner, G. H. F., 1980, 2D & 3D velocity estimation problems: Seismic Acoustics Laboratory Annual Progress Review, v. 6, M1–M22.

Sattlegger, J., 1975, Migration velocity determination—Part I, Philosophy: Geophysics, v. 40, p. 1–5.

———, 1980, Common offset plane migration: Geophys. Prosp., v. 28, p. 859–871.

Slotnick, M. M., 1959, Lessons in seismic computing: Tulsa, SEG.

Taner, M. T., and Koehler, F., 1969, Velocity spectra—Digital computer derivation and applications of velocity functions: Geophysics, v. 34, p. 859–881.

8. THREE-DIMENSIONAL MIGRATION VELOCITY ANALYSIS IN THE SPACE-FREQUENCY DOMAIN

Thomas R. Morgan and Fred J. Hilterman

INTRODUCTION. The ultimate objective of wavefield extrapolation, as it is used in this chapter, is to form images or reconstructions of either source image points or common reflection points. The images serve in two basic capacities: first, their locations yield information on strike and dip parameters for the reflector they represent; second, their coherence can be related to the correct media velocity down to the reflector. The truth of this last point will be discussed in detail shortly.

The purpose of the experiments presented here will be both to characterize the reconstruction algorithms under simple, controlled conditions and to simulate common situations that might be encountered in the field. This series of tests in no way exhausts the possibilities to be encountered in practice; rather, it is aimed at understanding some of the basic principles involved. Both theoretical and physical model data were employed.

IMAGE INTENSITY. Several methods can be used to determine relative coherences among a sequence of images of the same object. Here, the sequence will be created for a range of reconstruction velocities at a particular depth, $\bar{z}$, or normal incident traveltime, T_0, and a simple measure will be used to compare the absolute magnitudes. Other statistical measures could also be used, such as standard deviation and variance, but they will not be presented here. It will be shown that, for a single-point object, a reconstruction velocity equal to the correct media velocity will produce the image of highest magnitude or intensity at the location of that object. For data containing many objects, a particular object will then produce an image of at least relative maximum intensity.

The following development will apply strictly for images formed using the Rayleigh-Sommerfeld formula as the phase-propagation term. It will be true for the holographic method only when the reference position is the same as the desired image location, a fact that will be demonstrated in the first round of synthetic results and used to advantage in later experiments.

We begin by recalling the extrapolation equation in the frequency domain:

$$B(\bar{x},\bar{y},\bar{z};\omega) = \frac{-1}{2\pi} \iint_{\Sigma} \left(\frac{j\omega}{c} + \frac{1}{R}\right) B(x,y,z;\omega) \, \frac{e^{j\omega R/c}}{R} \cos\theta \, dx dy, \tag{8.1}$$

where

$$R = [(\bar{x} - x)^2 + (\bar{y} - y)^2 + (\bar{z} - z)^2]^{1/2},$$

$$\cos\theta = (\bar{z} - z)/R.$$

In general, $B(\bar{x}, \bar{y}, \bar{z}; \omega)$ is a complex number; that is, it has the form

$$B(\bar{x},\bar{y},\bar{z};\omega) = B(\mathrm{Re}) + jB(\mathrm{Im}). \tag{8.2}$$

This chapter was originally presented as a paper at the SEG convention in Los Angeles, October 1981.

The image intensity M can be defined as the complex spectral magnitude of $B(\bar{x}, \bar{y}, \bar{z}; \omega)$:

$$M \equiv [B(\mathrm{Re})]^2 + [B(\mathrm{Im})]^2. \tag{8.3}$$

For the recorded data from a single point scatterer, we have

$$B(x, y, z; \omega) = A(x, y, z) e^{-j\omega R_0/c_0}. \tag{8.4}$$

The amplitude term can be removed from the data, and equation (8.1) can be rewritten in terms of a phase-only reconstruction process:

$$B(\bar{x}, \bar{y}, \bar{z}; \omega) = \int\int_{\Sigma} \exp(-j\omega R_0/c_0) \exp(j\omega R/c) dx dy. \tag{8.5}$$

Let

$$\phi = -\omega \left(\frac{R_0}{c_0} - \frac{R}{c} \right), \tag{8.6}$$

and rewrite equation (8.5) in terms of sines and cosines:

$$B(\bar{x}, \bar{y}, \bar{z}; \omega) = \int\int_{\Sigma} \cos \phi \, dx dy - j \int\int_{\Sigma} \sin \phi \, dx dy. \tag{8.7}$$

Substitution of equation (8.7) into equation (8.3) gives

$$M = \left[\int\int_{\Sigma} \cos \phi \, dx dy \right]^2 + \left[\int\int_{\Sigma} \sin \phi \, dx dy \right]^2. \tag{8.8}$$

The first and second derivative tests will be used to evaluate equation (8.8) for maxima and minima. The derivatives will be taken with respect to c, $\bar{x}$, $\bar{y}$ and $\bar{z}$. For clarification, R and R_0 are restated as follows:

$$\begin{aligned} R &= [(\bar{x} - x)^2 + (\bar{y} - y)^2 + (\bar{z} - z)^2]^{1/2}, \\ R_0 &= [(x_0 - x)^2 + (y_0 - y)^2 + (z_0 - z)^2]^{1/2}. \end{aligned} \tag{8.9}$$

The first derivatives of M with respect to c and $\bar{x}$ are as follows (expressions for $\bar{y}$ and $\bar{z}$ are analogous to those for $\bar{x}$ and are not included):

$$\frac{dM}{dc} = \frac{2\omega}{c^2} \left[\int\int \sin \phi \, dx dy \int\int R \cos \phi \, dx dy \right.$$

$$\left. - \int\int \cos \phi \, dx dy \int\int R \sin \phi \, dx dy \right],$$

$$\frac{dM}{dx} = \frac{2\omega}{c} \left[\int\int \sin \phi \, dx dy \int\int \frac{x_0 - x}{R} \cos \phi \, dx dy \right.$$

$$- \int\int \cos \phi \, dx dy \int\int \frac{x_0 - x}{R} \sin \phi \, dx dy \Bigg] . \qquad (8.10)$$

Equations (8.10) are of the same form for all four parameters and have similar sets of solutions. A particular solution is $\phi = 0$ for all x, y; that is $\bar{x} = x_0$, $\bar{y} = y_0$, $\bar{z} = z_0$, $c = c_0$. Second derivatives evaluated at this point are listed next (again, expressions with respect to $\bar{y}$ and $\bar{z}$ are not given):

$$\frac{d^2M}{dc^2}\Bigg|_{\substack{c=c_0\\R=R_0}} = \frac{2\omega^2}{c^4}\left\{\left[\int\int R \, dx dy\right]^2 - \int\int dx dy \int\int R^2 \, dx dy\right\} \leq 0,$$

$$\frac{d^2M}{d\bar{x}^2}\Bigg|_{\substack{c=c_0\\R=R_0}} = \frac{2\omega^2}{c^2}\left\{\left[\int\int \frac{x_0 - x}{R} \, dx dy\right]^2\right.$$

$$\left. - \int\int dx dy \int\int \left(\frac{x_0 - x}{R}\right)^2 dx dy\right\} \leq 0. \qquad (8.11)$$

The comparison with respect to zero is a result of Schwarz's inequality (see, for example, Abramowitz and Stegun, 1968). Since all the second derivatives are never concave upward, the point $c = c_0$, $R = R_0$ is seen to be a maximum.

In order to test the extrapolation principles developed previously, a number of theoretical data sets are calculated and used as input for the Rayleigh-Sommerfeld and lensless Fourier transform algorithms (hereafter referred to as the convolutional and holographic methods). Point objects were employed to represent either source images or common reflection points. Only phase relations were considered, so no amplitude terms were included. Theoretical data frequency slices were calculated directly, thus avoiding the time-consuming task of Fourier-transforming time traces and reordering into frequency slices.

CONVOLUTIONAL METHODS: CSP DATA SETS. This section presents the results of the imaging tests made using the Kirchhoff integral cast in the temporal frequency domain, referred to as the convolutional method. All data arrays are 32 by 32, for a total of 1024 complex-valued input points in each frequency slice. The processing sequence involves a 2-D spatial convolution of a particular data set, a frequency slice, with an operator array based on a set of input parameters. The inputs are temporal frequency, velocity, source-to-source image path length, and the x,y location of the aperture center. The operator is calculated on a 63-by-63 grid as the input. Unless specified otherwise, the data arrays are sampled on 100-ft spacings in both the x and y directions, with the (0, 0) position in the lower left-hand corner, (1600, 1600) as the center, and (3100, 3100) in the upper right-hand corner. The convolution was designed and implemented so that the output image array directly overlays the lateral coordinates of the data arrays.

To build a velocity analysis output plot, a series of reconstructions was done of the object image point (which the theoretical test serves as either a common-source image point (CSP) or a common depth point (CDP) for flat reflectors) over a range of velocities. Except for a few initial monofrequency tests, a number of temporal frequencies were reconstructed at each velocity. These results were

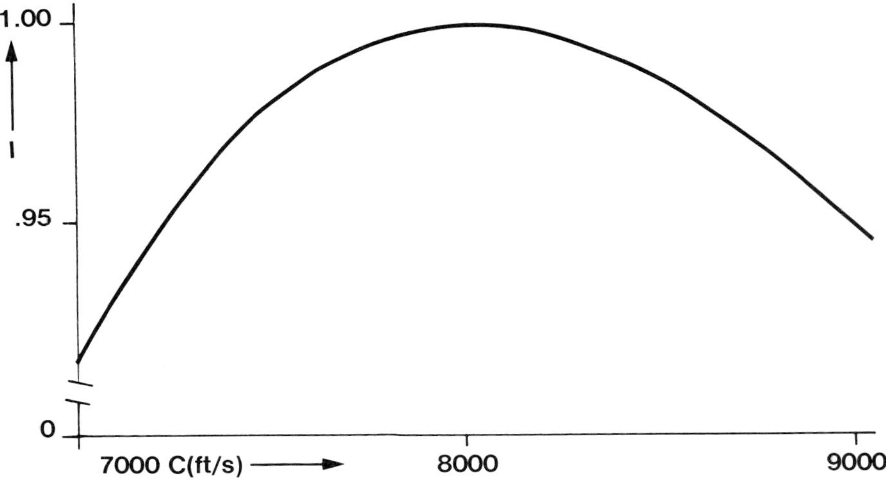

***Figure 8.1.** Convolutional reconstruction of single-object theoretical model. Data parameters: $x = 1600$ ft. $y = 1600$ ft, $z = 4000$ ft, $C = 8000$ ft/s, $f = 20$ Hz. Reconstruction parameters: $z = 4000$ ft, $f = 20$ Hz, $C = 7000$ ft/s to 9000 ft/s by 200 ft/s intervals.*

summed, real and imaginary parts separately, to produce a final image at one velocity. The entire image array, after calculating the complex spectral magnitude, was searched for the single maximum-intensity value. This value was plotted for each velocity in the scan on the top portion of the exhibits. The x and y coordinates of the maximum were plotted immediately below, except for tests in which no variation took place along one or both of the lateral axes.

THEORETICAL DATA SETS: SINGLE-LAYER CSP MODELS. The first examples are monochromatic reconstructions of an isolated point source. Since no other objects are present, it is possible to examine the results of the reconstructions over a range of velocities in an uninhibited manner. In the first two plots (Fig. 8.1 and 8.2), the object was placed directly beneath the middle of the recording aperture. A smoothly varying relationship between image intensity and reconstruction velocity resulted, with maximum intensity occurring for the correct media velocities (8000 ft/s and 7875 ft/s). This confirms the theory of the previous section, showing that the image intensity is, indeed, a maximum when the reconstruction parameters match the true parameters of the object. It should also be noted that the curves are asymmetric, with image intensity falling off less rapidly in one direction than it does in the other. This is because the fringe spacing on the operator does not change as rapidly for higher velocities as it does for equal lower-velocity increments. Therefore, the correlation at $C_0 + \Delta C$ is better than for $C_0 - \Delta C$.

Figures 8.3 and 8.4 show similar reconstructions for objects with the same parameters as those of Figures 8.1 and 8.2, except that they are located off the center of the array. This sequence was made to examine the effects resulting from arbitrary object locations and as a comparison to the physical model data sets, which were restricted to object points near the array border. The image intensities are again maximized at the correct media velocities, but the curves are much more irregular.

COMPARISON WITH CMP DATA SETS. In the common midpoint (CMP) method for each set of reconstruction parameters—strike, dip, velocity, normal incident

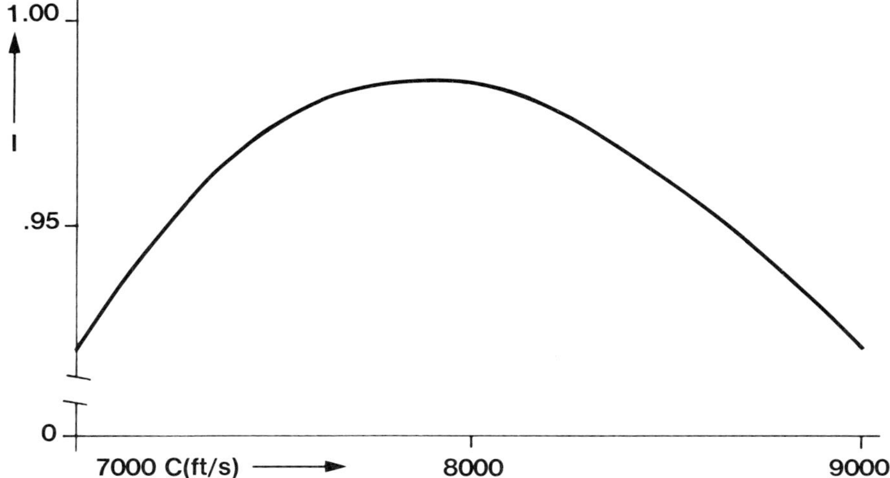

Figure 8.2. *Convolutional reconstruction of single-object theoretical model. Data parameters: x = 1600 ft, y = 1600 ft, z = 4000 ft, C = 7875 ft/s, f = 20 Hz. Reconstruction parameters: z = 4000 ft, f = 20 Hz, C = 7000 ft/s to 9000 ft/s by 200 ft/s intervals.*

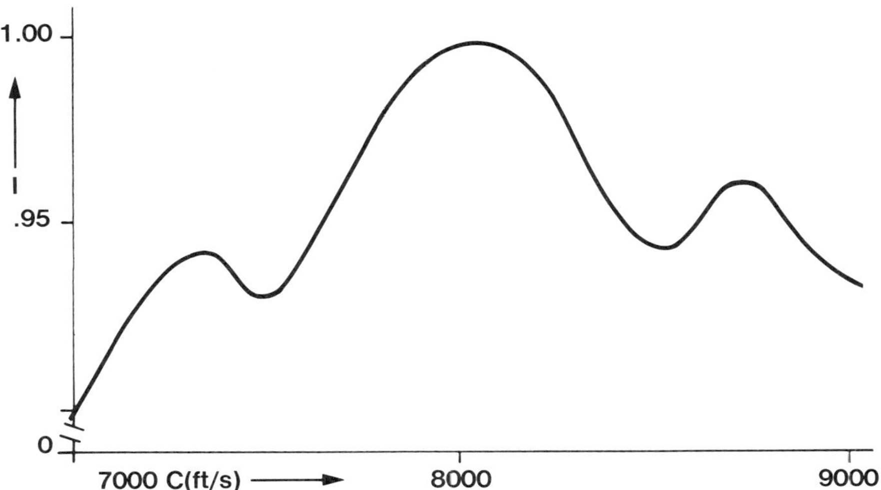

Figure 8.3. *Convolutional reconstruction of single-object theoretical model. Data: x = 2500 ft, y = 2500 ft, C = 8000 ft/s, f = 20 Hz. Reconstruction parameters: z = 4000 ft, f = 20 Hz, C = 7000 ft/s to 9000 ft/s by 200 ft/s intervals.*

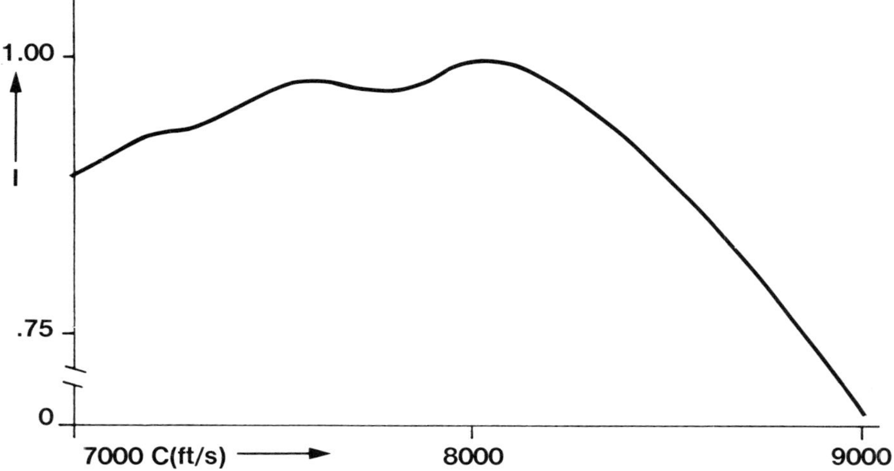

Figure 8.4. *Convolutional reconstruction of single-object theoretical model. Data: $\chi = 3200$ ft, $y = 1600$ ft, $z = 4000$ ft, $C = 8000$ ft/s, $f = 20$ Hz. Reconstruction parameters: $z = 4000$ ft, $t = 20$ Hz, $C = 7000$ ft/s to 9000 ft/s by 200 ft/s intervals.*

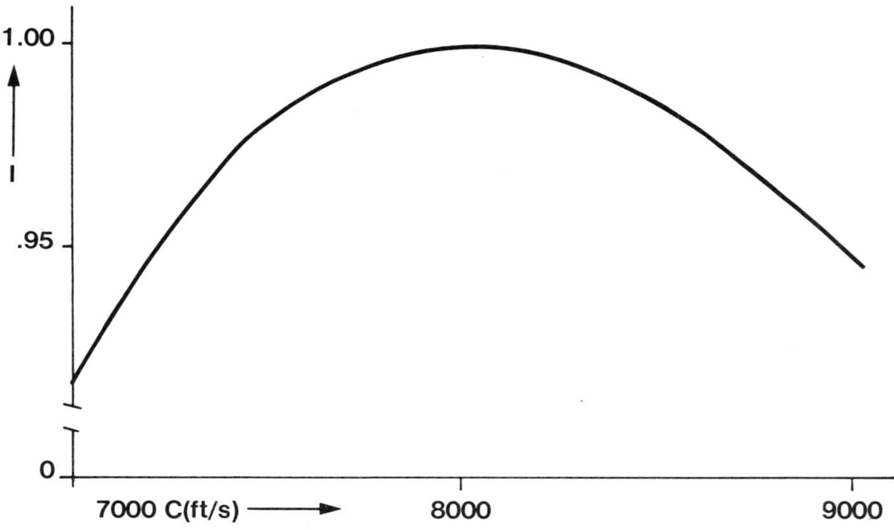

Figure 8.5. *Convolutional reconstruction of single-object theoretical model using CMP operator. Data: dip = 0°, strike = 0°, $z = 4000$ ft., $C = 8000$ ft/s, $f = 20$ Hz. Reconstruction: same as data, with $C = 7000$ ft/s to 9000 ft/s by 200 ft/s intervals.*

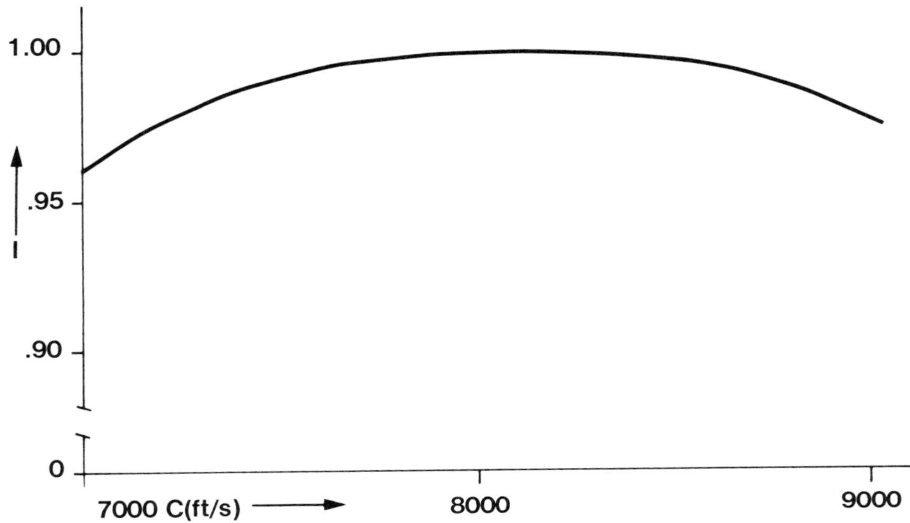

Figure 8.6. Convolutional reconstruction of single-object theoretical model using CMP operator. Data: dip = 33°, strike = 45°, z = 2371 ft, C = 8000 ft/s, f = 20 Hz. Reconstruction: same as data, with same velocity scan as Figure 8.5.

path length, and temporal frequency—a phase operator is calculated for a 32-by-32 grid. The input data array, a single 32-by-32 temporal frequency slice (but of traces that are CMP-oriented rather than CSP-oriented) is complex-multiplied by the operator, and an array summation is done on the output to obtain an image intensity value at the particular set of reconstruction values. A single parameter (usually velocity) is then varied while the others are held fixed.

The results of two velocity scans are presented in Figures 8.5 and 8.6. The input models were selected to duplicate the CSP models of Figures 8.1 and 8.3. The reconstructions were made using correct dip, strike, and normal incidence path length, and scanned over velocity. Figure 8.5 is a duplicate of Figure 8.1, as expected from theoretical considerations for plane horizontal reflectors. Differences can be noticed between Figures 8.6 and 8.3, however. The CMP intensity curve shows none of the irregularity of the CSP curve and, as a result, has a broader central peak area. These differences may be attributed to the fact that, in the CMP method, the time data are always centered about the common midpoint, and the midpoint was chosen to be the center of the array; whereas, in the CSP method, the data is centered vertically above the source image point, which is off the aperture center for dipping beds. In Figure 8.7 the test shown in Figure 8.6 is repeated, with the CMP surface location off center at $x = 2500$ ft, $y = 2500$ ft. This now yields a data array similar to that of Figure 8.3, except for the inherent differences in fringe spacing and shape as shown in Figure 8.8. Again, the CMP velocity scan gives a smooth concave downward curve.

The appearance of inflection points and multiple relative maxima in the CSP velocity scan output, as outlined earlier, can be traced to the action of the spatial convolution itself. For simple data sets, as presented here, the convolution can be viewed as a matching of two circular ring patterns on a flat plane, with the correlations being attempted for a range of offsets. The shape of the intensity-versus-reconstruction velocity relationships in Figure 8.3 indicates that, at some large offsets (that is, large strike and dip errors), the correlation is better than at some smaller ones under certain circumstances—when the data array (the fixed

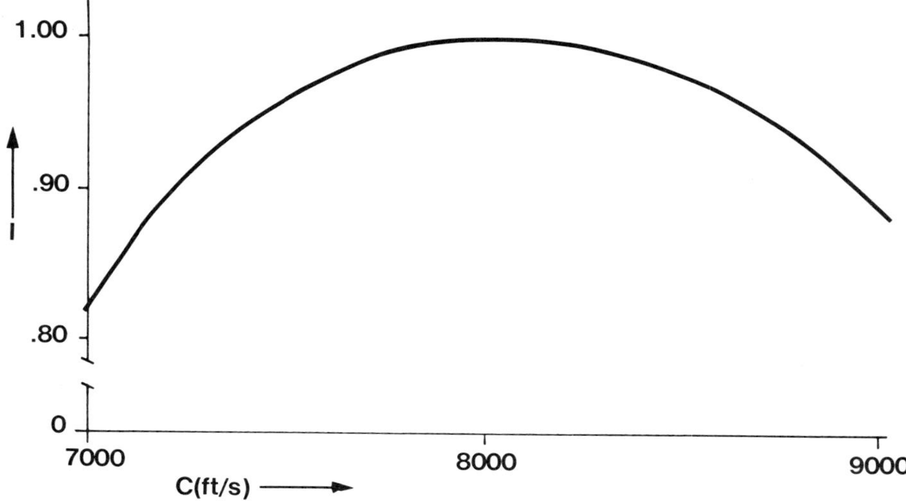

Figure 8.7. *Same model as Figure 8.6. Reconstruction with CMP located off center at $x = 2500$ ft, $y = 2500$ ft; all other parameters the same. (Note: Aperture center at $x = 1600$ ft, $y = 1600$ ft.)*

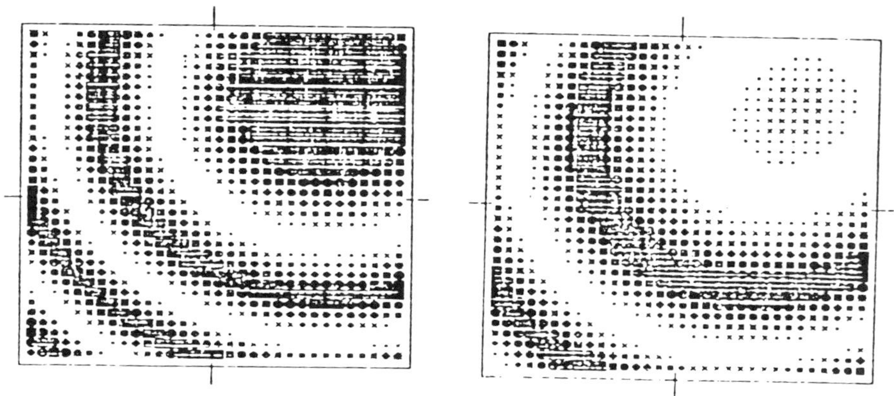

Figure 8.8. *CSP and CMP monochromatic data slices for models of Figures 8.3 and 8.6, respectively.*

circular fringe pattern) does not have its center of symmetry at the array center. The operator array (variable circular fringe pattern) always has its center of symmetry at the array center for each set of trial parameters and, hence, can have no direct bearing on whether the results appear as in Figures 8.1 or 8.3. Since the test using CMP data did not involve the equivalent of the multiple offsets between operator and data functions (that is, a search over strike and dip), similar effects did not appear in the output. In order to duplicate completely the dip and strike search of the convolutional method employed on the CSP data, 1024 strike and dip combinations would have to be tried at each velocity in the CMP approach. It is anticipated that similar relative maxima would then occur in the intensity-versus-reconstruction velocity curve. This would merely indicate that, again under some circumstances, erroneous strike-dip combinations would correlate better at some higher erroneous velocities than at lesser erroneous velocities. Such a test was not undertaken because of the prohibitive amount of computer time involved.

This does not mean, however, that 3-D parameter analysis on CMP data, as it is conducted here, is not practical. Normally, much fewer than 1024 strike-dip combinations are needed; as advocated previously, 11 strike and 11 dip positions, for a total of 121 strike-dip combinations, may be sufficient. The ultimate optimum choice is, naturally, highly data-dependent. The reason 1024 combinations can be used on the CSP-oriented data sets is simply the efficiency of the convolution-correlation process on modern array processor hardware and the need to calculate actually only one $2N$-1 square (in this case, 63 by 63) operator array at the outset. Even faster processes could be envisioned if only every other or every third output point were obtained for the convolution. This should not be attempted by spatially decimating the input data, however, because of aliasing considerations for steep dip.

THEORETICAL DATA SETS: MULTILAYER CSP MODELS. A multiobject data set containing four separate object points was used to test the convolutional algorithm under more realistic conditions and to check on depth resolution by frequency summation. The point objects were located at different depths and lateral positions and were computed using four different media velocities.

A sequence of reconstructions for the second object point (which is located directly under the center of the aperture), with a summation over eight frequencies from 10 Hz to 45 Hz, is displayed in Figures 8.9 and 8.10. Figure 8.9 shows the central image value—that is, the actual image intensity of the second object—as a function of velocity. It can be seen that the summation over frequency greatly improves velocity resolution. Figure 8.10 shows the maximum intensity image point found on the entire image plane and its x, y coordinates as functions of reconstruction velocity. For a given velocity, this maximum value may come from any of the four objects in the data set, but the maximum intensity for the whole scan is still associated with the second object, which is at the depth of interest. This is important because, in general practice, the correct lateral position of an image point at a particular depth is unknown, so the functions presented in Figure 8.10 are the output that must be evaluated for strike, dip, and velocity.

PHYSICAL MODEL RESULTS. The physical modeling system at the SAL was used to collect an areal CSP gather over an example of a simple geologic structure. Pertinent scaling factors are explained in Chapter 1.

A top view of the CSP data array is shown in Figure 8.11. For CSP data, the receiver scan is restricted to one side of the source, because the mechanical scanning arms of the modeling tank apparatus cannot cross one another. Fixing the receiver at the source position and scanning the source to get the second half of the array was not practical, because the coordinate systems of the two scanners could not be referenced closely enough to avoid producing a split image. As a

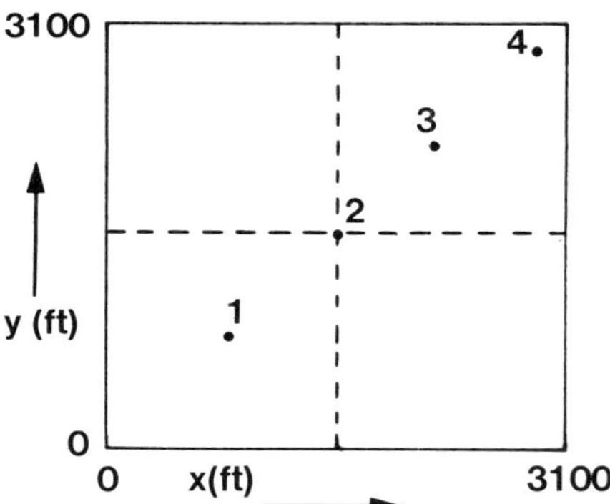

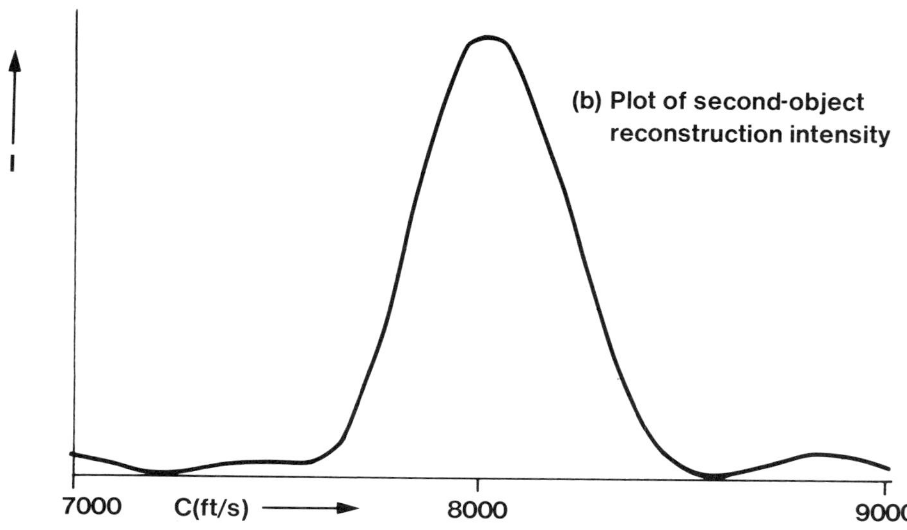

Figure 8.9. *Convolution reconstruction of four-object theoretical model. Data: (1) x = 900 ft, y = 900 ft, z = 3800 ft, C = 7750 ft/s; (2) x = 1600 ft, y = 1600 ft, z = 4000 ft, C = 8000 ft/s; (3) x = 2300 ft, y = 2300 ft, z = 4300 ft, C = 8250 ft/s; (4) x = 3000 ft, y = 3000 ft, z = 4400 ft, C = 8500 ft/s; f = 9.92 Hz to 35.55 Hz at 3.66 Hz intervals. Reconstruction: on second object at z = 4000 ft, with velocity scanned at 200 ft/s intervals.*

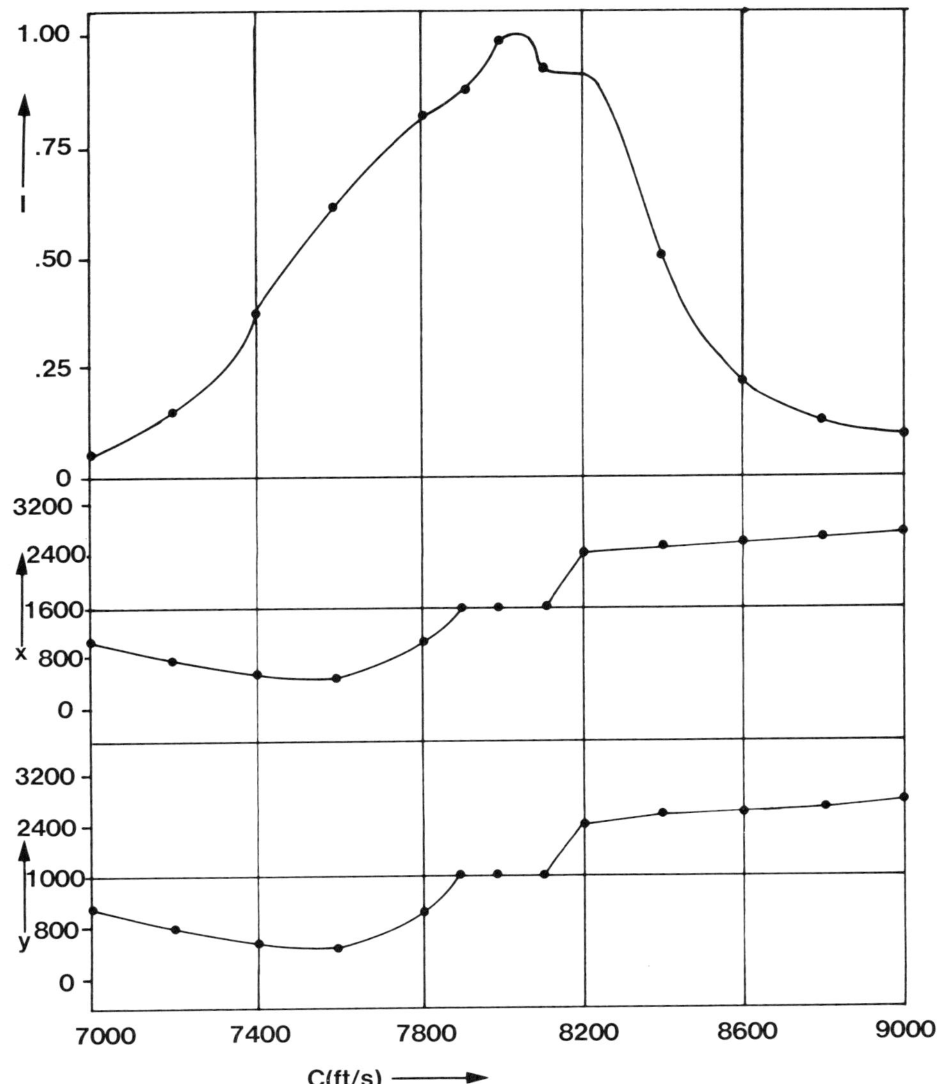

Figure 8.10. *Results from same reconstruction as in Figure 8.9, but with maximum intensity on output plane plotted. Note: These intensity values could result from any of the four objects. The x, y positions of maximum intensity value for each reconstruction velocity are plotted in the lower curves.*

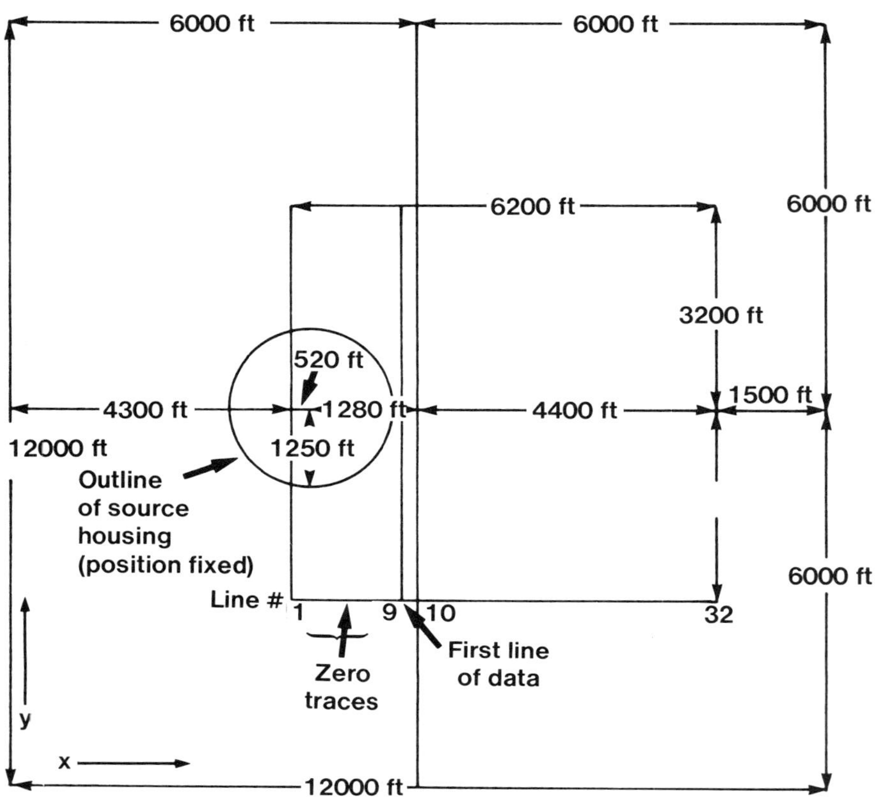

Figure 8.11. Top view of CSP physical model data scan. Source was fired and receiver was scanned starting with line 10. Lines 1 through 9 are all zero-valued traces. Complete data array is 32 lines of 32 traces each, with 200 ft line spacing and 200 ft trace spacing.

Table 8.1. Measured physical parameters for the physical model in Figure 8.12

Layer	Material	Lab bench velocity (ft/s)	Density (g/cm^3)	Acoustic impedance ($gft/s/cm^3$)
1	Water	4,970	1.00	4,970
2	RTV 184	3,385	1.04	3,744
3	Plexiglas	9,010	1.17	10,530

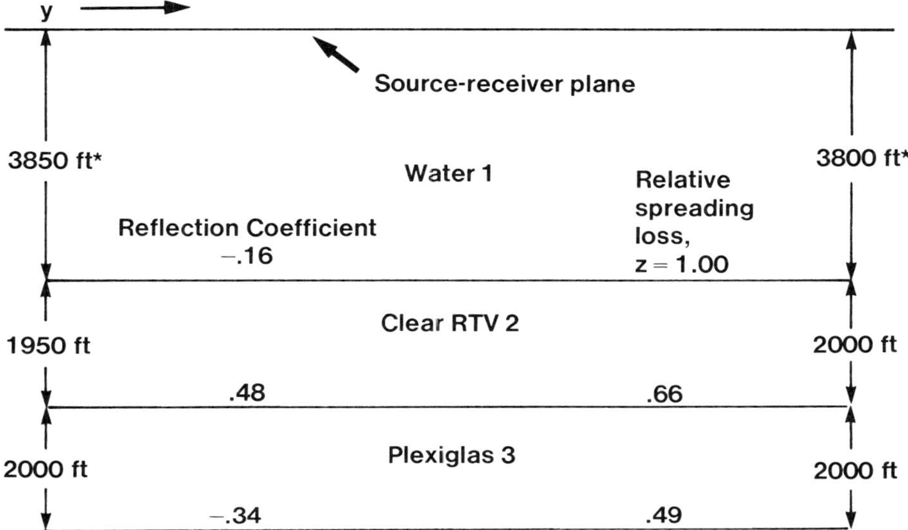

**Originally set at 4000 ft and adjusted from reconstruction results.*

Figure 8.12. *Side view of RTV-Plexiglas plane layer physical model.*

result, the fixed source position was restricted to the border of the receiver array.

The first model consisted of a two-inch-thick block of clear RTV silicone rubber above a two-inch-thick block of Plexiglas. A diagram of the physical model is given in Figure 8.12, and the physical parameters are summarized in Table 8.1; in those experiments, different scales from the normal were used. A time section of line 10, the first line of actual data traces, is shown in Figure 8.13. Events 1, 2, and 3 correspond to the water/RTV, RTV/Plexiglas, and Plexiglas/water interfaces, respectively. Events 4, 5, 6, and 7 have been identified as either direct-source receiver arrivals or trapped modes within the source-shaping apparatus. All seven events were picked and velocities were calculated using the $T^2 - x^2$ method. The results are listed in Table 8.2, and the expected values for the primary reflections, based on the physical parameters of Table 8.1, are presented in Table 8.3.

Reconstruction was then undertaken over a range of velocities for a series of depths. In order to provide vertical resolution, eight frequencies from 25 Hz to 42 Hz (200 kHz to 336 kHz) were summed at each velocity-depth combination (see Fig. 8.14). The final output, using the convolution algorithm described here with the CSP data as input, appears in Figure 8.15.

Two images have been formed at the depth corresponding to the top of the model. The lower-intensity image at 7500 ft/s (5000 ft/s physical) represents the true reflection. The higher-intensity image at 6500 to 6700 ft/s was determined to be a result of spatial aliasing. This conclusion resulted from the fact that the second image had the same lateral position as the first and that summation of 16 frequencies (see Fig. 8.16) over the same range suppressed the low-velocity image and shifted it away from the 7500 ft/s image. Eight frequencies were retained for subsequent runs, however, because of the large amount of additional computer time needed for 16 frequencies. The diagonal trend of the images in both the depth plot and the T_0 plot of Figure 8.17 was judged to be due to the ringing present in the reflection event (see the time section, Fig. 8.13). It can also be seen,

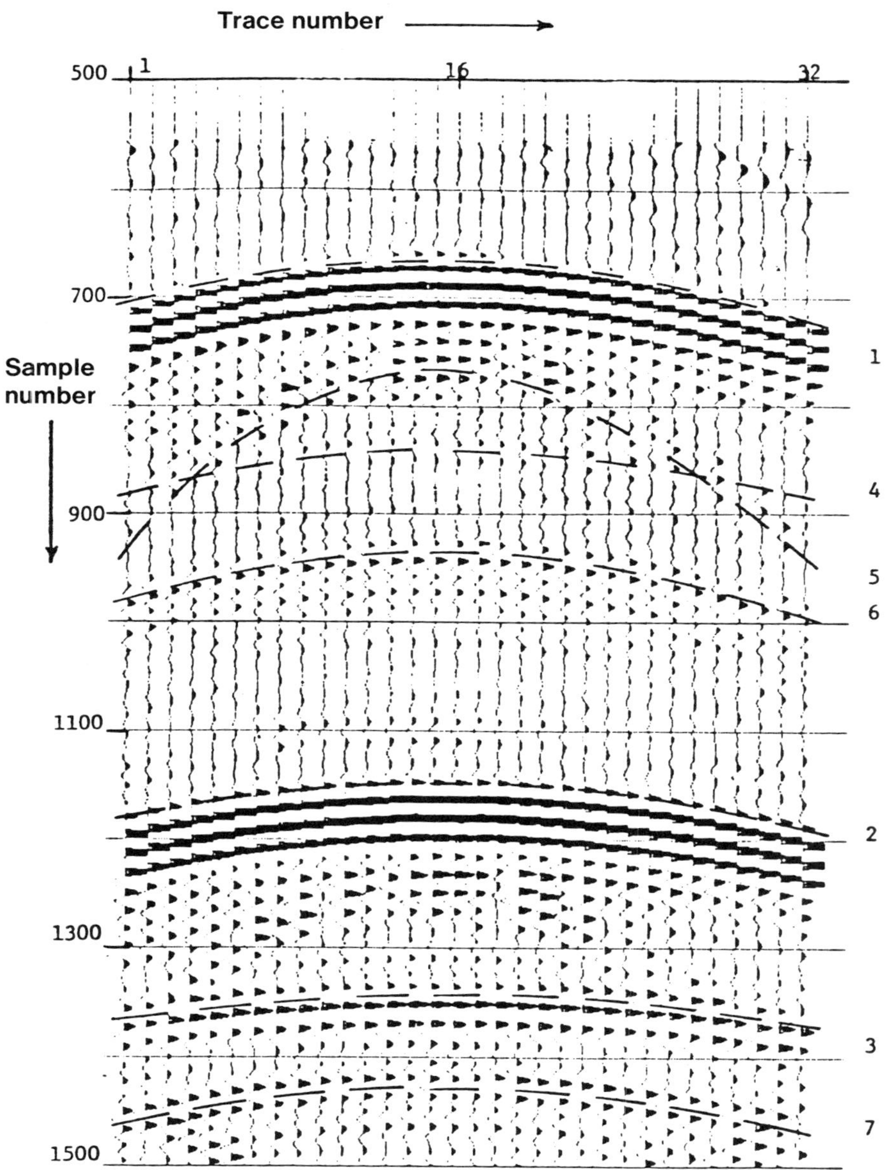

Figure 8.13. Time section of physical model data, line 10. Horizon numbers refer to Table 8.3.

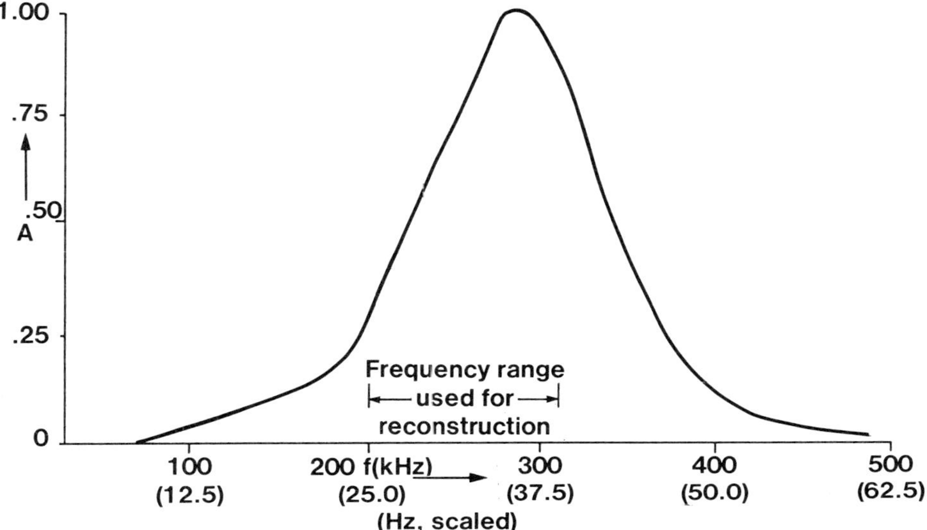

Figure 8.14. *Envelope of composite amplitude spectra; line 10, traces 15 and 16.*

Table 8.2. *Calculated parameters from arrival time data for physical model in Figure 8.12*

	Arrival time from first break on line 10 (in samples)			
Event	Trace 16	Trace 32	V_{rms}*	Interpretation
1	655	715	4970	Water/RTV interface
2	1140	1180	4390	RTV/Plexiglas interface
3	1330	1335	5215	Plexiglas/water interface
4	810	870	4480	Long tail of wavelet
5	720	895	2680	Reflection of source shaper
6	920	980	4220	First intersource mode
7	1412	1451	4240	First intersource mode

*Calculated using the $T^2 - x^2$ method and arrivals from line 10.

Table 8.3. *Physical properties of the model in Figure 8.12*

Interface	T_0 ($\times 10^{-4}$ s)	V_i* (ft/s)	Z_i (in.)	V_{rms} (ft/s)	Z (ft)	V_{rms}** (ft/s)	V_{rms}† (ft/s)
1	1.29	4970	3.85	4970	3850	7455	7455
2	2.26	3385	1.95	4360	5800	6540	6585
3	2.65	9010	2.00	5310	7800	7960	7825

V_i = interval velocity, ith layer; Z_i = thickness, ith layer; V_{rms} = root-mean-square velocity.
*From physical measurements.
**Scaled to prototype units.
†Calculated using the $T^2 - x^2$ method and arrivals from line 10, scaled to prototype units.

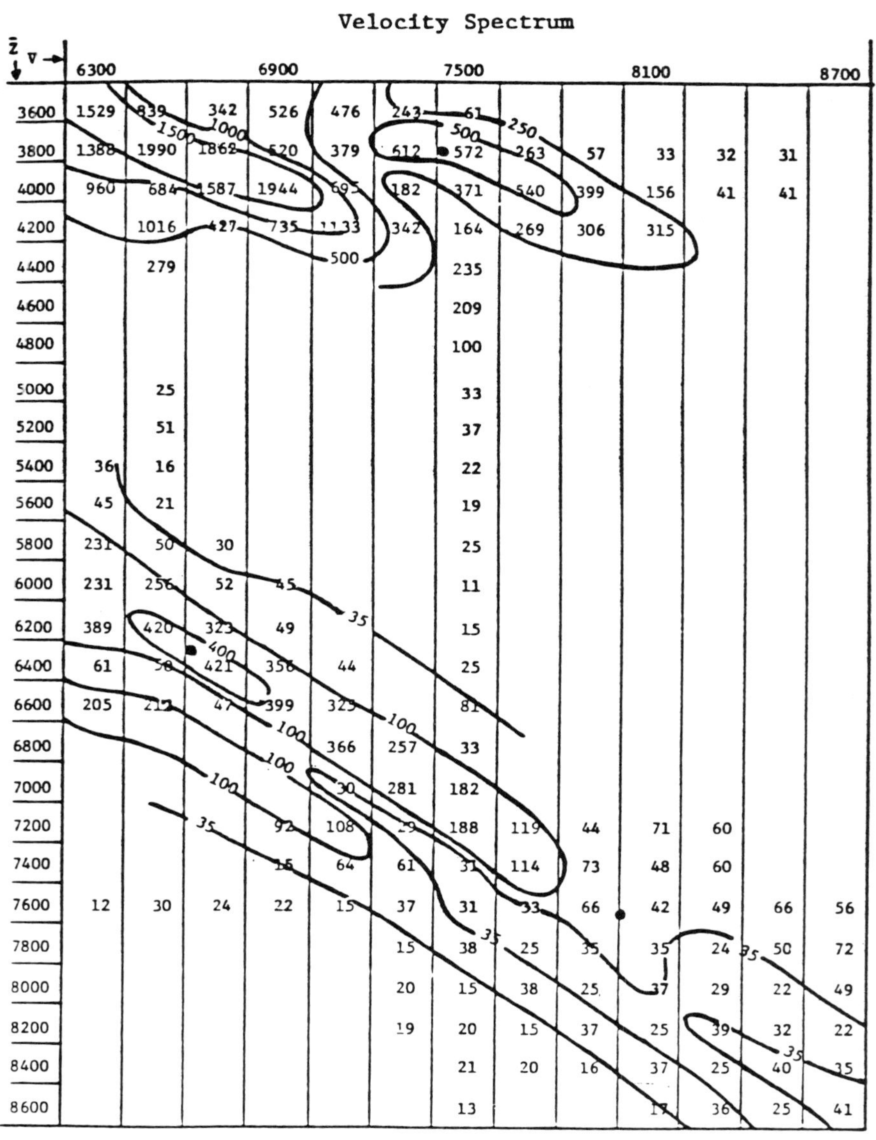

Figure 8.15. *Image intensity values as a function of depth,* $\bar{Z}$, *and velocity,* V, *for convolutional reconstruction of physical model data. Eight frequencies, from 25.02 to 43.33 Hz, were summmed.*

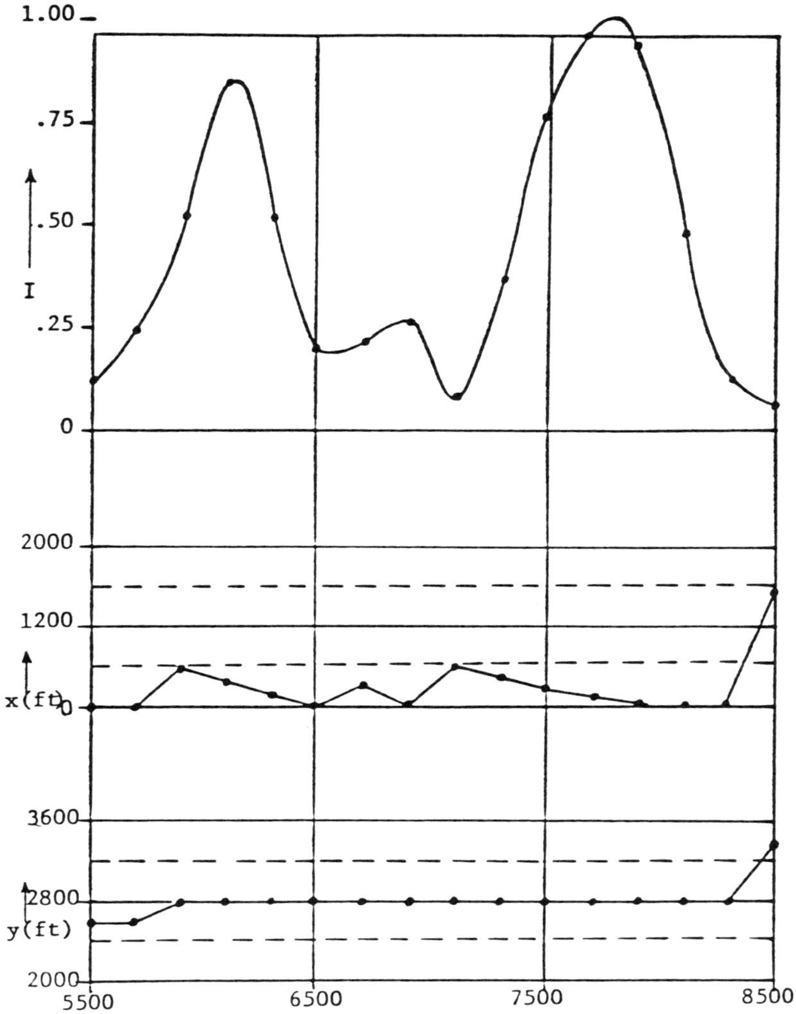

Figure 8.16. *Line at 3800 ft depth in Figure 8.15 filled in with 16 frequency points in the same interval. Lower plots indicate x, y positions of intensity maxima.*

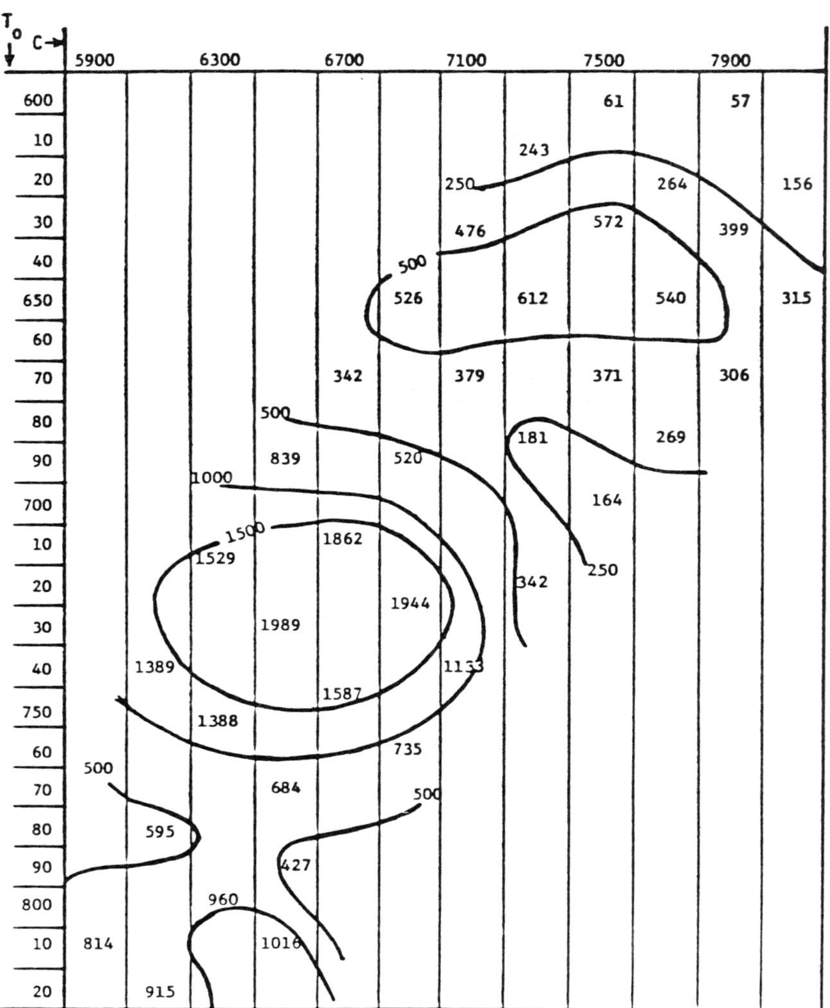

Figure 8.17. *Top portion of Figure 8.15 replotted as image intensity as a function of velocity and normal incidence traveltime,* T_0.

in the lateral position plots of Figures 8.18 and 8.19, that the intensity for the adjacent grid point is nearly equal to the maximum. This indicates that the actual image is located between two lines.

The image for the second interface forms at 6200 to 6400 ft and 6500 to 6700 ft/s. This agrees well with the expected velocity of 6555 ft/s for the RTV/Plexiglas boundary, although it is slightly deeper (400 to 600 ft) than anticipated. This may be a result of the straight ray approximation or, possibly, a combination of that and the ringing mentioned earlier. The lateral position of this image point is offset approximately one grid point from that of the top surface. This is because the RTV layer is not of precisely uniform thickness. It actually varies from 1950 to 2000 ft from one edge to the other in the y direction. This represents a dip of about 2.4°,

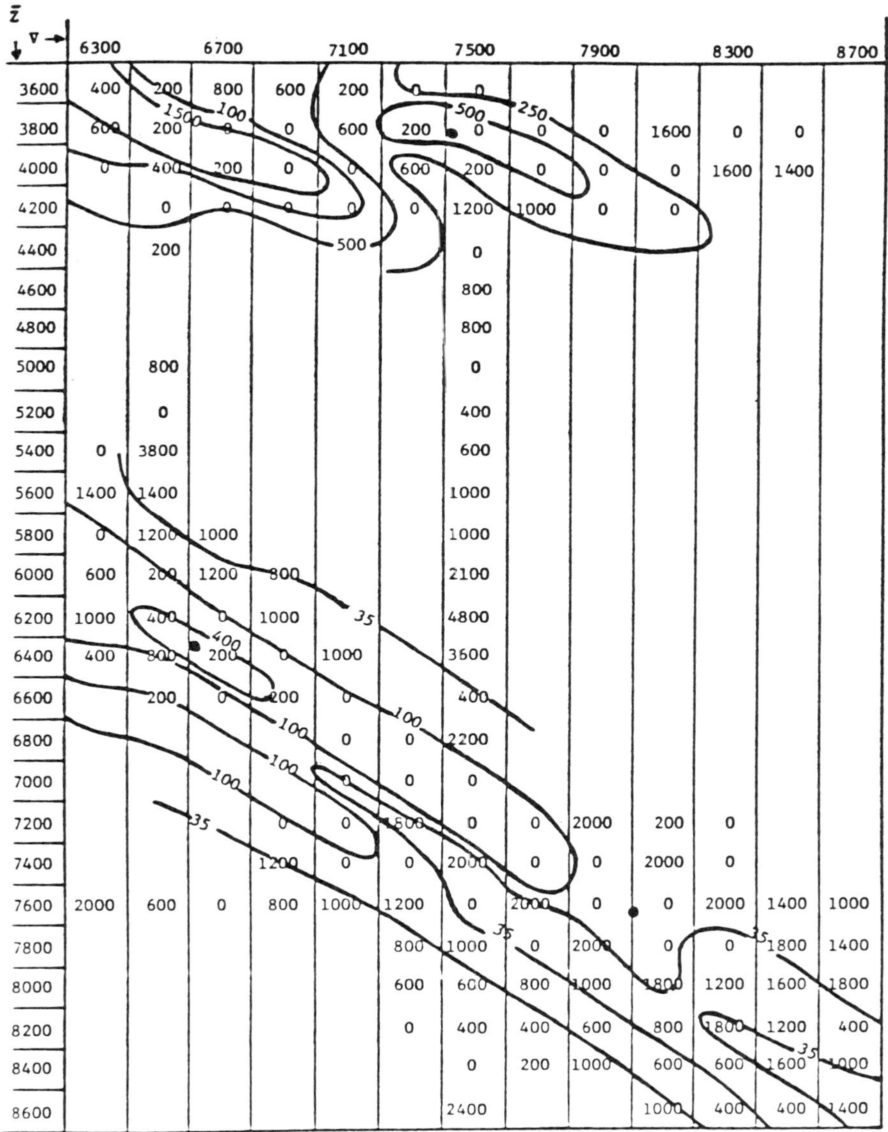

Figure 8.18. *Values of x coordinate as a function of depth and velocity for convolution reconstruction of physical model data. Intensity contours from Figure 8.15 are overlaid.*

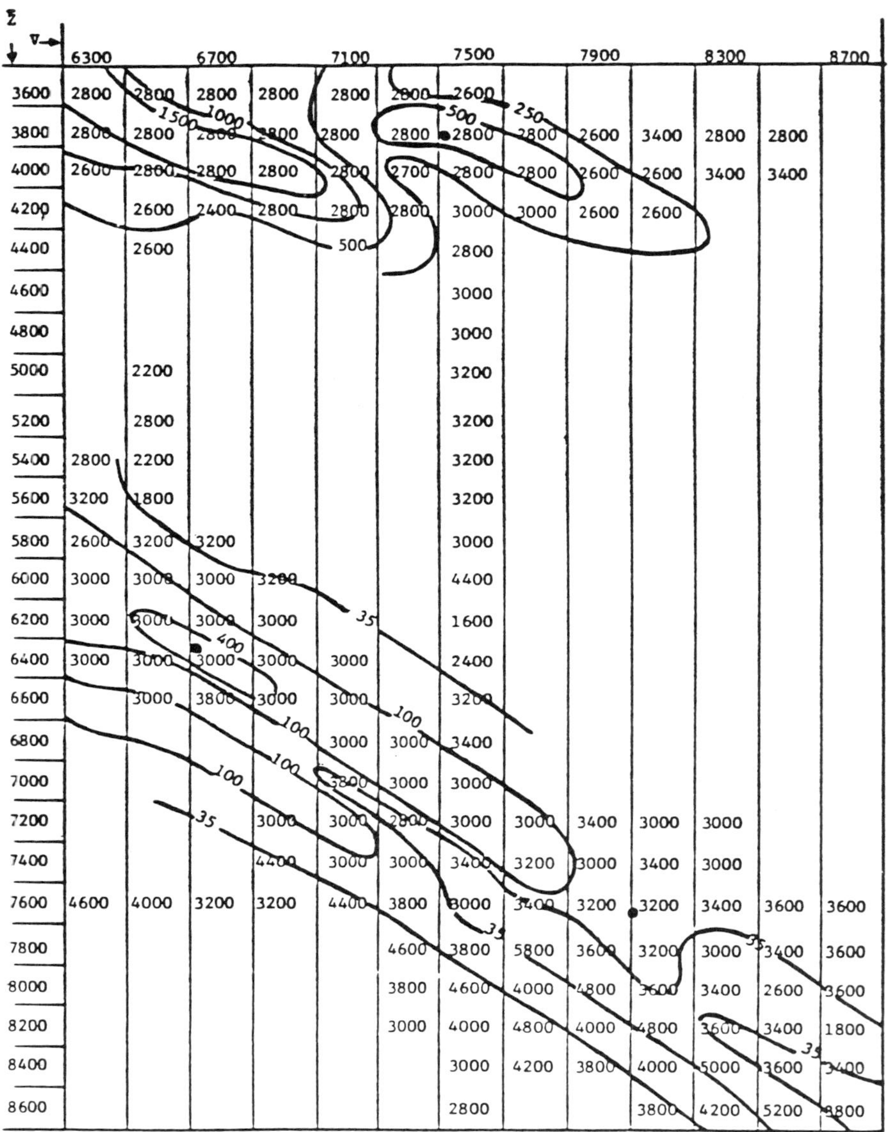

Figure 8.19. *Values of y coordinate as a function of depth and velocity for convolution reconstruction of physical model data. Intensity contours from Figure 8.15 are overlaid.*

which causes a lateral displacement of the source image for the second interface of about 0.8 grid point, or 160 ft. This result provides a good check on the quality of depth resolution and dip discrimination obtained.

Results for the third interface indicate a velocity of 7900 to 8100 ft/s and a depth of 7500 to 7700 ft. The velocity is in close agreement with the actual model, but the depth is 10 to 300 ft shallower than expected. Again, this may be a result of the straight ray approximation. Some confusion is caused by the plunging contours from the image of the second event (caused by ringing) plus an aliased image in

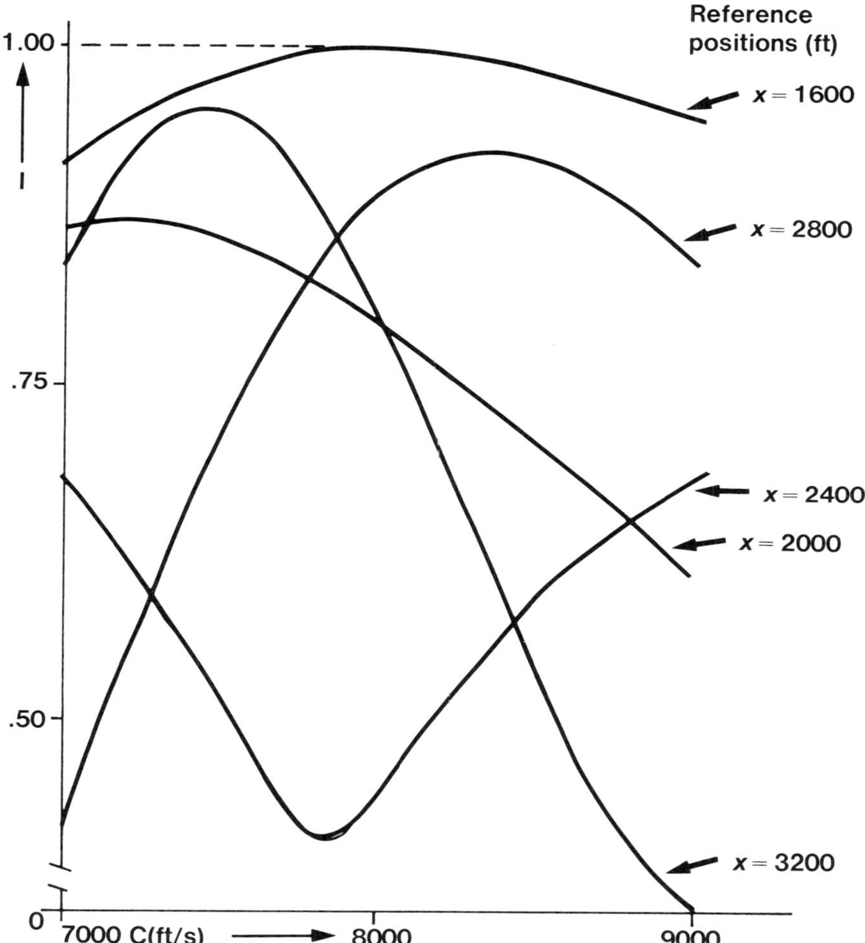

Figure 8.20. *Holographic reconstructions of single-object theoretical model. Data: x = 1600 ft, y = 1600 ft, z = 4000 ft, C = 8000 ft/s, f = 20 Hz. Reconstructions at y = 1600 ft, z = 4000 ft, f = 20 Hz, C at 200 ft/s intervals for x position indicated.*

the 8700 to 9100 ft/s range (Z = 8600 ft), and the pick for this event is based mainly on the lateral coordinate plots. The image for the Plexiglas/water interface is displaced an additional grid point in the *y* direction, again because of the nonuniform thickness of the RTV layer. The variation in the *x* direction is also caused by thickness changes across the model layers, although they are less systematic than those along the *y* axis.

LENSLESS FOURIER TRANSFORM METHOD. The same methods as outlined previously were used to calculate theoretical data sets for input to the lensless Fourier transform imaging algorithm. A similar set of tests was conducted, with the same objectives in mind, as well as several additional tests needed to evaluate certain characteristics peculiar to the lensless technique.

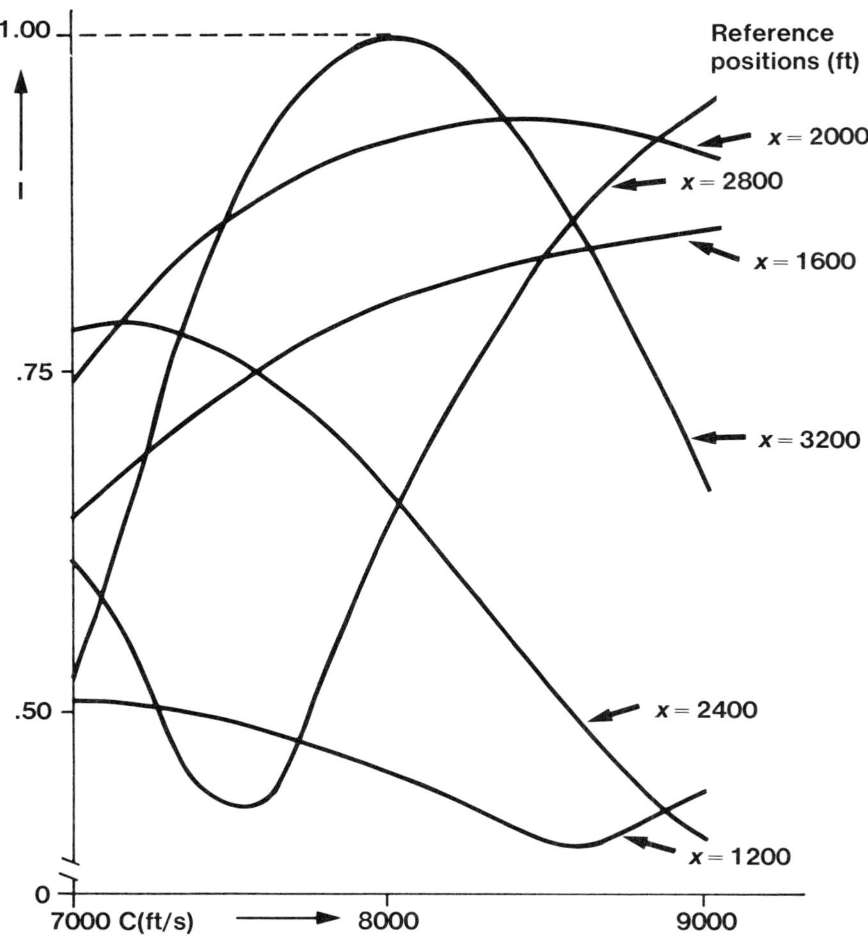

Figure 8.21. *Holographic reconstructions of single-object theoretical model. Data: x = 3200 ft, y = 1600 ft, z = 4000 ft, C = 8000 ft/s, f = 20 Hz. Reconstructions at y = 1600 ft, z = 4000 ft, f = 20 Hz, C at 200 ft/s intervals for x positions indicated.*

Imaging by the lensless Fourier transform method uses the same input data as outlined earlier for the convolutional method. The calculated operator, however, is only 32 by 32 and is determined for the same grid points as the input. The imaging portion of the algorithm involves a point-by-point complex multiplication of the input data array with the operator, followed by a two-dimensional (2-D) Fourier transform to obtain the image. The image array is also 32 by 32, but the sample spacing is variable, depending on the reconstruction parameters. Generally, the only point of the output or image array that directly corresponds to an input array position will be the reference location.

The first plot (Fig. 8.20) is for a single, isolated point object located beneath the center of the recording array. A series of velocity scans for various reference locations is displayed. Figure 8.21 shows a similar series of reconstructions, but with the single object located under the middle of one edge of the aperture. It can be seen that, with the reference position in the same location as the object, a curve similar to the convolution method results. In fact, they are exactly the same, as indicated by the theory.

When the reference is moved laterally off the coordinates of the object, the curve varies considerably, and the maximum intensity occurs at widely erroneous velocities. It should be noted, however, that the absolute maximum intensity occurs only for the correct x, y positions at the correct media velocity. The erratic behavior is attribútable to image scaling considerations; that is, the output sample spacing is a function of the reconstruction parameters C, z, f, and (P, δx) or (Q, δy), and will be different for each velocity in a particular reconstruction sequence. As a result, the same positions on the image are not sampled at all velocities, as is done in the convolution algorithm. This difference is shown diagramatically in Figure 8.22. The major contributing factor is the large sample spacing and hazardous undersampling that result when using a 32-by-32-point input to the 2-D spatial Fourier transform. For the parameters of Figure 8.20 at a velocity of 8000 ft/s, the output spacing is 500 ft for an input spacing of 100 ft.

A direct solution to this problem would be to pad zeros onto the fast Fourier transform (FFT) input to 128 by 128 or even 256 by 256, but computer memory limitations and computation time constraints made this option unattractive. An alternate solution considered was to resample the input data in order to obtain a constant output spacing. Unfortunately, even after interpolation, the spatial sampling mismatches are bothersome. Higher-order effects for incorrect velocities were too great, as confirmed by the results shown in Figures 8.23 and 8.24. A successful solution was achieved by simply summing over temporal frequency without regard for the output spatial sampling. Since the output spacing is a function of frequency, if a sufficiently large range and number of frequencies is included, the aforementioned effects will be smoothed out, though at the expense of lateral resolution. This is demonstrated in Figure 8.25 for the single-object data set. Eight frequencies from 10 Hz to 45 Hz were summed. Some deviation from the correct medium velocity is still evident for incorrect lateral positions, but it is much more systematic, and the image intensity is much more sensitive. Similar behavior is exhibited in Figures 8.26 through 8.29 for the four-object data sets (Fig. 8.26 to 8.28) and the CSP physical model data (Fig. 8.29). Again, the absolute maximum intensity for the desired object is obtained at the correct x, y, C coordinates. An important result is that the incorrect x, y locations chosen over any of the other objects do not produce an image intensity greater than the correct coordinates.

The foregoing results indicate a potential for a recursive search over velocity and lateral reference location for the lensless Fourier transform holography (LFTH) method. However, the problem of slow convergence or nonconvergence for a poor choice of initial reference parameters (x, y, C) must be addressed. In order to evaluate the size of this potential difficulty, a series of reconstructions were done using single-object theoretical data. A later position of $z = 2400$ ft, $y = 1600$ ft was used as the first reference, the results of which are shown in Figure 8.30. Maximum image intensity occurred at 8800 ft/s. Using the x, y coordinates of this image as a new reference position, Figure 8.31 was obtained. The image intensity maximum was again used as a new reference location to obtain Figure 8.32. At this stage, the image intensity maximum occurred at the reference position used for the reconstructions, so the correct position had been found. The velocity at the image maximum was 8000 ft/s, which agrees with the model parameters.

A similar series of reconstructions was also conducted using the four-object models previously described. It was discovered that the procedure would not always converge on the correct object point for certain choices of the initial reference position. For example, one set of tests was aimed at determining the parameters of the fourth object point ($x = 300$ ft, $y = 30$ ft, $z = 4400$ ft, $C = 8500$ ft/s). When the initial reference position was selected too near the second or third object (for example, at $x = 1200$ ft, $y = 2800$ ft), the recursion did not converge on the fourth object, even though the reconstructions were done at the depth of the

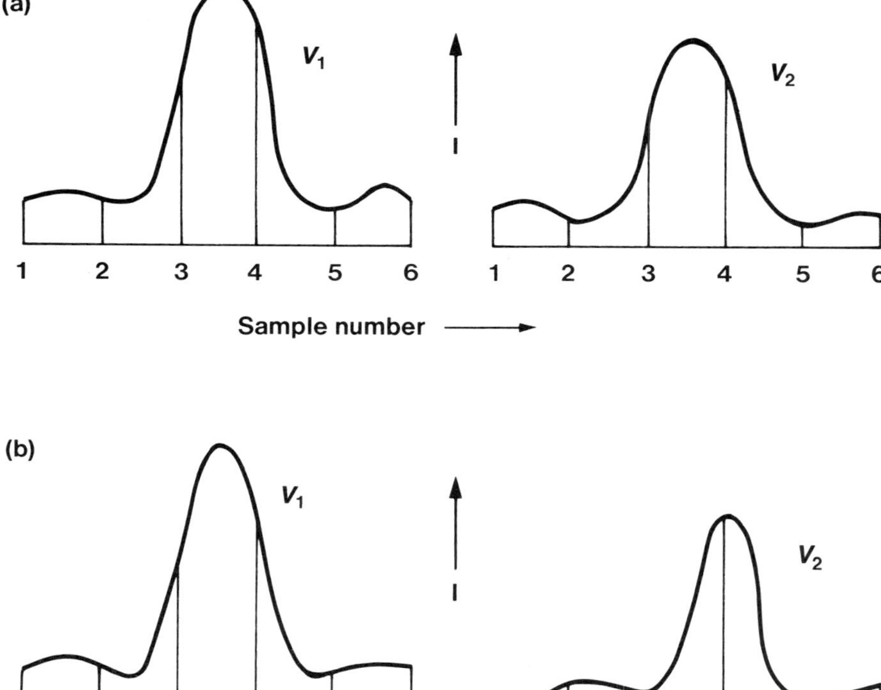

Figure 8.22. *Image sampling comparison: (a) In the convolution method, the same output spacing is maintained throughout, and the image function is sampled consistently. (b) Since the output sampling is variable in the lensless Fourier transform holography (LFTH) method, inconsistencies arise, as in the example here. The magnitude of sample 4 at V_2 is greater than that of samples 3 or 4 at V_1, which would lead to the mistaken conclusion that the image at V_2 is of greater magnitude than that at V_1.*

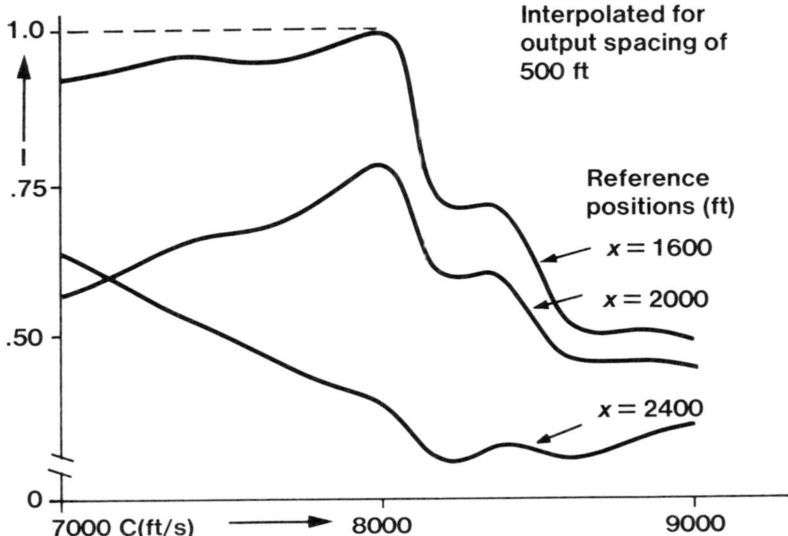

Figure 8.23. Holographic reconstruction with interpolation. Model is single-object theoretical. Data: $x = 1600$ ft, $y = 1600$ ft, $z = 4000$ ft, $C = 8000$ ft/s, $f = 20$ Hz. Reconstructions at $y = 1600$ ft, $z = 4000$ ft, $f = 20$ Hz, C at 200 ft/s intervals for z positions indicated.

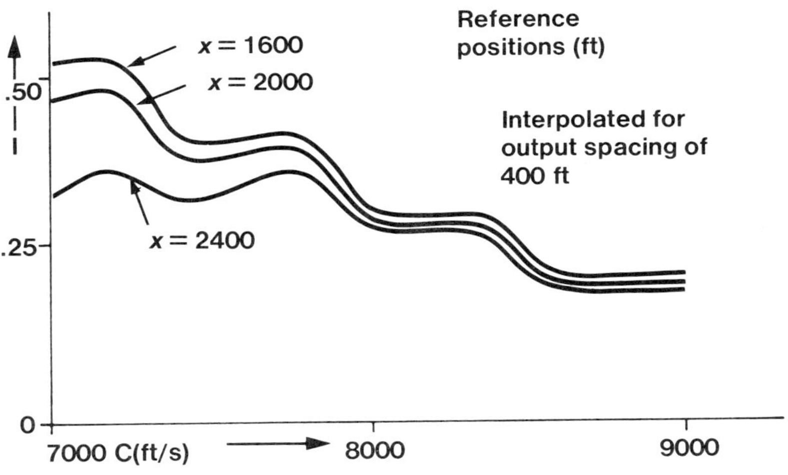

Figure 8.24. Holographic reconstruction with interpolation. Same model and reconstruction parameters as in Figure 8.23.

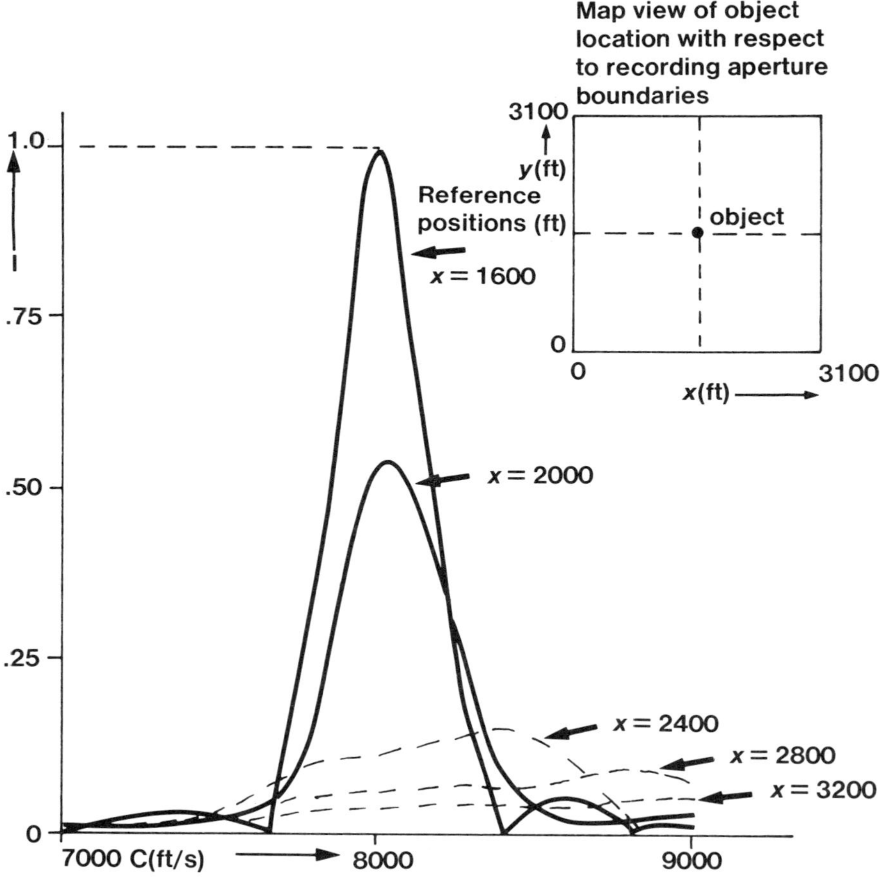

Figure 8.25. *Holographic reconstruction of single-object theoretical model. Data: $x = 1600$ ft, $y = 1600$ ft, $z = 4000$ ft, $C = 8000$ ft/s, $f = 9.9$ to 45.2 Hz by 5 Hz intervals. Reconstructions at $y = 1600$ ft, $z = 4000$ ft, with 200 ft/s velocity increments, and eight frequencies summed; x positions as indicated.*

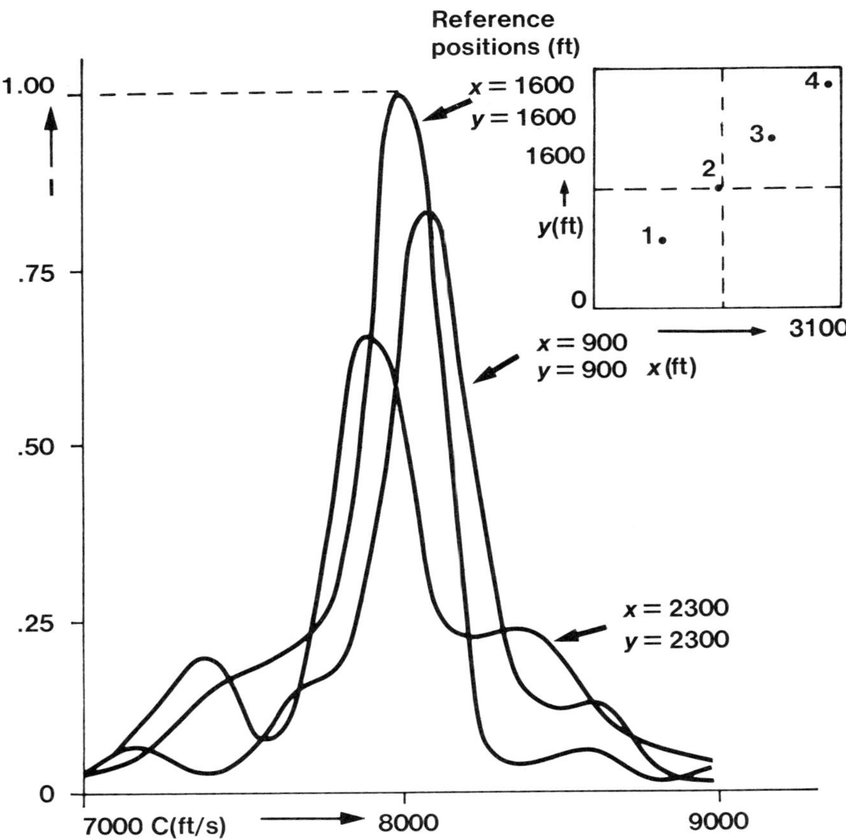

Figure 8.26. *Holographic reconstruction of four-object theoretical model. Data: (1) x = 900 ft, y = 900 ft, z = 3800 ft, C = 7750 ft/s; (2) x = 1600 ft, y = 1600 ft, z = 4000 ft, C = 8000 ft/s; (3) x = 2300 ft, y = 2300 ft, z = 4200 ft, C = 8250 ft/s; (4) x = 3000 ft, y = 3000 ft, z = 4400 ft, C = 8500 ft/s; f = 10 to 45 Hz at 5 Hz intervals. Reconstruction at z = 4000 ft for 200 ft/s velocity increments, with eight frequencies summed from 10 Hz to 45 Hz; x, y positions as indicated.*

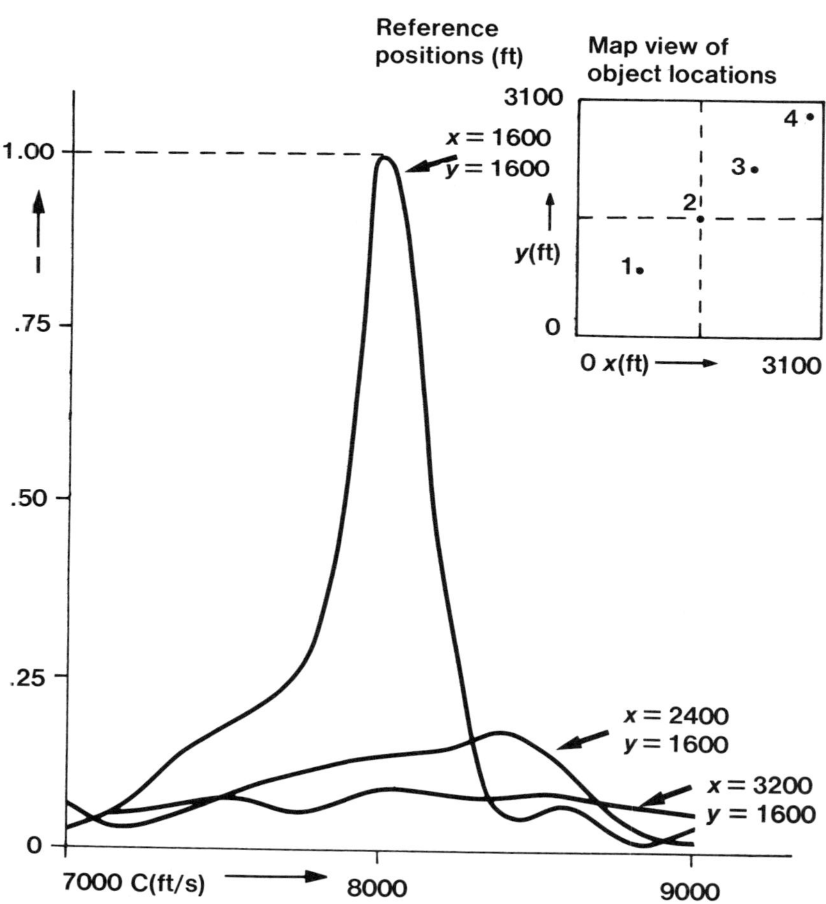

Figure 8.27. Holographic reconstructions using same data set as in Figure 8.26. Reconstructions at $z = 3800$ ft for 200 ft/s velocity increments, with eight frequencies summed; x, y positions as indicated.

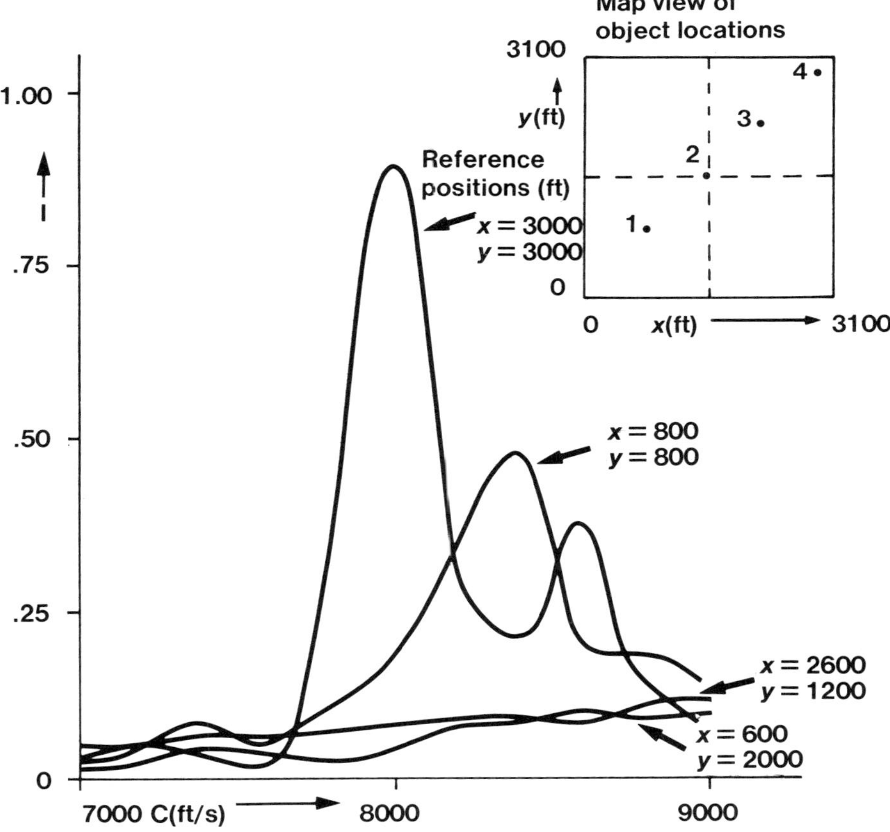

Figure 8.28. *Holographic reconstructions using same data set as in Figures 8.26 and 8.27. Reconstructions at z = 4400 ft for 200 ft/s velocity increments, with eight frequencies; x, y positions as indicated.*

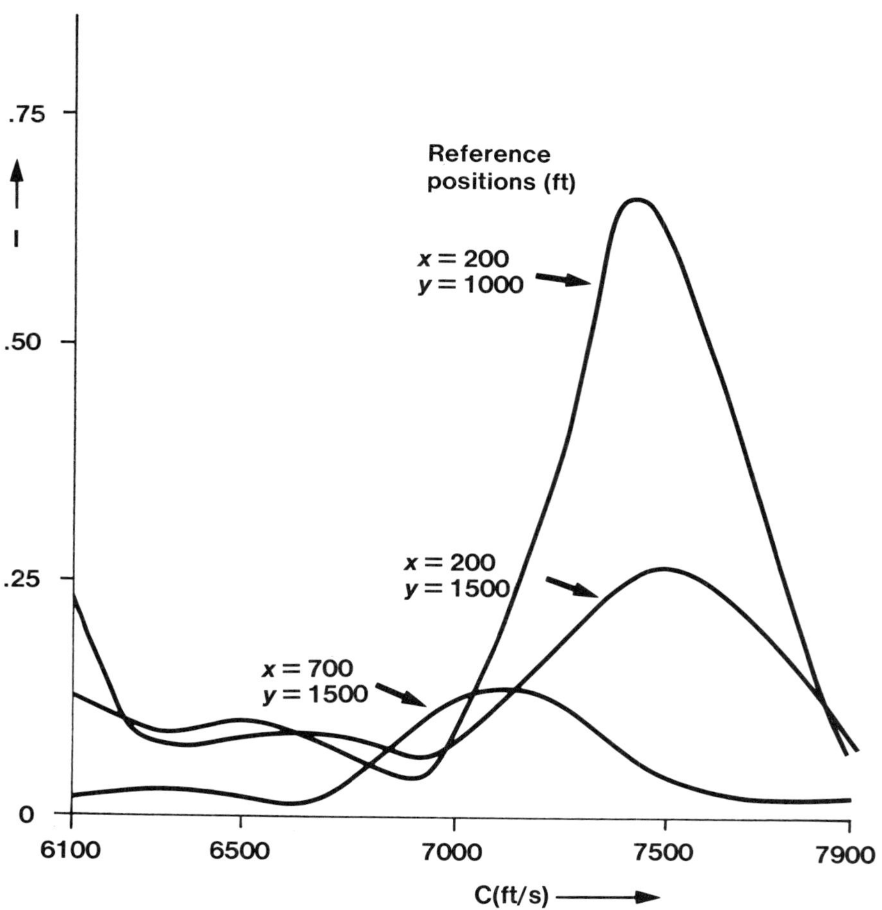

Figure 8.29. Holographic reconstruction of physical model data. Reconstructions at z = 3800 ft for 200 ft/s velocity increments, with 16 frequencies summed from 25 to 43 Hz; x, y positions as indicated.

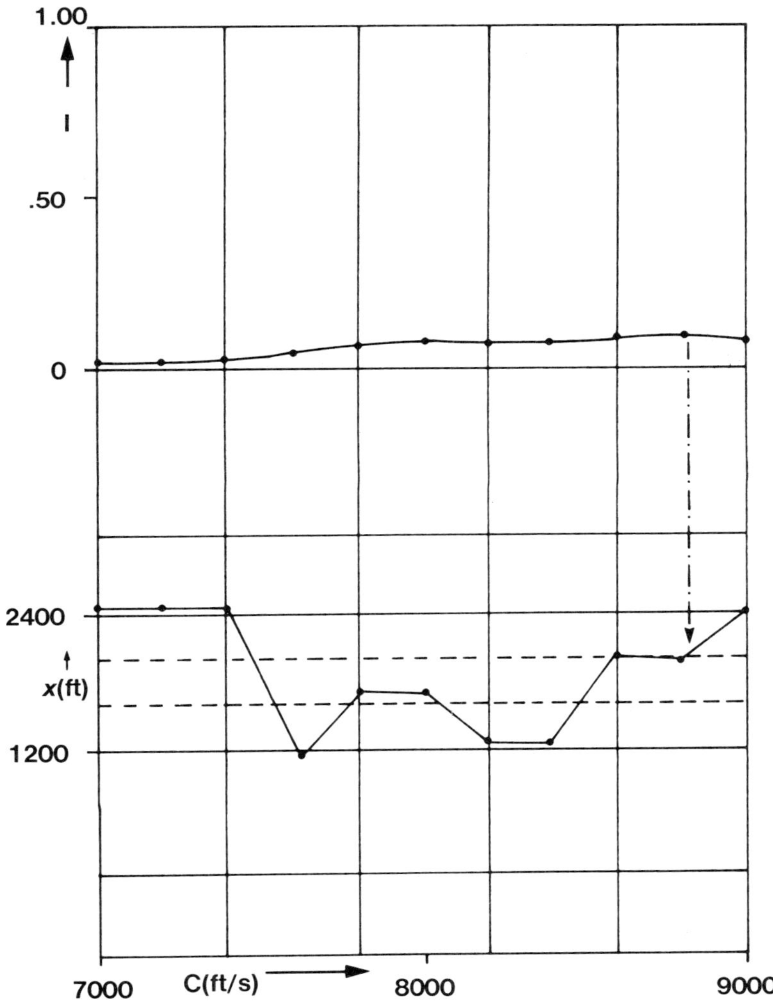

y position = 1600 ft for all reconstructed images.

Figure 8.30. *Holographic reconstruction of single-object theoretical model. Data: x = 1600 ft, y = 1600 ft, z = 4000 ft, C = 8000 ft/s. Reconstructed at x = 2800 ft, y = 1600 ft, z = 4000 ft, for 200 ft/s velocity increments, with eight frequencies summed (10 to 45 Hz).*

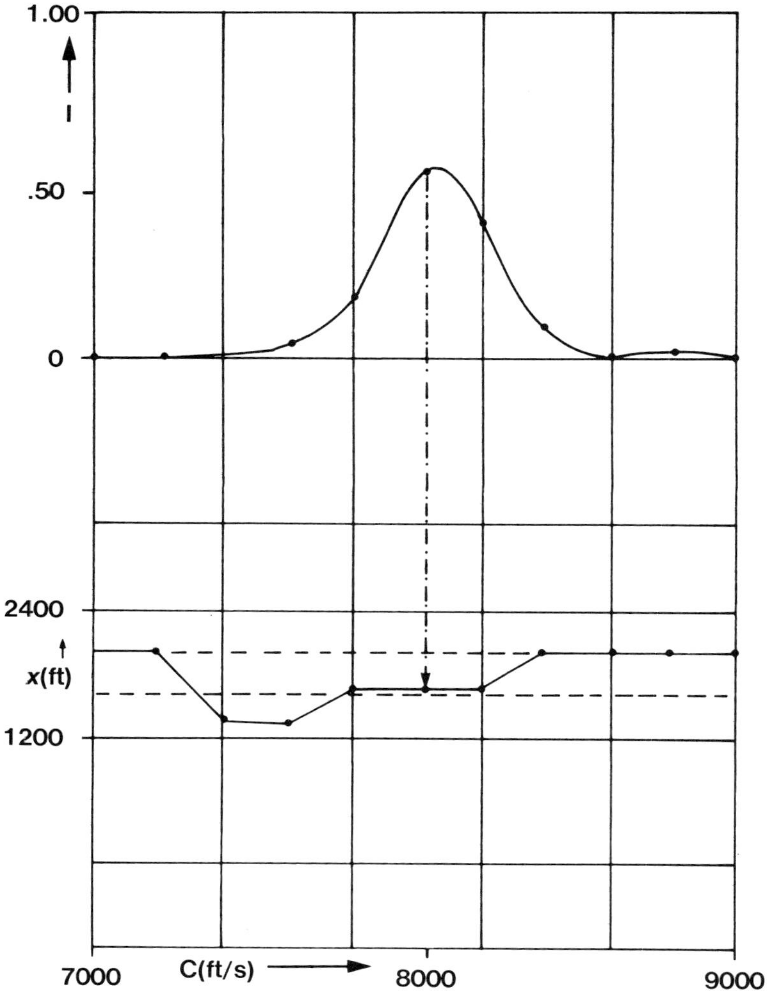

y position = 1600 ft for all reconstructed images.

Figure 8.31. *Holographic reconstruction of same data set as in Figure 8.30. Reconstructed at z = 4000 ft for 200 ft/s velocity increments, with eight frequencies summed; x, y positions of 2000, 1600 used from results of Figure 8.30.*

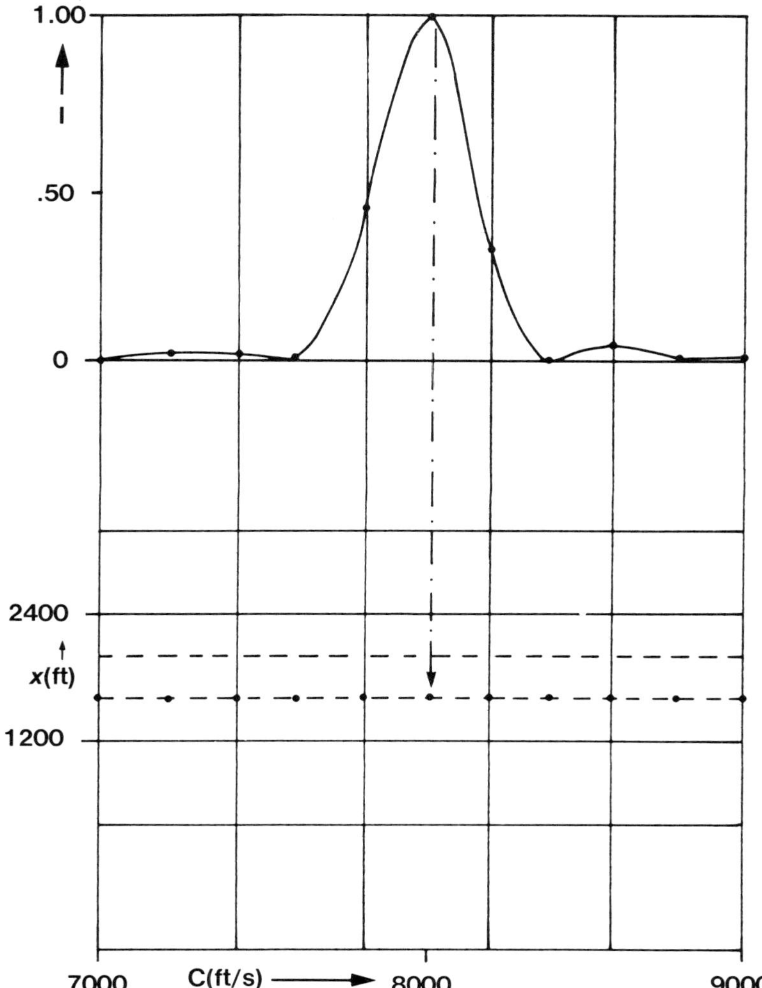

y position = 1600 ft for all reconstructed images.

Figure 8.32. *Holographic reconstruction of same data set as in Figures 8.30 and 8.31. Reconstructed at z = 4000 ft for 200 ft/s increments, with eight frequencies summed; x, y positions of 1600, 1600 used from results of Figure 8.31.*

fourth object. An initial selection of $x = 2400$ ft, $y = 4000$ ft, which is closer to the fourth object than any of the others, produced the desired convergence, and at an intensity level higher than the false convergences previously obtained. The use of additional frequencies (16 instead of 8) did not change these results significantly. The end result is that, to assure convergence on the correct object point, the recursion must be started at several widely spaced initial positions, the correct object parameters being those for the recursion sequence that converges to the highest-intensity maximum.

A last series of tests on the recursion method was performed on the CSP physical model data. These results are presented as a top-view mapping of

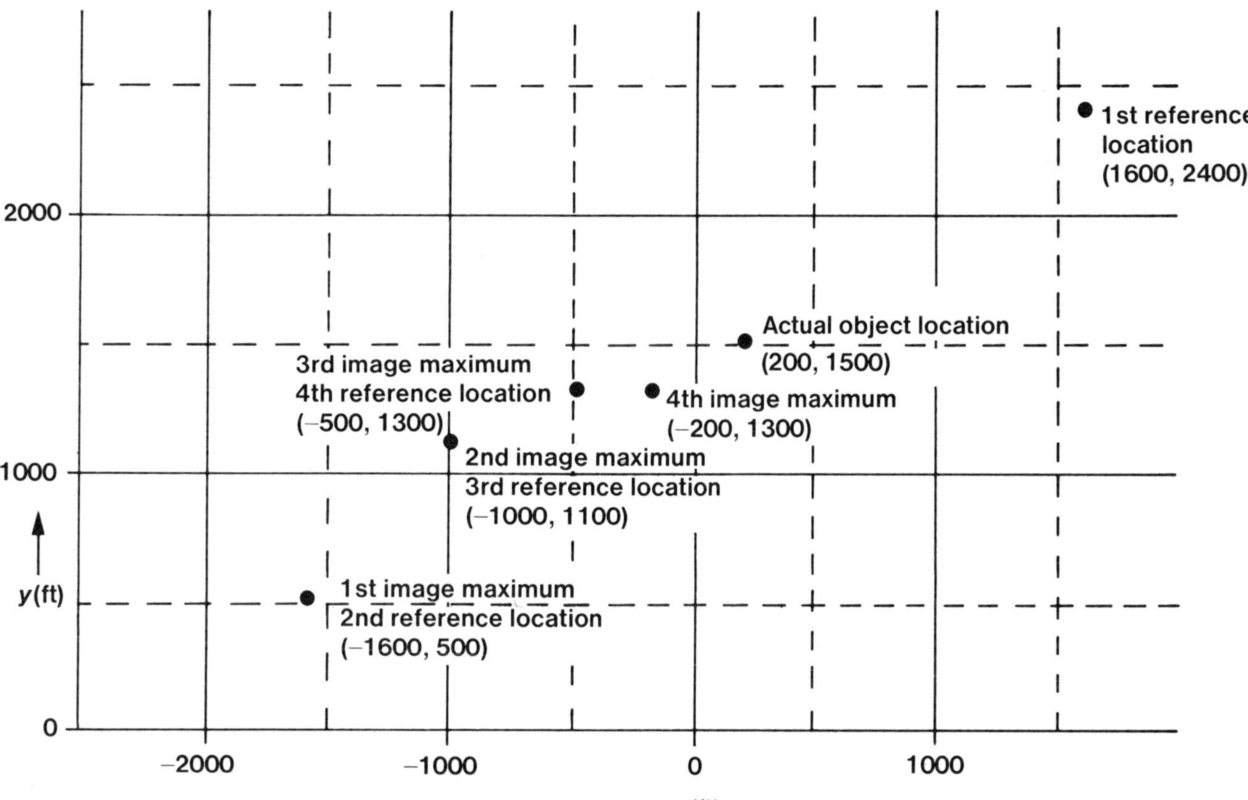

Figure 8.33. Holographic search results using physical model data.

reference positions and reconstructed image locations in Figure 8.33. Again, the location of the image intensity maximum for a particular reference position was used as a new reference location for the next step in the process. Convergence on the proper object point was assured after five iterations for an initial position of $x=1600$ ft, $y=2400$ ft (the actual object location was $x=200$ ft, $y=1500$ ft). Although this was a multiobject or multilayer model, no false convergences resulted. The reason for this is that the model was of essentially horizontal boundaries, and the object points were stacked vertically, one above another, and spaced widely in the z direction. In the previous theoretical example, the objects, all of unit reflection coefficients, were arranged typically, with respect to problems of parameter analaysis in sedimentary basins, and spaced closely together in the z direction. These factors undoubtedly contributed to the severity of the false convergence problem, although the existence of the problem should be considered inherent in the method.

CONCLUSIONS. The viability of 3-D earth parameter determination by frequency domain imaging has been demonstrated for data sets representative of simple geologic sections. Theoretical aspects of two imaging techniques have been discussed in detail: the Rayleigh-Sommerfeld diffraction formula, which is a modified version of the Kirchhoff integral cast in the temporal frequency domain; and an approximation of the Rayleigh-Sommerfeld equation, lensless Fourier transform holography. Both algorithms were used to reconstruct image points for synthetic and physical model data sets.

During the model experiments, a number of issues were raised and resolved:

(1) Reliable velocity and position data were obtained from data sets, using the concepts and methods described.
(2) Velocity spectra were more easily interpreted when plotted with time, T, as the vertical axis rather than depth, z. This was especially true for physical model data, which was very ringy.
(3) Spatial aliasing and wavelet ringing could be alleviated by summing additional frequencies.
(4) For the same size array, computation time for the lensless Fourier transform holography method was approximately one-tenth the time for the convolution (Rayleigh-Sommerfeld) technique.
(5) For monochromatic solutions, the LFTH method gave correct results only for the right position parameters.
(6) Simple bilinear interpolation of the input data could be used to produce an LFTH image with a predetermined output spacing, but this method was not completely successful when used to alleviate the sampling difficulties encountered during frequency summation. It was found that this problem could be modified by summing frequencies and ignoring the effects on image scaling.
(7) The possibility of using the LFTH technique in an interactive search algorithm was demonstrated.
(8) Phase-only reconstructions must be done.

REFERENCE

Abramowitz, M., and Stegun, I. A., 1968, Handbook of mathematical functions: National Bureau of Standards.

ADDITIONAL BIBLIOGRAPHY

Abramowitz, M., and Stegun, I. A., 1972, Handbook of mathematical functions: Dover Publications, Inc. New York.

Acoustical holography, Plenum Press. v. 1–9, 1967–1979: New York.
Claerbout, J. F., 1970, Coarse grid calculations of waves in inhomogeneous media with applications to delineation of complicated seismic structure: Geophysics, v. 35, p. 407–418.
Dohr, G. P., and Stiller, P. K., 1975, Migration velocity determination. Part II, applications: Geophysics, v. 40, p. 6–16.
Duffy, R. E., 1980, Seismic coal seam modeling constrained by depositional environment: Master's thesis, Department of Geology, University of Houston.
French, W. S., 1974, Two-dimensional and three-dimensional migration of model-experiment reflection profiles: Geophysics, v. 39, p. 265–277.
Gabor, D., 1948, A new microscope principle: Nature, v. 161, p. 777–778.
———, 1949, Microscopy by reconstructed wavefronts: Proc. Roy. Soc., v. A197, p. 454.
Gardner, G. H. F., French, W. S., and Matzuk, T., 1974, Elements of migration and velocity analysis: Geophysics, v. 39, p. 811–825.
Gardner, L. W., 1947, Vertical velocities from reflection interpretation: Geophysics, v. 12, p. 221–228.
Gates, J. P., 1957, Descriptive geometry and the offset seismic profile: Geophysics, v. 22, p. 589–609.
Goodman, J. W., 1968, Introduction to Fourier optics: New York, McGraw-Hill Book Co., Inc.
Green, C. H., 1938, Velocity determinations by means of reflection profiles with special reference to geophysical prospecting: Geophysics, v. 3, p. 295–305.
Hagedoorn, J. G., 1954, A process of seismic reflection interpretation: Geophys. Prosp., v. 2, p. 85–127.
Hoover, G. M., 1972, Acoustical holography using digital processing: Geophysics, v. 37, p. 1–19.
Leith, E. N., and Upatnieks, J., 1962, Reconstructed wavefronts and communication theory: J. Opt. Soc. Am., v. 52, p. 1123–1130.
Leith, E. N., and Upatnieks, J., 1967, Recent advances in Holography, *in* Progress in optics, v. VI: E. Wolf, Ed., Amsterdam, North Holland.
Levin, F. K., 1971, Apparent velocity from dipping interface reflections: Geophysics, v. 36, p. 510–516.
Mayne, W. H., 1962, Applications of the expanding reflection spread: Geophysics, v. 27, p. 981–993.
Meir, R. W., 1965, Magnification and third-order aberrations in holography: J. Opt. Soc. Am., v. 55, p. 987.
Morgan, T. R., Fitzpatrick, G. L., Hilterman, F. J., and Wang, K., 1979, Lensless Fourier transform holography as applied to imaging of reflection seismic data, *in* Acoustical holography, v. 9: K. Wang, ed., New York, Plenum Press, p. 761–796.
Musgrave, A. W., 1952, Wave-front charts and raypath plotters: Quarterly of the Colorado School of Mines, v. 49.
———, 1961, Wave-front charts and three-dimensional migrations: Geophysics, v. 26, p. 738–753.
———, 1962, Application of the expanding reflection spread: Geophysics, v. 27, p. 981–993.
Rieber, F., 1936, A new reflection system with controlled directional sensitivity: Geophysics, v. 1, p. 97–106.
Sattlegger, J. W., 1975, Migration velocity determination. Part I, philosophy: Geophysics, v. 40, p. 1–5.
Schneider, W. A., 1978, Integral formulation for migration in two and three dimensions: Geophysics, v. 43, p. 49–76.
Schneider, W. A., and Backus, M. M., 1968, Dynamic correlation analysis: Geophysics, v. 33, p. 105–126.
Slotnick, M. M., 1959, Lessons in seismic computing: Tulsa, SEG.
Stolt, R. H., 1978, Migration by Fourier transform: Geophysics, v. 43, p. 23–48.
Stoke, G. W., 1968, An introduction to coherent optics and holography: New York, Academic Press.
Taner, M. T., and Koehler, F., 1969, Velocity spectra-digital computer derivation and applications of velocity functions: Geophysics, v. 34, p. 859–881.
Wylie, R. C., 1975, Advanced engineering mathematics: New York, McGraw-Hill Book Co., Inc.

INDEX

MARATHON
RESEARCH CENTER
LIBRARY
LITTLETON, COLORADO